Elasticity, plasticity and structure of matter

Elasticity, plasticity and structure of matter

edited by
R. Houwink and H. K. de Decker

third edition

Cambridge at the University Press 1971

Published by the Syndics of the Cambridge University Press
Bentley House, 200 Euston Road, London N.W.1
American Branch: 32 East 57th Street, New York, N.Y.10022

© Cambridge University Press 1971

Library of Congress Catalogue Card Number: 72–154515

ISBN: 0 521 07875 X

Printed in Great Britain
at the University Printing House, Cambridge
(Brooke Crutchley, University Printer)

Preface

To introduce the new edition of a book first written 34 years ago (1937), it may be interesting to quote part of the foreword to the original edition:

There are three important groups of scientists and engineers who have shown a particular interest in the study of the elastic and plastic phenomena which take place upon deformation of matter. The physicist from the earliest times has observed these phenomena, and has attempted to express his observations by means of formulae. The chemist has carried on this work in his own way, and has attempted especially to discover a relation between physical observation and the structure of matter. By a cooperation of these two groups a certain insight into the constitution of matter has thus been obtained, which furnishes an explanation of various observations on the elastic and plastic behaviour of certain substances. Finally, an insight into the problems under discussion is particularly important to the engineers since it increases the possibility of the correct choice of material.

It has been one of the purposes of this book to bring the three groups of workers into closer contact with each other, so that they may understand one another more easily, and by the aid of an insight into the structure of matter may seek to improve existing materials and to discover new ones.

One of the tasks which the author has set himself is the attempt to trace a unity of thought as far as it is possible. In order to achieve this purpose the observations on very different substances (crystalline substances, glass, resins, asphalt, textile fibres, natural and synthetic rubber, proteins, paints, clay, sulphur) had to be compared with each other.

The result is often disappointing, either because of the lack of sufficient observations, or because of the lack of insight into the structure of the substance. It appears possible however in many cases, and with some degree of certainty, to make certain prophecies about the behaviour of a material upon deformation on the basis of its structure. Herein, according to the opinion of the writer, lies the key to the deliberate synthesis of materials with definite desired elastic and plastic properties, which will be carried out in the coming decades. In this connection one has only to think of synthetic rubbers, fibres and plastics.

That the expectations expressed in 1937 were fulfilled, may be proved by the fact that a second edition appeared in 1952 (incorporating some modernizations), followed by a paperback edition (Dover Publications, 1958). Moreover, translations appeared in German (two editions), Italian and Russian.

Seeing that the time is ripe for a new, totally modernized version, the two editors and the publisher agreed to maintain the spirit of the original book, as explained in the 1937 Foreword. Taking into account, however, the spectacular developments in rheology and materials technology of the last 25 years, it was obvious that the objective could only be reached by bringing together a number of eminent experts to write the specialized chapters (thermoplastics, metals, etc). But, in order to maintain the original idea of coherence, the general surveys (structure, rheology) were mainly composed by the editors, in close cooperation.

As experience with the earlier editions shows, the book should be useful for science students in their third or fourth year in college. On the other hand it is also an 'idea' book that shows the expert in one field how to relate his experience to the data in another field.

Starting from the simple well-known principles that atoms and molecules have certain volumes and exert certain forces upon one another, the book tries to show how the mechanical properties of materials can be understood as a result of their structure.

In recent years there has been a growing realization that scientists can only help to solve a complicated problem if they broaden their knowledge and look at the entire problem, not just at their particular area of interest in it. In other words, we are rediscovering the drawbacks of narrow specialization and see the need for broader understanding. For these general reasons, our book may be particularly useful.

Another significant development has taken place since the first edition of 1937. The subjects of 'Materials science' and 'Materials technology' (a special name 'Materiology' has been suggested) have developed rapidly as important fields of activity, and in a number of Universities have obtained their own departments, even sometimes including doctoral studies.

We are convinced that the present book not only supplements existing textbooks, but provides instructive additional background reading in Materials technology.

For example, it will satisfy a student's curiosity aroused by the strange fact that rubber stretches elastically to 600 % of its length, but glass only 1 %. It will guide him to see the 'intrinsic' possibilities in a material: the ductility by crystal slippage in a metal wire, the brittleness due to the rigid irregularity of a glass, the ductility due to long molecules in a plastic bag. It will enable him to simplify the concepts of deformation, and rheology, and visualize them as everyday phenomena, sometimes called 'strength', or 'creep', or 'stiffness'. Finally it will help him to re-define any deformation problem into simple parameters, such as modulus, viscosity, temperature, and time, and to relate them naturally with structure: places of atoms in space and the forces between them. In turn, differences between order (crystals) and disorder (liquids or glass), or between polymer (plastics) and monomer (simple liquids) will acquire an automatic meaning in terms of mechanical properties.

Wassenaar, The Netherlands　　　　　　　　　　　R. HOUWINK
New York (N.Y.), U.S.A.　　　　　　　　　　　H. K. DE DECKER
January 1971

Contents

List of most frequently used Symbols

C	concentration
C_V	volume concentration
D	velocity gradient
E	modulus of elasticity (Young's modulus)
E_v	activation energy
e	energy
e_r	rupture energy
e_s	energy at melting point
f	yield value
G	shear modulus
K	coefficient of compressibility
M	molecular weight
M_B	bulk modulus
p	pressure
T	temperature
T_g	glass transition temperature
T_m	melting temperature
T_s	melting temperature
γ	deformation
ϵ	elongation
η	viscosity coefficient
η^*	viscosity coefficient for quasi flow
λ	relaxation time
μ	Poisson's ratio
σ	tension or stress
τ	shearing stress
ψ	fluidity

Contributors

R. D. ANDREWS, Department of Chemistry and Chemical Engineering, Stevens Institute of Technology, Hoboken (N.J.), U.S.A.

A. H. BLOKSMA, Research Chemist at the Institute for Cereals, Flour and Bread, T.N.O. Wageningen, The Netherlands.

M. N. M. BOERS, Research Chemist at the Paint Research Institute, T.N.O. Delft, The Netherlands.

B. B. BOONSTRA, Associate Director of Rubber and Plastics Research, Cabot Corporation, Billerica, Mass., U.S.A.

U. DAUM, Central Laboratory, T.N.O. Delft, The Netherlands.

H. K. DE DECKER, Uniroyal International, New York, U.S.A.

R. W. DE DECKER, Pan American World Airways, New York, U.S.A.

J. L. DEN OTTER, Central Laboratory, T.N.O. Delft, The Netherlands.

G. E. DIETER JR., Dean of the Department of Metallurgical Engineering, Drexel University, Philadelphia, Penn., U.S.A.

D. R. HAY, Associate Professor of Metallurgical Engineering, Drexel University, Philadelphia, Penn., U.S.A.

A. HESLINGA, Plastics and Rubber Research Institute, T.N.O. Delft, The Netherlands.

R. HOUWINK, Honorary Director at Euratom, Brussels.

R. H. PETERS, Professor of Polymer and Fibre Science, University of Manchester Institute of Science and Technology, Great Britain.

D. UHLMANN, Assistant Professor of Ceramics, Department of Metallurgy, Massachusetts Institute of Technology, Cambridge (Mass.), U.S.A.

M. VAN DEN TEMPEL, Unilever Research Laboratory, Vlaardingen, The Netherlands.

Acknowledgements

Grateful thanks are due to the following copyright holders for permission to make use of figures.

Scientific American for Figs. 2.2, 3.6, 7.10, 7.12, 7.13; J. Wiley & Sons, Inc. for Figs. 9.6, 10.1, 10.8, 10.10, 15.2; Edward Arnold (Publishers) Ltd. for Fig. 9.16; American Chemical Society for Figs. 9.33, 11.2; Textile Institute for Figs. 10.2–10.4; Elsevier Publishing Co. for Figs. 10.2, 10.5, 10.9; Butterworth & Co. (Publishers) Ltd. for Figs. 10.3, 10.4, 10.12, 10.13–10.27; Chapman & Hall Ltd. for Fig. 10.4; *La Chimica e l'Industria* for Fig. 10.6; Dr Dietrich Steinkopff Verlag for Fig. 10.7; Institute of Physics and the Physical Society for Fig. 10.11; Textile Research Institute (Princeton, N. J.) for Fig. 10.15; Shirley Institute (Manchester) for Fig. 10.26; Royal Society for Fig. 10.27, Plate 13.3; American Ceramic Society for Figs. 11.1, 11.4; Metallurgical Society of AIME for Figs. 11.8, 13.28. 13.29; Pergamon Publishing Co. for Fig. 11.9, Plate 13.4, Fig. 13.23; Academic Press for Figs. 11.14, 11.15; Addison-Wesley Publishing Co. for Fig. 13.11; McGraw-Hill Book Co. for Figs. 13.13, 13.27, 13.31.

1 Rheological behaviour of matter

R. Houwink

1.1 Rheology and its parameters[1]

The study of elasticity, plasticity, viscosity and other types of deformation in matter is called rheology. Thus rheology is the science of deformation (shape changing) of materials.

When a force is applied to a body, the two extremes of behaviour which may occur from a rheological viewpoint are the pure elastic deformation of a solid and the pure viscous flow of a liquid. Pure elasticity means that the body returns to its original form after release of stress; pure viscosity means that the liquid flows even under the smallest stress applied and does not retake shape after release of stress.

The main parameters related to these deformations are as a first approximation:

The resistance against deformation, expressed (in an extension experiment) by the modulus of elasticity E for an elastic body and (in laminar flow) by the viscosity coefficient η for a liquid.
The ultimate stress and elongation which the elastic body can bear without flowing or breaking.
The ultimate shearing stress which can be applied to a liquid before turbulent flow sets in.

The elastic deformation (Fig. 1.1 a) is given by

$$\mathrm{d}l = \frac{\sigma}{E},\qquad(1.1)$$

where σ is stress; and the viscous flow (Fig. 1.1 b) by

$$D = \frac{\mathrm{d}v}{\mathrm{d}y} = \frac{\tau}{\eta},\qquad(1.2)$$

where τ is shearing stress. The full differential equation from which (1.2)† has been derived is

$$v = \frac{\mathrm{d}x}{\mathrm{d}t};\quad \frac{\mathrm{d}v}{\mathrm{d}y} = \frac{\partial^2 x}{\partial y\,\partial t}.$$

† The definition of equation (1.2) is possible only in laminar flow as illustrated in Fig. 1.1(b). Laminar flow occurs only as long as a characteristic number, R, the

HEM

In Fig. 1.1 these two types of deformation are presented schematically. The modulus E which is a physical constant as long as the deformation is proportional to the stress (Hooke's law), is defined as the stress per unit of *deformation*. The viscosity coefficient η, also only a physical constant in the case of proportionality, is the shearing stress per unit of *deformation rate*. E is expressed in kgf/cm² (lbf/in²), η in poise.†

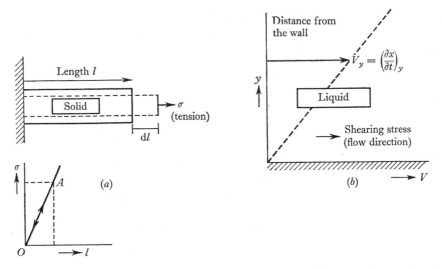

Fig. 1.1. The two main types of deformation. (*a*) Elastic deformation (stress–strain diagram). (*b*) Viscous flow (shear-flow diagram).

The viscosity of water at room temperature is about 1/100 poise, 37 times that of air. The maximum possible value of η is somewhat over 10^{13} poise; then the material involved is so hard and brittle (e.g. glass) that breaking occurs without any signs of flow.

E accounts for the energy stored in elastic deformation and η for the energy

Reynolds number, remains below a certain limit. Above the limit, turbulence occurs and equation (1.2) cannot be used. The Reynolds number is defined by $R = \rho v l / \eta$, in which ρ and η are the density and viscosity of the fluid, respectively, and v and l are a velocity and a length characteristic of the flow system under examination.

† The dimensions of viscosity are ($ML^{-1}T^{-1}$). When the shearing force is 1 dyn cm⁻², the distance between two planes 1 cm and the velocity 1 cm s⁻¹, the value of η is said to be 1 *poise*. 1/100 part of a poise is the centipoise. Besides the dynamic viscosity one distinguishes a *kinematic viscosity*, indicated by η/d, where d is the density. It is expressed in stokes. The reciprocal of η the *fluidity* Ψ, is expressed in *rhes*, so that

$$1 \text{ rhe} = \frac{1}{1 \text{ poise}}.$$

dissipated in flow, which is due to transfer of momentum (motion in direction of flow) perpendicular to this direction.

Where both effects are simultaneously involved, the deformation is said by certain schools to be visco-elastic. Visco-elastic, also frequently referred to as plastic behaviour, can be characterized by the additivity of stress according to

$$\sigma = E\,dl + \eta D, \qquad (1.3)$$

$$\text{where } D = \frac{dv}{dy}.$$

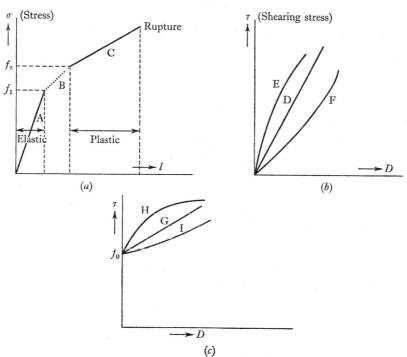

Fig. 1.2. Main types of elastic and plastic behaviour. (*a*) Stress–strain curves. (*b*) Shear-flow curves (viscous). (*c*) Shear-flow curves (plastic).

This formula may therefore be considered as a general equation of rheological behaviour. In cases where $E\,dl = 0$, the deformation is sometimes called anelastic.

In Fig. 1.2 these principles are worked out in somewhat more detail. Fig. 1.2 *a* represents the approach from the elastic point of view. Part A of the curve represents purely elastic deformation and part C on the other hand represents plastic deformation. Between these two there is usually a transition region, B, where both phenomena occur together. In such a case flow begins at the yield value f_1, called the lower yield value, and the elastic part finishes at the upper yield value f_2.

For substances with a predominant viscous of plastic behaviour an approach from that particular point of view is more adequate. In viscous flow (Fig. 1.2 *b*) even the smallest force will cause displacement and so there is no yield value. The general form of the flow curve is

$$D = \frac{\tau^n}{\eta^*}. \tag{1.4}$$

Curve D presents the case where $n = 1$; this is called pure viscous or Newtonian flow,

$$D = \frac{\tau}{\eta}, \tag{1.5}$$

where η is a constant, expressed in poise.

In the cases E and F, on the contrary, $n \neq 1$ and consequently η is no longer a simple constant expressing the relationship between deformation and applied stress. η^*, called the apparent or quasi-viscosity and marked by an asterisk, has no proper dimensions and so cannot be expressed in poise.

Fig. 1.2 *c* represents cases where plastic deformation predominates over the elastic above a finite yield value f_1. Curve G represents *true plastic flow*, and is described by the formula.

$$D = \frac{1}{\eta}(\tau - f). \tag{1.6}$$

Curves H and I are very similar to those for quasi-viscous flow discussed above and can be represented by the formula

$$D = \frac{1}{\eta}(\tau - f)^n. \tag{1.7}$$

This is called *quasi-plastic flow* and the remarks already made concerning η hold here.

We will encounter materials having a temporary structure which can be altered by shearing and which is restored upon relaxation. This phenomenon is called *thixotropy*. A thixotropic material is stiff and does not flow under relatively small stresses. It can be made to flow with less and less viscosity as the shearing stress is increased, until the low viscosity persists even under reduced stress. But when the material is stress-free, it will, with time, regain its original stiffness. An example is drilling mud.

More or less contrary to thixotropy is the phenomenon of *dilatany*, that is, the stiffening of a liquid when it is brought into motion. A well-known example is beach sand.

The importance of a well-established nomenclature in rheology is obvious from the following list of layman terminology: chalky, crisp, doughy, firm, flaky, fleshy, floury, flabby, greasy, hard, juicy, lean, limp, lumpy, mucky, sloppy, slushy, slimy, soft, syrupy.

1.2 Elastic after-effects, relaxation, hysteresis

Most elastic and plastic deformation phenomena are influenced by factors such as time, previous history of the materials (internal tensions due to manufacturing conditions, etc), and temperature.†

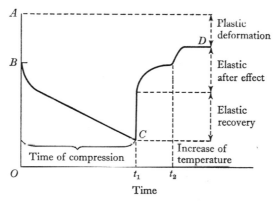

Fig. 1.3. Stages in which elastic recovery can take place. On compressing a cylindrical specimen of height OA between two parallel plates, first a spontaneous (elastic) compression AB can take place, then a slow (only partly elastic) compression BC; on release of external stress at t_1 the recovery CD may take place as indicated. In this scheme it is supposed that temperature is raised at t_2.

The slow recovery after some previous deformation which produced residual internal tensions is called the *elastic after-effect*. Fig. 1.3 shows the recovery stages after a compression test. In this example the elastic after-effect occurred in two stages, one in the normal way, the other after a short period of heating (t_2).

† For these reasons all definitions must be based on a precise specification of the experimental conditions, especially regarding time. So it can happen, that in a quick experiment a yield value is found, whereas in the same experiment of a longer duration, however, no yield value is detected. A substance which is purely elastic in quick experiments (unvulcanized rubber or polyethylene) may show a viscous or plastic behaviour in slow deformations. These are often neglected in technical literature, but will be carefully considered in this book.

Residual tensions in a material tend to dissipate; a phenomenon called *relaxation*. The time necessary for the tension to decrease to $1/e$ of its original value, the so-called relaxation time λ, is expressed by equation (1.8):

$$\tau = \tau_0 e^{-\frac{t}{\lambda}}, \tag{1.8}$$

where τ_0 is the original tension, t is time.

Another phenomenon connected with relaxation is the gradual increase in length of a specimen when it is kept under constant stress; this is called *creep*.

During elastic deformation – especially when it is accompanied by plastic phenomena – a certain amount of the work done may be dissipated as heat.

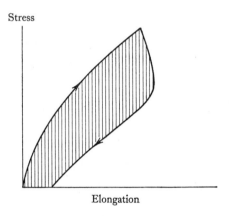

Fig. 1.4. Elastic hysteresis.

This fact will be manifested in the stress–strain curve, where a somewhat different form is obtained for the loading of the specimen from that recorded for its unloading. This point is shown in Fig. 1.4 where the cross-hatched area is a measure of the work lost. The phenomenon is called *elastic hysteresis*.

1.3 Methods of measuring

Although testing methods are often very different in different countries, the basic principles are in many cases the same. For this reason emphasis will be given to the principles only, and to methods in which the results can be expressed in terms of physical instead of technological units.

1.3.1 Testing in the solid state

Tensile strength, modulus of elasticity and permanent set unfortunately are not physical constants independent of the form of the test piece and the testing rate, due to relaxation during the testing operation. A rough comparison is usually possible if the forms and speeds under comparison do not differ too much. Test specimens (Fig. 1.5 *a*) usually consist of a bar with a reduced section in the centre (dumbell): for rubber, rings are also used, giving noticeably lower values. Typical curves are shown in Fig. 1.1. When the initial part is a straight line through the origin, its gradient is a measure of the *modulus of elasticity*.

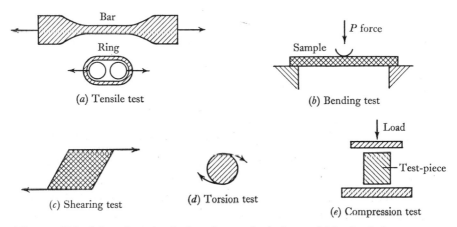

Fig. 1.5. Principles of mechanical testing methods from which physical constants can be derived.

The permanent set can easily be determined by measuring the length of the test piece after release of the stress.

Instead of measuring the stress–strain relationship by extension, other methods like bending (Fig. 1.5 *b*), shearing (Fig. 1.5 *c*), torsion (Fig. 1.5 *d*) and compression (Fig. 1.5 *e*) can be used for obtaining purely physical parameters.

Hardness tests register the depth of penetration of a pointed or rounded object into the plane surface of a solid test-piece. The material around the penetrating object undergoes shear deformation. The geometry of this deformation is complicated, and therefore it is difficult to correlate hardness with more elementary properties such as modulus or tensile strength. Since the deformation in a hardness test is mostly permanent, the basic property underlying this test is the yield strength.

Deformation may be accompanied by a change in volume as, for instance, molecules take new positions closer to each other – for example when the lattice constant in crystals is changed, or when rubber molecules crystallize by stretching (which may cause 20% volume decrease).

It is often useful to recalculate the elastic parameters for the actual cross-section by means of the equation:

$$\text{actual tensile stress} = \text{nominal stress} \times \frac{l+\Delta l}{l}. \tag{1.9}$$

This formula needs correcting, however, when the volume changes during deformation.

The coefficient of compressibility (K) is:

$$K = -(1/V)(dV/dp), \tag{1.10}$$

in which V is the volume and p the hydrostatic pressure. The relation between the three elastic constants for isotropic materials is:

$$E\frac{3(1-2\mu)}{K} = 2(1+2\mu)G, \tag{1.11}$$

where $\mu = \dfrac{\text{relative lateral contraction}}{\text{relative longitudinal strain}}$ under unilateral stress (Poisson's ratio).

Since μ measures the volume contraction on deformation, it follows that μ will be equal to 0.5 when no contraction occurs. This will be understood when it is recalled that, in liquids where E equals zero and where the molecules are so mobile that under the influence of shearing stresses no contraction occurs, the term $(1-2\mu)$ must also be zero. The closer the character of a substance approaches that of a liquid, the closer the value of μ approaches 0.5. The following figures are illustrative.

Steel	$\mu = 0.27$	Ebonite	$\mu = 0.39$
Glass	$\mu = 0.25$	Soft rubber	$\mu = 0.49$
Lead	$\mu = 0.40$	Gelatin gel	$\mu = 0.50$

Examples of further technological tests, which may be of great practical value although it is often difficult to express the results in terms of pure physical parameters, are presented in Fig. 1.6. In both tests one measures the resistance against brittle fracture (by rupture under stress). In the tear test this is done under slow conditions, and in the impact test under fast conditions.

1.3.2 Plasticity measurements

The main principles used in plastometers are presented in Fig. 1.7.

The rotary principle rests on a safe theoretical basis and can be applied to a wide range of materials with a great variety of viscosity or plasticity. A further advantage is that the test can be continued with the one and same specimen so that the influence of other factors (temperature, additions) can be measured. In the extrusion plastometers, based on capillary flow, the theory is difficult but the practical value of these instruments is sound.

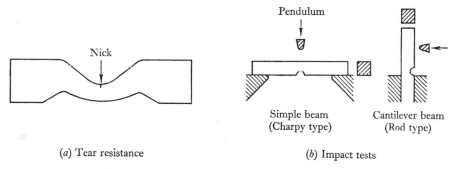

(a) Tear resistance   (b) Impact tests

Fig. 1.6. Principles of some technological tests.

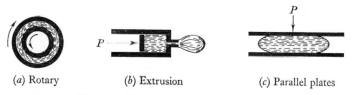

(a) Rotary   (b) Extrusion   (c) Parallel plates

Fig. 1.7. Main types of plastomers.

In the compression type of plastometer the theory is difficult. Application is restricted to rather hard materials because in soft materials the deformation is soon complete. Therefore, use of the compression type is limited.

Table 1.1 demonstrates the order of magnitude of the rates of shear obtainable in these instruments.

1.3.3 Viscosity measurements

In most viscometers the fluids flow through a capillary under their own weight. Many plastometers can also be used for the examination of viscous fluids, and then there is no fundamental difference between the two sorts of instruments. Table 1.2 presents the range covered by such equipment.

TABLE 1.1. *Comparison of the rates of shear obtainable in plastometers*

Name	Type of instrument	D_{av} in s^{-1}
Griffith	Extrusion, piston type	60
Marzetti	Extrusion, air pressure type	1–2
Dillon and Johnson	Extrusion, screw type	50
Dillon and Johnson	Extrusion, piston type	500
Williams	Parallel plate	1–2

TABLE 1.2. *Measuring range of some plastometers and viscosimeters*

Name	Principle	Viscosity range in poise
Redwood 11	Capillary	1–50
Vogel Ossag	Capillary	10^{-2}–10^{2}
Sinker	A cylinder sinks in a tube	1–10^{5}
Capillary pressure	The material flows out under a pressure of 10 atm	up to 10^{6}
Couette	Concentric cylinders	10^{4}–10^{8}
Pochettino	Two concentric cylinders move in opposite directions along the axis	10^{7}–10^{10}
Williams	Parallel plate	10^{3}–10^{8}
Ungar	The material is sheared between oppositely rotating half globes	10–10^{8}
Weissenberg	Rheogoniometer. The material is sheared between a cone and a flat plate	10^{-3}–10^{10}
Brookfield	Synchro-lectic. A cylinder rotates in a relatively large sample of fluid	10^{-2}–10^{6}

1.3.4 Others

For asphalts and resins extensive use is made of a *penetration test* (A.S.T.M. D5–25), in which a sharp needle is forced by a given weight (100 g for example) during 5 seconds into a specimen of the material. The viscosity is expressed by:

$$\eta = \frac{5.5 \times 10^9}{\text{pen}^{1.93}} \text{ poise}, \qquad (1.12)$$

where pen = penetration.

The ring and ball softening point, mostly used for asphalts, measures the temperature at which a steel ball sinks through a cylindrical sample as the temperature is increased at the rate of 5° C per min.

References to chapter 1

1   General literature on rheology:

T. Alfrey, *Mechanical behaviour of high polymers* (New York, 1948).
F. Eirich, *Rheology* (New York, 1956 (4 volumes)).
M. Reiner, *Deformation, strain and flow* (New York, 1960).
K. Weissenberg, *The testing of materials by means of the rheogoniometer* (Farol Research Engineers Ltd, Bognor Regis (Sussex), England, 1964). Also *Proc. Oxford Meeting Faraday Society* 1959 (Pergamon Press, 1960). This is an excellent treatise of general rheology as well as of measuring technique, plus literature references
J. R. van Wazer *et al.*, *Viscosity and flow measurement* (*a laboratory handbook of rheology*) (New York, 1963). This is a complete survey of measuring methods.

2 Structure of matter in relation to its elastic and plastic behaviour and failure[1]

R. Houwink, H. K. de Decker (§2.3.2) and M. van den Tempel (§2.4)

2.1 Cohesive and repulsive forces

The rheological properties of materials are primarily dependent on the energy content of the bonds between their constituent atoms (ions) and molecules although, as will appear throughout this book, there are many structural factors which over-rule completely the sort of rheological behaviour that might be expected on the basis of the nature of these bonds.

In every solid the structural units exert attractive forces upon each other. Such forces must always be conceived as the resultant of an attractive and a repulsive force. If this were not the case it would be impossible to explain why resistance is experienced, not only during elongation, but also during the compression of a solid material.

The following schematic treatment of the forces making up the atomic bond may, with the necessary changes, be applied in the cases of the various types of bonds to be encountered in the following sections. Fig. 2.1 a shows the variation of the attractive and repulsive forces σ with changing distance between the two atoms 1 and 2 according to the following formulae:

$$\text{for the attractive force:} \quad \sigma_A = \frac{mA}{r^{m+1}}, \tag{2.1}$$

$$\text{and for the repulsive force:} \quad \sigma_r = -\frac{nB}{r^{n+1}}, \tag{2.2}$$

where n, m, and A and B are constants. For the resultant force one may therefore write

$$\sigma = \sigma_A + \sigma_r = \frac{mA}{r^{m+1}} - \frac{nB}{r^{n+1}}. \tag{2.3}$$

In this equation the value of n is always much greater than that of m.

At the interatomic distance r_0 in Fig. 2.1 a, that is, at the separation at which the two atoms exist uninfluenced by external forces, the resultant of σ_A and σ_r is zero. At the positions 1 and 2 the atoms execute heat vibrations. In the displacement of atom 2 toward the right, a continually increasing is experienced until the distance r' is reached. From that moment the resistance decreases. The force σ_{max} is therefore a measure of the yield value (= tensile strength in the case of an ideally elastic body), and the distance $r' - r_0$ is a measure of the elongation at the yield value. The modulus of

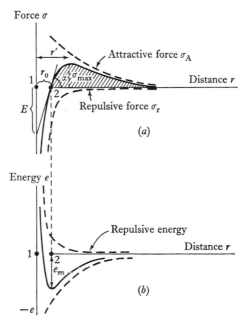

Fig. 2.1. Causes of cohesion in matter. (a) Schematic representation of the forces between two atoms 1 and 2. (b) Potential trough between two atoms 1 and 2.

elasticity is given by the distance designated by E in the figure, for according to the equation

$$\frac{\Delta r}{r} = \frac{\sigma}{E},$$ (2.4)

one may write:

$$E = r\frac{\sigma}{\Delta r} = r\tan\alpha.$$ (2.5)

The potential energy as a function of the separation of the atoms is represented in Fig. 2.1 b. The minimum is reached when the separation is r_0, and the atom may then be said to be at the bottom of a *potential trough*.

Fig. 2.1 a enables us to calculate the rupture energy e_r per pair of atoms. This energy is

$$e_r = -\int_{r_0}^{\infty} \sigma\, dr,$$ (2.6)

and is represented by the shaded area in Fig. 2.1 a.

From consideration of the oscillations of the atoms, the mean interatomic energy may be deduced as $3kT$ ($3RT$ per gram atom), where k is the Boltzmann constant, R the gas constant (1.98) and T the absolute temperature. At the melting point T_s of such a system one may thus, as a first approximation, write

$$e_s = 3kT_s. \tag{2.7}$$

From this it follows, also as a first approximation, that at any given temperature T lower than T_s the energy content of the bond will be

$$e_s - 3kT = 3k(T_s - T). \tag{2.8}$$

As a result of the heat vibrations, which take place with alternating impulses in such a way that the energy $3kT$ is continually being exceeded momentarily, local migrations will take place before the temperature has reached T_s, the actual melting point of the material (cf. p. 21). Consequently the resultant rupture energy or 'separation' energy, leading to flow is

$$e_r = 3k(T_s - T). \tag{2.9}$$

2.2   Types of bonds

2.2.1   Cohesion in general

The five types of bonds holding matter together may, broadly speaking, be classified in two main groups. The stronger ones, often denoted as *primary bonds*, have an energy content of the order of 100–200 kcal/mol; for the weaker ones, the *secondary bonds*, the energy content is of the order of 0.1–10 kcal/mol, or somewhat higher.

Table 2.1 shows some examples: the distances between the centres of the atoms are shorter for the stronger bonds.

Fig. 2.2[1] gives a schematic picture of the five types of bonds. See also Table 3.2 on p. 43.

2.2.2   Primary bonds

In the *ionic or heteropolar bonds* the atoms have completed their outer electron shell either by gaining or by loosing an electron. They are thus electrically charged in an opposite sense and consequently attract each other. In a heteropolar lattice, present in many inorganic crystals, the values of m and n in equation (2.3) are, respectively 1 and 9–11.

In the *covalent or homopolar bonds*, pairs of atoms fill their outer shells by sharing pairs of their outer electrons between both atomic nuclei. The

TABLE 2.1. *Energy content of bonds*

Bond	Primary Distance between the atomic nuclei (Å)	Energy equivalent (kcal)	Group	Secondary Distance between neighbouring groups (Å)	Molecular cohesion increment per group (kcal)
C—C (aliph)	1.54	81	—CH₃	3–4	1.78
C—C (arom)	1.36	—	=CH₂	3–4	1.78
C=C	1.33	145	—CH₂—	3–4	0.99
C≡C	1.20	200	=CH—	3–4	0.99
C—O	1.42	87	—OH	3–4	7.25
C=O	1.21	174	=CO	3–4	4.25
C—H	1.08	100	—CHO	3–4	4.70
C—Cl	1.76	78	—COOH	3–4	8.97
C=S	1.60	129			
H—H	0.75	103			

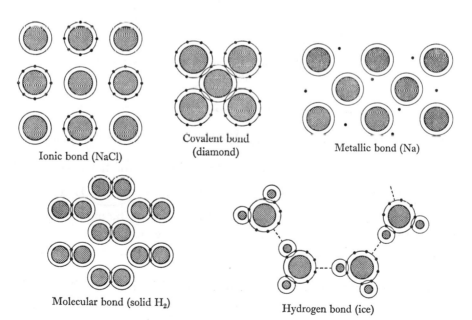

Ionic bond (NaCl)

Covalent bond (diamond)

Metallic bond (Na)

Molecular bond (solid H₂)

Hydrogen bond (ice)

Fig. 2.2. The five types of bonds (ref. 1).

C—C bond and the C—H bond, so preponderant in organic chemistry, are examples of this type.

In the *metallic bonds* all atoms share all the outer electrons and these free electrons are the reason for the electrical conductivity of metals. This 'gas' of electrons in a metal can be very dense, more than 10^{22} per cubic centimetre, or of the order of magnitude of the number of molecules in the same volume of a liquid.

2.2.3 Secondary bonds[2]

The secondary or intermolecular bonds are mostly responsible for weaker structures that can easily be ruptured, giving rise to viscous and plastic deformations.

The intermolecular bonds, usually called van der Waals' bonds,† arise from the displacement of charge within electrically neutral atoms or groups of atoms in various ways.

(a)   (b)

Fig. 2.3. Two different possibilities in the mutual attraction of dipoles. (a) Parallel. (b) Non-parallel.

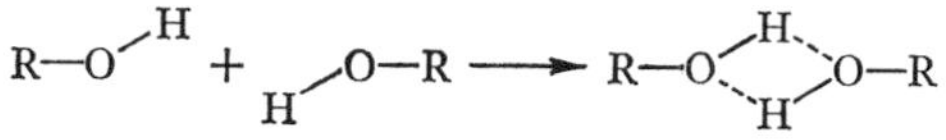

$$R—O\overset{H}{\diagdown} + \diagup O—R \longrightarrow R—O\overset{H}{\diagdown}\diagup O—R$$
$$H\qquad\qquad\qquad\qquad H$$

Fig. 2.4. The formation of hydrogen bonds between two alcohol molecules.

1. *Permanent dipoles* occur if the centres of gravity of the positive and negative charges do not coincide, for instance H_2O, $\left(\overset{+}{\underset{+}{}}\ \ -\right)$. A sort of permanent magnet is formed (see Fig. 2.3) whose magnetic moment (dipole moment) is equal to the charge times the distance between the centres of the charges. For this bond, the values of m and n in equation (2.3) are 3 and > 3 respectively.

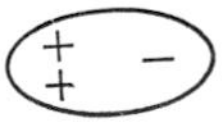

The *hydrogen bond* is a special type of dipole which is responsible for the association of molecules as in water, alcohols and carboxylic acids. Fig. 2.4

† There exists a confusion in nomenclature. Some people comprise all secondary bonds together under the name of van der Waals' bonds, others restrict this name to the dispersion forces.

shows the formation of such a hydrogen bond, due to the interaction of OH groups. With an energy content of between 4 and 10 kcal/mol the hydrogen bond is the strongest of the secondary bonds.

2. *Induced dipoles*, which can be generated in neutral molecules by permanent dipoles, also lead to mutual attraction. They may be permanent or alternating, according to the character of the inducing source. If this is a permanent dipole for instance, the attraction may change during deformations. If the induced dipole is due to an ion, the values of m and n are 4; if due to another dipole the values are 6.

3. The *dispersion forces*. During the movement of the electrons around the positive nuclei, continuously changing dipoles are formed which have an inducing effect on neighbouring atoms resulting in a mutual attraction. These forces often play a considerable part in the cohesion of matter (see Fig. 2.3 *b*).

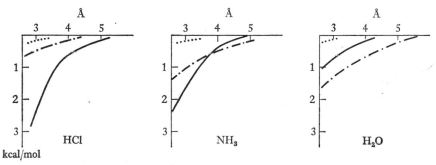

Fig. 2.5. Contributions of the various kinds of effects to the secondary bonding energy (see ref. 2). ... induced dipole, -·-·-·- permanent dipole, — dispersion.

2.2.4 Interaction of the various forces

Interactions between these various kinds of forces lead to the overall van der Waals' attraction. Fig. 2.5 shows how the contributions of each of the various effects to the total secondary bonding energy can vary considerably in magnitude for different substances.

In Fig. 2.6 is shown the attraction of diethyl ketone by water. The dipole–dipole attraction is indicated by an arrow. The OH and CO dipoles induce dipoles, whereas each particular atom contributes its share to the dispersion forces.

The importance of the strength of the secondary bonds for polymers is discussed on pp. 22 and 43. It will appear that – in absence of complications due to crystallizing phenomena – the following rough classification can be made:

rubbers have a cohesion increment of 1–2 kcal/mol per chain unit

plastics have a cohesion increment of 2–5 kcal/mol per chain unit
fibres have a cohesion increment of > 5 kcal/mol per chain unit

Summarizing the role played by various bonds in the cohesion of a solid material, we can examine the example of nylon (polyamide). This material hangs together by the following forces:

1. Inside the long molecules of the polyamide, homopolar bonds link the atoms together.

Fig. 2.6. Dipole–dipole attraction of diethyl ketone and water (parallel adjustment).

2. Between the different molecules, a number of secondary bond types operate: (*a*) the dispersion effect between any group in one molecule and all parts of other molecules that are close to it; (*b*) the dipole effect between C=O or N—H groups belonging to different molecules; (*c*) the hydrogen bonds in which the H atom belonging to an N—H group is in 'resonance' with (or partially belongs to) an adjoining C=O group of a neighbouring molecule.

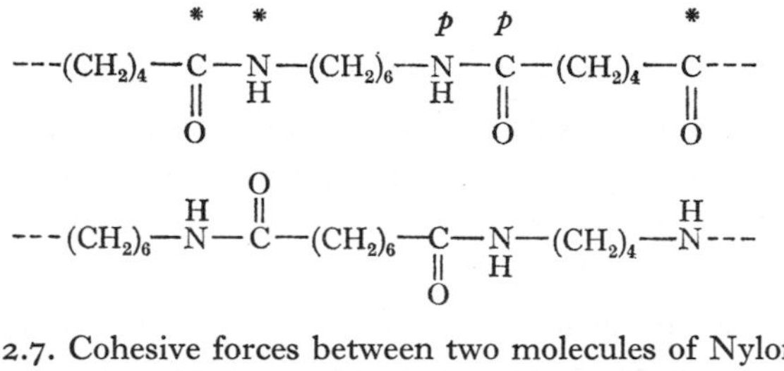

$$---(CH_2)_4-\overset{\ast}{C}-\overset{\ast}{\underset{H}{N}}-(CH_2)_6-\overset{p}{\underset{H}{N}}-\overset{p}{C}-(CH_2)_4-\overset{\ast}{C}---$$
$$\qquad\qquad \underset{O}{\|}\qquad\qquad\qquad \underset{O}{\|}\qquad\qquad \underset{O}{\|}$$

$$---(CH_2)_6-\underset{H}{N}-\overset{O}{\overset{\|}{C}}-(CH_2)_6-\overset{\|}{C}-\underset{H}{N}-(CH_2)_4-\overset{H}{N}---$$
$$\qquad\qquad\qquad\qquad\qquad\qquad\underset{O}{\|}$$

Fig. 2.7. Cohesive forces between two molecules of Nylon 66.
* = Hydrogen bond; *p* = dipole effect.

Fig. 2.7 shows small sections of two molecules of Nylon 66 in close proximity and allows the reader to locate all the cohesive forces mentioned.

It has been calculated that for large ionized particles the minimum of potential energy may lie at a distance which is of the same order of magnitude as the radius of the particle. On this basis it may be seen that the resultant attractive energy would be still of the order of kT, even when the distance between the particles is of the same magnitude as the radii of the particles. This means that at this distance the attraction might be strong enough to maintain a coherence between particles which would otherwise be separated by Brownian movement, an important conclusion which will for example be applied in the case of clay (p. 417).

The secondary bonds play an especially interesting part in colloid chemistry, since all kinds of group formation recognized in that science must be traced back to the action of these bonds. We mention merely *solvation* (the 'gluing' of a liquid to a solid substance), *association* (the junction of certain kinds of molecules to form larger groups, e.g. $nH_2O \dashrightarrow (H_2O)_n$) and the formation of *swarms*, very loose groups in which the cohesion is disturbed even by the process of flowing.

2.3 Cohesion and rheological properties

As pointed out already, the rheological parameters of modulus (and viscosity), yield value and ultimate strength can basically be derived from the potential trough of Fig. 2.1 (p. 13). In practice however, many complications occur due to inherent imperfections in matter or to structural changes during deformation. Crystals can be strengthened by interstitial atoms but they are weakened by glide planes. Polymers can be strengthened by crystallization due to the orientation of the molecules but the viscosity of polymer solutions can be lowered by such orientation. These complications will be considered in Chapter 3. Although their influence often over-rules the cohesive forces within matter, a short treatment neglecting these complicating factors will be given here in order to make it clear where we stand.

2.3.1 The modulus and the yield value

A potential trough (or energy trough) may be drawn for each pair of atoms. Considering together all the troughs of the atoms surrounding a given atom A, one may speak of the resultant potential trough with a depth e_R. The atom vibrates in all directions in this trough with a frequency of the order of 10^{13} per second due to thermal agitation, but it is unable to escape from the trough as long as e_R is greater than the energy impulse corresponding to the largest amplitude of vibration occurring. If, however, any energy impulse due to thermal vibration becomes greater than the energy e_R at a certain moment, the atom will attempt to migrate into a neighbouring trough. By a multiplication of this process a plastic deformation may take place. The mechanism of this deformation is quite analogous to that described on p. 6 in the case of relaxation.

In a crystalline material one can expect congruent potential troughs in each crystallographic direction (Fig. 2.8 a); accordingly in a crystal the yield value can be different in various directions.

If, however, we consider the distribution of potential energy along a random section of an amorphous body, we will encounter troughs of differing depths, e_{R_1}, e_{R_2}, etc., as shown in Fig. 2.8 b. For certain amorphous substances one may even expect to find troughs where $e_R = 0$. The conclusions to be drawn from this very schematic picture are obvious. The tension necessary

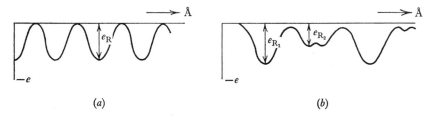

(a)   (b)

Fig. 2.8. (a) Potential troughs along a lattice plane of a crystalline body. (b) Potential troughs in a random section of an amorphous body.

to draw the various atoms out of their troughs will vary considerably; for troughs where $e_R = 0$, it will be zero, so that such a material will flow spontaneously. Every increase in tension will cause new atoms to migrate and a state of equilibrium will be attained only when a further increase of tension causes no further increase in the number of 'flowing atoms'.

Because of the congruence of the potential troughs for a crystalline body, and the fact that the energy necessary to remove each atom from its trough is about the same, the yield value of such a body is rather sharp, and by comparison the yield value of an amorphous body is unsharp. It will be seen that the stresses necessary to remove the atoms from their potential troughs are much lower than might be expected from theoretical considerations.

The resistance to deformation is governed by the curvature of the energy trough and is expressed by the modulus of elasticity in tensile tests and by the shear modulus in shear tests. As the curvature at the base of the troughs is different from the slope elsewhere, one expects a modulus which is constant over small deformations. Since in practice the modulus is a statistical average over a great number of potential troughs in which the displacements of all the atoms are not exactly simultaneous, the modulus can nevertheless be constant over a rather large deformation.

Since the vibrations of an atom are irregular, that is vibrations with small amplitudes may be followed by those with large amplitudes, it is evident that the rate of testing will have an influence on the magnitude of the yield value (and also of the tensile strength). The longer a given tension is maintained, the greater the chance will be that the atom will carry out a vibration with an amplitude sufficiently great to cause it to leave its potential trough. This influence of the rate of testing will be observed especially with amorphous materials, since in such materials there will always be atoms for which e_R is of about the same magnitude as the mean energy impulse. The chance that such an atom will escape depends therefore upon whether or not an energy impulse greater than e_R occurs, a factor which is governed by the laws of probability. At first sight one would be inclined to conclude from this consideration that the yield value of crystalline bodies with deep potential troughs is independent of the rate of testing. Actual observations seem however to point to the opposite conclusion. It is well known that under the influence of time even the atoms of such a crystal can change their places without any stress being applied (migration of atoms, diffusion of two metals into each other). The atoms can evidently leave their troughs under the influence of thermal agitation alone.

For silver, with a melting point of 960° C, the mean time necessary before the migration of a certain atom under ordinary condtions is 100 million years; at 927° C, however, this time becomes 0.001 s.

Since this migration must be taken into account, and since it will be increased by any shearing stress applied, the conclusion must be drawn that, *strictly speaking, no yield value can exist except at absolute zero.* Since, however, in practical (rapid) experiments a yield value can be observed, a *practical* definition of the yield value has to take into account the conditions of the test (see p. 45).

2.3.2 Tensile strength (by H. K. de Decker)

The tensile strength of a solid can basically also be derived from the potential trough of Fig. 2.1. As appears, however, from Table 2.2, in many cases the experimental values are 1/500 to 1/1000 of the theoretical strength. It will be shown (p. 36) that this is attributed to glide planes and/or faults within the materials.

Table 2.2 shows theoretical and observed strength for some materials.

TABLE 2.2. *Discrepancies between theoretical and experimental strength of some materials*

Substance	σ_{max} calculated in kg mm^{-2}	σ_{max} observed in kg mm^{-2}	Type of bonds involved
1. NaCl crystal, normally	200–400	0.6	NaCl (ionic)
NaCl crystal, under special condition		160	
2. Glass, normally		3.5–8.5	Si—O—Si (heteropolar)
Glass, under special condition	1100	360–630	
3. Phenol-formaldehyde resin (C stage) at −195° C	4300	7.8	C—C (homopolar)
4. Cresol-formaldehyde resin (C stage) at −195° C	3800	3.8	C—C (homopolar)
5. Paraffin wax	50–100	0.05	$>$CH$_2$---CH$_2<$ (secondary)
6. Polyethylene oriented fibre	50–100	50	$>$CH$_2$---CH$_2<$ (secondary)

In cases 1 to 4 inclusive, *primary* bonds (ionic or heteropolar or homopolar) are taken as the basis for strength. As we shall see in Chapter 12, the special conditions under which NaCl or glass can show a strength close to the theoretical values are mostly connected with the total absence of flaws in the crystal structure (NaCl) and in the surface (glass). The complete uniformity of the applied stress is also vital.

The reason that the thermosetting resins show a strength so very much below the theoretical calculation is simply the assumption used in this calculation. The C—C bond occurring between adjoining benzene rings ($-\phi-$CH$_2-\phi-$) was used as the basis; the calculation simply adds up all the bonds of this type occurring in a plane perpendicular to the pulling force, using equation (2.10), in which N = Avogadro's number.

$$\sigma_{max} = \frac{(\text{bond strength}) \times (\text{bonds per mm}^2)}{(\text{density}) \times (\text{bonds per mole}) (\text{elementary displacement distance}) \times N}$$

$$(2.10)$$

One assumes that all the bonds in the unit cross-section participate equally and that they all have to be displaced one elementary distance (about half the atomic distance) simultaneously to make the sample yield. In the case of the resins such assumptions are not realistic. It is clear that these amorphous materials will yield (and break) because very few of the relevant bonds are in an exposed position; they carry the total load and their failure initiates the failure of the entire sample. A comparison of the theoretical and practical values shows that probably only one out of every 1000 ϕ—CH_2—ϕ bonds carries the load. Another theory (p. 335) ascribes the low strength to 'Lockerstellen'.

As follows from Fig. 2.9, longer molecules have a larger cohesive surface and consequently produce a greater strength than shorter or globular ones. This will be the more true, the better they are oriented.

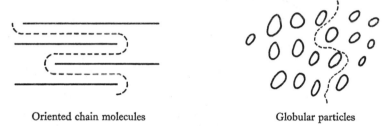

Oriented chain molecules   Globular particles

Fig. 2.9. Cohesive surface (dotted) in materials with oriented linear and with globular molecules.

As a first approximation the following formula may be postulated for materials with linear molecules:

$$\sigma_{max} = kn\alpha, \qquad (2.11)$$

where k = constant, n = number of chain elements, α = molecular cohesion per monomer group.

Although the proportionality according to equation (2.11) is not confirmed experimentally, the increase of strength with increasing chain length is clearly detectable in Fig. 2.10(a). It is seen in Fig. 2.10(b) that similar behaviour is found on a macro scale for spun yarn.

Materials number 5 and 6 of Table 2.2 offer an interesting comparison. They both have pure linear hydrocarbon molecules and their cohesion is through secondary bonds, but paraffin wax has a molecular weight of about 200, whereas the polyethylene has an average molecular weight of 200000 and can be strongly oriented and crystallized by drawing into a fibre.

Using equation (2.9) a theoretical strength of 50–100 kgf/mm² is found for both, assuming the same secondary bonds between —CH_2— or —CH_3 groups. Material 5 thus has a real strength equivalent to the effect of only one in every 1000 secondary bonds available; rupture is initiated by the yielding of these few 'load-carrying' bonds.

On the other hand, it appears that material number 6 has a strength virtually equal to the theoretical one, and that the strength of one secondary

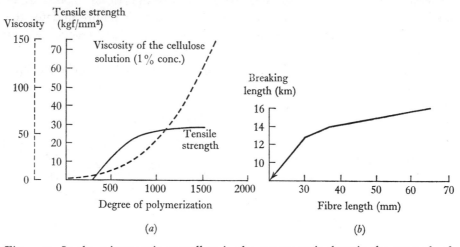

Fig. 2.10. In the microscopic as well as in the macroscopic domain the strength of substances built up with oblong particles increases with the length of these particles. (*a*) The strength of cellulose increases with increasing molecular length. (*b*) The strength of a thread increases with increasing fibre length.

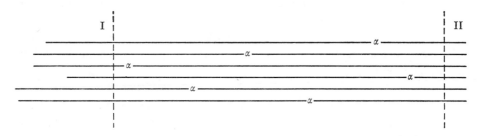

Fig. 2.11. Diagram of a fully oriented polyethylene fibre. Cohesion between regions I and II is by secondary bonds at α between end groups.

bond per oriented macromolecule in a cross-section is participating. One has to assume that this one participating bond is that between end groups (α in Fig. 2.11).

Thus the maximum strength obtainable from secondary bond cohesion can be practically realized by high molecular weight and maximum orientation in the direction of pulling.

2.3.3 Viscosity

The molecular picture of viscous flow is obtained by remembering that the thermal motion of individual molecules can make them relatively independent of one another. This is, of course, particularly true above the melting point temperature, where a large number of molecules have thermal kinetic energy exceeding the amount needed to escape each other's attraction. This has already been mentioned in §2.1 and equation (2.7) showed that the 'critical' energy a molecule must surpass to move freely is $e_s = 3kT_s$, where T_s is the melting temperature.

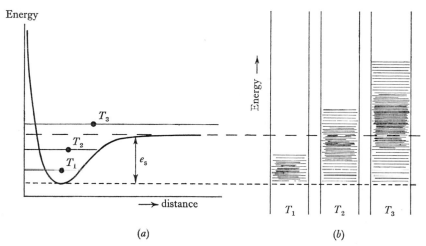

(a) (b)

Fig. 2.12. (a) Average position and (b) thermal energy distribution of molecules at temperatures T_1: far below the melting temperature; T_2: slightly below the melting temperature; T_3: above the melting temperature.

Although molecules in a 'true' liquid (above the melting point) are randomly moving around and do not fall into any ordered pattern from the *static* point of view, it is nevertheless quite possible to picture their *average* arrangement in a *statistical* manner. This also has the advantage of showing how the same type of arrangement exists at *all* temperatures, and how some degree of viscous flow will always occur under stress, be it sometimes so little that it can hardly be measured.

In Fig. 2.12(b), the density of cross-hatching represents the numbers of molecules at that particular energy level. At the low temperature T_1 the vast majority of molecules has kinetic energy below the critical level e_s, and viscous flow will be practically unnoticeable. At the temperature T_2, almost at the melting point (amorphous solidfication is assumed), a substantial number of molecules will be able to move, and viscous flow becomes quite evident. At T_3, well above the melting point, free flow occurs because almost all molecules move freely. The *average* distance between molecules at these three temperatures is indicated in Fig. 2.12(a).

It becomes clear that viscous flow is dependent on the fraction of molecules that is freely moveable, or has a kinetic energy above the critical level e_s. That fraction is known to be proportional to $\exp{(-e_s/kT)}$. It can be seen, therefore, why the viscosity, or resistance against flow, will show a temperature dependence in the form of equation (2.12).

$$\eta = A \exp{(E_V/RT)} = A \exp{(E_V/NkT)}. \tag{2.12}$$

The absolute levels of e_s in Fig. 2.12 and that of E_V/N in equation (2.12) are different, because Fig. 2.12 is idealized and the mutual interaction of molecules changes e_s.

In Chapter 4 a somewhat different approach for the derivation of equation (2.12) will be discussed; there it will also become clear that this reasoning holds for any molecule, including the very large ones, provided one looks not at the entire molecule but at 'kinetic units', those segments of large molecules that move relatively freely.

It will further be shown that, for small molecules, E_V in equation (2.12) is about one-third of the heat of vaporization. This indicates that the elementary step in viscous flow is a one-dimensional event related to the removal of a molecule from its surroundings and the formation of a 'hole'. Thus, E_V is also correlated with the cohesion of matter or the specific heat.

For polymers in solution or in the melt condition, the movement of kinetic units depends on a jump frequency J and a statistical factor F. One can write $\eta = F/J$. The statistical factor F depends on the molecular weight and the entanglements, and expresses the co-ordinated restriction of each kinetic unit by its primary bonding. The jump frequency J depends on the temperature (mobility) and the free volume (space to move).

Fig. 2.13 shows for polystyrene that the 'constants' A and E_V of equation (2.12) are somewhat variable; it also shows the influence of a diluent (dibenzyl ether) on the mobility and free volume of the polystyrene kinetic units, for example, the viscosity at 127° C (400° K) is about 300000 if 5 % diluent is present, but it sinks to 30000 for 12 % diluent.

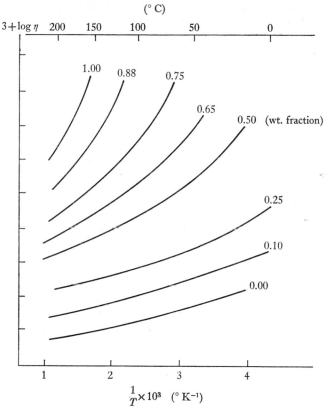

Fig. 2.13. Plot of log η versus $1/T$ for solutions of polystyrene in dibenzyl ether at the weight fractions of polymer shown on the curves (mol. wt. — 70000).

2.4 Structure of multiphase systems (by M. van den Tempel)

Many materials of practical importance consist of an intimate mixture of at least two different phases. A phase is defined as a region in which all the properties (such as composition, pressure, etc) are *uniform*, and which is bounded by a closed surface on which these properties change abruptly to those of the neighbouring phase. The difficulty in this definition is connected with the term 'abruptly'; the change in properties is never restricted to a mathematical surface of zero thickness, but occurs over a distance of at least a few molecular diameters. In most cases, if the properties which we shall require to remain uniform in a phase include the electrical state (charge and potential), the transition across the phase boundary is then even more gradual. If the dimensions of the regions constituting a phase become smaller than the transition region, the phase concept breaks down. Examples of such systems are solutions of macromolecules and micellar solutions of soap and detergents. The structure of these systems cannot be described in terms of a mixture of phases; the formalism required in such cases is discussed in §7.2.

It is evident that the minimum size of a particle constituting a separate phase, depending on the properties of the material, varies between less than 100 Å† and several microns. Most multiphase systems contain particles in this range of sizes, dispersed in a continuous phase. Recent developments in construction materials have resulted in two-phase systems in which both phases are continuous. Such materials are elastic solids; their properties will be discussed in §7.2. The remaining part of this section is restricted to a discussion of dispersions.

The rheological properties of dispersions vary between those of Newtonian liquids and those of brittle solids, depending upon a number of factors which we shall now discuss. Many dispersed systems show a typical rheological behaviour, which is loosely called 'plastic' and which is often of considerable importance in their applications. As was discussed in §1.1, plasticity means that the material behaves more or less like an elastic solid at small deformations (i.e. generally under the influence of small forces), whereas the properties become more liquid-like at larger deformations. Even if a dispersion does not show pronounced 'plastic' behaviour, it will generally exhibit properties like non-Newtonian flow, thixotropy, false-body or pseudo-plastic behaviour. The difference between these properties and plasticity is not a difference in kind but in degree (see §7.2), and therefore the following considerations are also valid for such dispersions.

The solid-like behaviour, that is, the possibility of elastic deformations, shows that a plastic dispersion must possess a definite structure which determines its form in the absence of external forces. The relatively small stress needed to produce moderate deformation suggests that the structure is much more open than in ordinary solids, in which strong bonds connect the atoms into a rigid three-dimensional structure. On the other hand, the small region of *elastic* deformation suggests that the bonds between the structural elements are of the same type as in an ordinary solid, that is, due to chemical interaction, to electrostatic or van der Waals' forces. There is no long-range elasticity showing more or less complete elastic recovery even after very large deformations.[3] It follows that the structure of dispersed systems, in relation to the rheological properties, must be described in terms of certain structural

† This is twice the shortest distance over which the presence of a phase boundary is noticeable.

elements and the forces acting between those elements. The structural elements are, evidently, the dispersed particles themselves or aggregates of these particles formed under the influence of forces acting between them. The various types of forces will now be briefly discussed.

2.4.1   Van der Waals'–London interaction

The van der Waals'–London attractive force between two isolated atoms decreases with a high power of the interatomic distance (7 or 8, depending on conditions). The force between two particles, each containing a very large number of atoms, is to a good approximation made up of the sum of the attractions between each pair of atoms. The result is that, for such particles, the attractive force decreases very much less rapidly with increasing distance: for spherical particles the force is inversely proportional to the *square* of the interparticle distance (for not too large distances). This is the same dependency as for electrostatic (Coulomb) interaction, and it is therefore incorrect to postulate that electrostatic forces have a longer range than the van der Waals'–London interaction.

The magnitude of the van der Waals'–London interaction depends on the nature of the particles and of the material present in the gap between the particles.[4] It is surprising, however, that this dependency is not very strong; for most materials where the values of the necessary parameters are known it is found that the magnitude of the interaction does not vary by more than about one order of magnitude. It is helpful to remember that the interaction *energy* between two spherical particles, placed at a mutual distance equal to their diameter, is of the order of one kT unit (i.e. comparable to the energy of thermal motion).

2.4.2   Electrostatic interaction

At any interface between two phases there will in general exist an electric field, due to unequal distribution of charges (electrons or ions) over the two phases. Moreover, many dispersions contain ionized surface-active materials, certain ions of which will be absorbed on the particle surface whereas other ions remain in solution. In any case, the dispersed particles will usually carry a net charge, and an equal but opposite charge must be built up in the continuous phase because the whole dispersion is electrically neutral. The ions in the continuous phase are attracted by the opposite charges on the particle, but due to their thermal motion they form a kind of diffuse cloud extending from the particle surface over a certain distance into the continuous phase. The system consisting of the charges on the particle together with the diffuse layer of opposite charges in the solution is called the *diffuse electrical double layer*, see Fig. 2.14. The thickness of the diffuse layer is mainly determined by the ionic concentrations in the continuous phase; it varies between less than 10^{-6} cm in aqueous electrolyte solutions to more than 10^{-3} cm in non-polar liquids with very low concentrations of ions.†

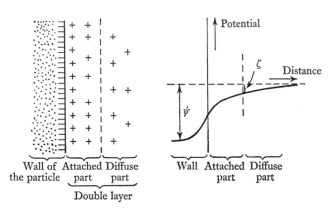

Fig. 2.14. Constitution and potential of the double layer for a negatively charged particle.

Each particle, together with its diffuse double layer, forms an electrically neutral entity, and there is no electrostatic interaction between such entities if their mutual distance is larger than twice the thickness of the diffuse double layer. Electrostatic interaction starts when the particles come so close together that partial overlapping of the diffuse double layers occurs. It has been shown[5] that this must result in a repulsive force in nearly all cases; only under very special conditions can the electrostatic interaction between *non-identical* particles become attractive over a certain range of interparticle distances.[5] This assumes that the charge of a particle is distributed uniformly over its surface. Nature has, however, provided us with colloidal particles with a non-uniform distribution of surface charge. Crystalline platelets present in certain types of clay (e.g. montmorillonite) may carry a negative charge on their flat surfaces and a positive charge on the edges. The resulting strong electrostatic attraction gives rise to a rigid aggregate structure.[6] In most

† In many cases, the Bingham equation (1.6) (p. 4) describes the rheological behaviour of such ionic suspensions and the yield value explains why such materials as clay are suitable for modelling (see also p. 419).

other systems, however, the surface charge must be assumed to be distributed uniformly.

If two particles are sufficiently close together to cause some overlapping of their diffuse double layers, the charges on the particles are still partly screened off from one another by the presence of opposite charges in the gap between the surfaces. This causes the electrostatic interaction between dispersed particles to be very much less than the Coulomb interaction between similarly charged particles in vacuum. Moreover, the interaction depends more strongly on the interparticle distance than the Coulomb force; to a good approximation the force decreases exponentially with increasing distance instead of inversely proportional to the square of the distance as in Coulomb interaction.

It is important to notice that the electrostatic repulsion is not strongly dependent upon the particle charge, or on the potential difference between the particle surface and the continuous phase. A potential difference of something like 30 or 40 millivolts is sufficient to cause appreciable electrostatic interaction, and a particle charge as low as 10^{12} elementary charges (i.e. monovalent ions) per cm^2 of surface area will give rise to a potential difference of that magnitude in many systems of practical importance.

An apparent paradox arises when the electrostatic repulsion between similar particles is compared in aqueous and non-polar media. Both the lower dielectric constant and the low ionic concentration in the non-polar solvent would result in enhanced electrostatic repulsion, compared with an aqueous system where most of the particle charge is screened off by the ions in the diffuse part of the double layer. Nevertheless, electrostatic repulsion between particles is nearly always negligible in non-polar media, whereas it is usually the main stabilizing factor in aqueous systems. The reason is that, in a dispersion of reasonable concentration, the particles are already so close together that further decrease of their distance has hardly any affect on their potential energy.[7]

2.4.3 Other types of interaction

If a dispersion contains dissolved macromolecules in the continuous phase, these will have a strong effect on the particle interaction and hence on the rheological properties. Although this effect is of considerable practical importance, our understanding of the phenomena involved is still very limited. Absorption of part of the macromolecule on the surface of the dispersed particles will generally occur. If the solution of macromolecules is very dilute, many of these molecules may be absorbed on two or more particles. In this way a continuous network of chain molecules can be formed, in which dispersed particles act as junctions. Such a network would possess elastic properties.† At higher polymer concentrations the tendency seems to be for each molecule to be absorbed on a single particle. This would prevent mutual approach of the particles to distances where an elastic network structure might be formed by aggregation, and is the reason why fat globules in milk are prevented from coagulating by an absorbed layer of protein.

This stabilization against aggregation due to absorbed macromolecules is connected with the tendency of each molecule to have a maximum number of possible configurations available. Mutual approach of particles carrying such absorbed molecules would decrease the number of possible configurations, and hence the entropy of the system. This is equivalent to the existence of a repulsive force between the particles. It is of interest to observe that 'entropic stabilization' requires only the absorption of flexible chain molecules, which need not be very long. Even absorbed chains containing not more than 20 carbon atoms may prevent mutual approach to distances less than about twice the chain length, that is, about 5×10^{-7} cm. Stability against aggregation, however, requires that the particles are prevented from approaching each other to distances where the van der Waals' attraction becomes appreciable. Since the magnitude of the van der Waals' attraction is proportional to the particle size, and becomes noticeable at a distance of the order of the particle size, this means that entropic stabilization against aggregation requires the absorption of macromolecules having a chain length comparable to the particle size.

The literature contains many references to another type of stabilizing mechanism, which relies on the change of the properties of a liquid in the vicinity of a surface. It has been held that such changes may even be perceptible at distances as far as 10^{-5} cm from the surface, especially in aqueous systems. In particular, the properties of aqueous clay suspensions have sometimes been explained in terms of thick 'hydration' layers around each clay particle. More recent evidence shows that the structure and properties of a liquid near an interface are only affected in a layer having a thickness of a few molecules (*ca.* 10^{-7} cm), and it is not known how such a structural change could affect the stability of a dispersion. It is probably safe to postulate that all interaction forces between dispersed colloidal particles extending beyond a few times 10^{-6} cm must be either van der Waals'–London or electrostatic interaction.

† Its porous but coherent structure would facilitate filtration. The cleaning of sewage effluent and the structure of soil can be improved by adding small amounts of materials such as polyacrylamides.

There is one other type of particle interaction which can have an appreciable effect on the mechanical properties of dispersions, but which acts only in flocculated systems. The addition of a very small amount of liquid which is immiscible with the liquid forming the continuous phase may give rise to the formation of 'necks' around the contact region of particles. The condition for this to occur is that the surface of the particles is wetted preferentially by the liquid present in small amounts. This results in stronger bonds between the particles, and therefore in a higher yield value of the dispersion.[8] The effect is used in modifying the consistency of putty, by the addition of a trace of water.

2.4.4 Aggregation of particles

The total energy of interaction of two dispersed particles, as a function of their distance, is made up of the sum of all the contributions of the various types of interaction. A more quantitative treatment can only be given for van der Waals' and electrostatic interaction, because our knowledge of the other types of interaction is still insufficient.

Electrostatic repulsion decreases more rapidly, with increasing interparticle distance, than van der Waals' attraction (exponentially and proportional to the second or third power of distance respectively). This means that, even if repulsion is dominant at some (small) distance, attraction will usually prevail at larger distances. A typical example of the shape of the curve showing interaction energy as a function of distance is shown in Fig. 2.1. The deep energy minimum at very small distances between the particles is called the 'primary minimum', whereas the shallow trough at larger distances is the 'secondary minimum'. When the distance between two particles corresponds with that of the 'secondary minimum', the system is in mechanical equilibrium in the sense that a force must be applied in order to increase or decrease the distance. Recent measurements of this equilibrium distance are in very good agreement with predictions from theory.[9] Although these measurements were carried out on soap films, they provide convincing evidence for the presence of a secondary minimum in the potential energy curve in many aqueous dispersions.

The amount of energy needed to separate particles flocculated in the secondary minimum (i.e. the depth of the minimum) has not been measured directly. If this energy becomes less than several times the energy of thermal motion (say $< 5kT$), only a fraction of the particles will be in a flocculated condition, and a 'dynamic equilibrium' will be set up between free and flocculated particles. Such a case has been studied[10] and the results were used for estimating the magnitude of van der Waals' forces between hydrocarbon particles in water. The aggregation is, in this case, easily reversible. The depth of the secondary minimum can, however, be much larger; it is obvious that this will affect the strength of the aggregates and hence the force necessary for their disruption on deformation of the system.

The amount of energy needed to induce aggregation in the primary minimum is determined by the height of the maximum in the potential energy curve. In many aqueous systems electrostatic repulsion is sufficient to produce an energy barrier which completely prevents flocculation in the primary minimum (i.e. with actual particle contact), but we have seen that even in these systems reversible aggregation will usually occur. In non-polar liquid attraction prevails to very small interparticle distances because repulsive forces are either absent or very short-range.

It follows from these considerations that the normal state of a dispersion is one in which the particles are aggregated more or less reversibly. For reasons that are not well understood, these aggregates usually elongate to form chain-like structures. In a completely flocculated suspension these chains may form a highly interlinked network, extending through the entire volume. The network of particle chains will spontaneously transform to a state where its volume is as small as possible (i.e. the maximum number of contact points between particles is formed), but in many systems this appears to be a very slow process manifesting itself in the phenomenon called syneresis.

If attractive forces predominate, an aggregate will be formed in a quiescent suspension under the influence of Brownian motion of the dispersed particles. The rate of the aggregation process depends mainly on the concentration of particles. To give an idea of the time scale involved, it may be noted that flocculation is fairly complete after 1 second in an aqueous dispersion containing 10^{11} particles/cm^3, if there are no repulsive forces to slow down the aggregation process. In more concentrated suspensions aggregation proceeds even more rapidly.

References to chapter 2

1   Reference is made to a series of excellent papers on 'Structure of matter and rheological behaviour' in *Scientific American* **217**, 619–176 (1967). The authors are: C. Stanley Smith, A. H. Cottrell, J. J. Gilman, R. J. Charles, H. F. Mark and A. Kelly. Figures and concepts were taken from these papers.
2   G. J. van Amerongen in *Elastomers and plastomers* **1**, 194 (Ed. R. Houwink; Elsevier, 1949).

3 H. M. James & E. Guth, *J. Chem. Phys.* **11**, 455 (1943).
4 E. M. Lifshitz, *J. Exper. Theor. Phys.* **29**, 94 (1955).
5 V. B. Sandomirskii & V. P. Smilga, *Soviet Phys., Solid State* **1**, 275 (1959).
6 H. van Olphen, *J. Colloid Sci.* **19**, 313 (1964).
7 W. Albers & J. Th. G. Overbeek, *J. Colloid Sci.* **15**, 489 (1960).
8 P. G. Howe, D. P. Benton & J. E. Puddington, *Can. J. Chem.* **33**, 1189 (1955).
9 J. Lyklema & K. J. Mysels, *J. Amer. Chem. Soc.* **87**, 2539, (1965).
10 S. N. Srivastava & D. A. Haydon, *Trans. Far. Soc.* **60**, 971 (1964).

3 Specific structures and rheological properties

R. Houwink

3.1 Solid matter

It has been shown in Table 2.2 that the experimentally calculated tensile strength does not confirm the theoretical calculations. These discrepancies which are often found in the rheological behaviour of solids can, as a first approximation, be traced back to irregularities or 'faults' in the structure of the material.

3.1.1 Crystalline materials

In crystalline materials (like metals), plastic flow is most likely to occur along the closest packed planes. The observation that slip mostly occurs under stresses far below that which might theoretically be expected, is connected with the presence of definite structural imperfections, called dislocations

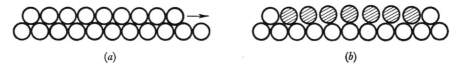

(a)   (b)

Fig. 3.1. Slip along crystallographic planes. (a) Two close-packed layers and (b) with a dislocation.

(Fig. 3.1 b). A dislocation, as shown here, is characterized by the fact that the number of atoms (actually: lattice rows perpendicular to the plane of the drawing) over a certain length of the glide plane is one row less than the number in the same length of the underlying plane, a kind of 'nonius' arrangement. As a result the dislocation can move through the lattice under a relatively low shear force, as the atoms at the 'front' of the dislocation (to the right in the drawing) are, as it were, helping the atoms at the 'back' to overcome the potential barriers formed by the atoms in the underlying layer.

If lattice distortions have the dimensions of flaws or cracks, their weakening effect will be increased by notch action at the edges of such cracks, resulting in a concentration of tension at that point. For an elliptical flaw (half axes a and b)

under a force, perpendicular to the long axis a, the tension at the edge of the flaw will reach a maximum

$$\sigma_{max} = \sigma\left(1 + \frac{2a}{b}\right). \tag{3.1}$$

The presence of impurity atoms may hinder the movement of dislocations and may explain why, in general, impure metals and alloys have a higher modulus and also a larger yield value than pure materials.

The effect of 'irregularities' on the strength of crystalline materials is underlined by tests on single crystal filaments. The filaments, grown like thin 'whiskers', are virtually free of imperfections and the experimentally determined tensile strength is very close to the value calculated from the bond energies.

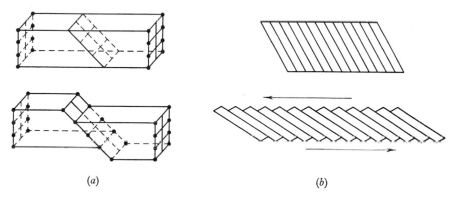

(a)   (b)

Fig. 3.2. Deformation in crystal structures. (a) Slip along a crystallographic plane. (b) Change in orientation of glide lamellae.

Most crystalline substances exhibit *strain hardening* (or shear hardening), that is, under constant load the deformation rate decreases. As deformation proceeds the orientation of the glide planes makes the angle under which the forces act less effective (Fig. 3.2). Another important cause of shear hardening is the interaction of dislocations moving along different glide planes, and in particular, in polycrystalline test-pieces, the 'build-up' of dislocations at grain boundaries. The different orientation of the atomic planes in adjoining crystallites prevents the movement of a dislocation as shown schematically in Fig. 3.1. For this reason shear hardening is much more pronounced in polycrystalline metals (as are normally used in metallic construction materials) than in single crystals.

3.1.2   Partly crystalline materials

Ceramics are compounds containing combinations of metallic and non-metallic elements (atoms or ions). An important group of ceramics takes a position between crystalline metals and amorphous glass in so far as they contain small, thin platelets (e.g. 5000 Å wide, 300 Å thick), bonded together by an amorphous glass formed from the clay with silica in the firing operation. This structure explains the lack of glide planes and why dislocations, if present in the small crystals, are easily stopped. Consequently the ductility, a characteristic of metals, is lacking: ceramics do have a high modulus and a high yield value, but they are brittle. The rheological properties of such ceramics largely depend on the irregularities at the crystal boundaries.

Efforts directed toward modifying these factors, and these effects, and other improvements like pre-stressing particularly at the surface, have led to the development of unbreakable earthenware.

3.1.3   Amorphous materials

Glasses are a combination of one or more ions with a non-metallic element, usually oxygen. The comparatively large oxygen atoms serve as matrix embedding the small metals atoms. The firm ionic and covalent bonds produce a high modulus and yield value. The ultimate strength might be much higher if glasses were not brittle.

Glass, and ceramics in general, have a rigidly fixed network of highly saturated bonds. These bonds all have slight deviations from the ideal geometry, and greatly strained angles and extended bond-lengths occur with statistical frequency. External stress will aggravate some of these strained bonds. Unless there is a 'relief mechanism' for these bonds, when a crack is initiated, brittle fracture will follow.

It is typical of ceramics and glass that they do not have a relief mechanism for internal stress, and are therefore brittle. In metals and other crystals, the movement of glide planes operates to relieve stresses. In plastics, a relaxation occurs through the viscous flow of molecular kinetic units. Neither of these mechanisms is available in ceramics or glass.

3.1.4   Linear polymers

Linear polymers have opened a wide range of possibilities for building up different sorts of molecules, to produce materials ranging from a putty to a hard infusable 'organic glass', with rubbers, leathery materials and soft or hard thermoplastics in between.

Broadly speaking the main structural possibilities are: different monomers (building blocks); the molecular size and its distribution; the building up of stereo-polymers; the building up of complex structures.

(a) The molecular size and its distribution

In a particular reaction, the macromolecules are formed with a range of molecular weight and length; the same is encountered in nature. Table 3.1 shows the range of the molecular size of three natural and three synthetic substances.

TABLE 3.1. *Molecular size of some polymers*

Substance	Molecular weight	Degree of polymerization
Cellulose (wood)	300 000–500 000	2000–3000
Cellulose, regenerated (rayon)	75 000–100 000	500–600
Rubber (native)	150 000–200 000	2000–3000
Polystyrene (for moulding)	60 000–500 000	1200–6000
Polystyrene (for coating)	80 000–120 000	800–1200
Nylon	16 000–32 000	150–300

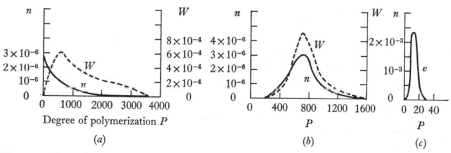

Fig. 3.3. Distribution curve of the degree of polymerization P by weight (W) and number (n) for different polymers. (*a*) Polyisobutylene. (*b*) Nitrocellulose. (*c*) Phenol formaldehyde resin (estimated).

Fig. 3.3 shows, for three different materials, the distribution by weight in grams (W) and by the number (n) of gram molecules of every fraction existing on one gram of the material.

In polyisobutylene the small molecules are most frequent. The curve for nitrocellulose is of an entirely different character; here the large molecules with a polymerization degree of 700 are the most numerous.

(b) The building-up of stereo-polymers and complex structures

Fig. 3.4 shows nine different sort of stereo-polymers and Fig. 3.5 gives an idea of the variety of complex structures that can be built up from tailor-made

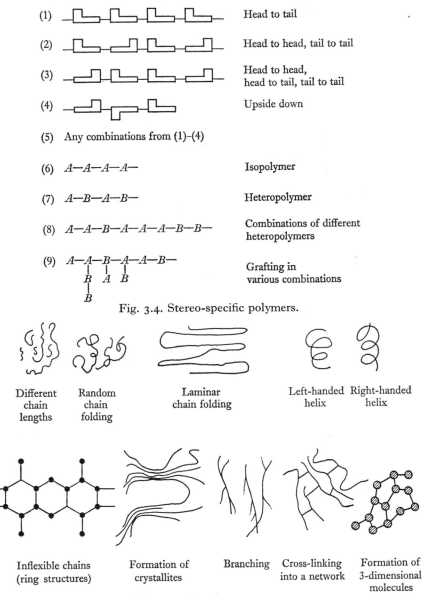

(1) Head to tail

(2) Head to head, tail to tail

(3) Head to head, head to tail, tail to tail

(4) Upside down

(5) Any combinations from (1)–(4)

(6) $A-A-A-A-$ Isopolymer

(7) $A-B-A-B-$ Heteropolymer

(8) $A-A-B-A-A-A-B-B-$ Combinations of different heteropolymers

(9) $A-A-B-A-A-B-$
      $B \quad A \quad B$
      B Grafting in various combinations

Fig. 3.4. Stereo-specific polymers.

Different chain lengths   Random chain folding   Laminar chain folding   Left-handed helix   Right-handed helix

Inflexible chains (ring structures)   Formation of crystallites   Branching   Cross-linking into a network   Formation of 3-dimensional molecules

Fig. 3.5. Complex structures.

molecules. Since many of the large number of possibilities can now be effected, the rapid growth of the polymer industry can be understood.

Fig. 3.6, from Mark,[1] demonstrates how particular combinations of polymer structures are reflected in the rheological behaviour of matter.

Although these steric factors play an important role in determining the rheological behaviour of polymers, the cohesive forces due to the secondary bonds (energies between 0.1 and 10 kcal/mol) still have an important part as well.

From Table 3.2 it appears that for rubbers, the specific molar cohesion is between 1 and 2 kcal compared with more than 5 kcal for fibres. Polymers with a cohesion of between 2 and 5 kcal are more or less intermediate; they form the group of plastics, although some of them are somewhat rubbery and others can be used as fibres. An exception is polyethylene. Although the cohesion is only 1 kcal, the fact that it is not a rubber is explained by its ready crystallization, due to the ideal symmetry of the monomer and to the easy rotational isomerism of the chain elements with respect to one another. A further factor, determining whether a material is rubbery or not, is the entropy of fusion (for rubbers about 1 cal/° C/mol), see Chapter 9.

It seems that there is a ceiling in the development of polymers with new properties, although doubtless 'new' polymers, in the sense that they will differ slightly from existing ones, will be made. Fundamentally promising prospects, however, are open to the younger family of inorganic polymers.[3]

3.2 The transition solid ⇄ fluid. Visco-elastic behaviour

The transition solid ⇄ fluid can occur:

(*a*) by changing in a two component system the concentration $\frac{\text{solid}}{\text{fluid}}$.

(*b*) by changing the temperature.

3.2.1 The influence of concentration

Small additions of solvents can often have considerable rheological consequences. The addition of plasticizers to polymers is a technique widely used to bring about a gradual change in the visco-elastic behaviour.

Ionic dispersions can, even with a high solvent content (gelatine), still show perfect elastic behaviour. The modulus E changes according to the equation

$$E = KC^n \tag{3.2}$$

where C = concentration, n = a constant.

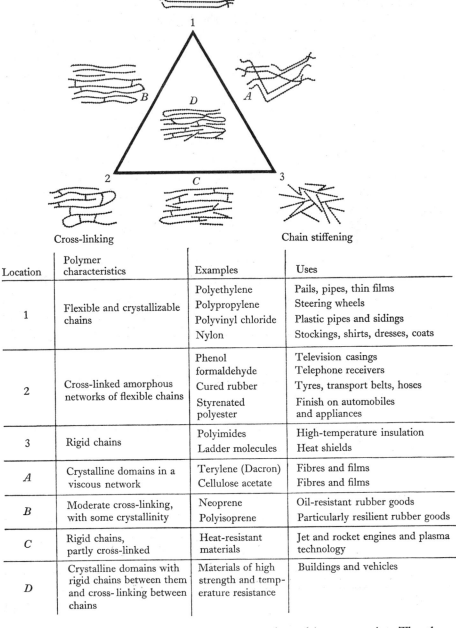

Fig. 3.6. Combination of polymer structures and resulting properties. The three basic principles are presented at the corners of the triangle. *A, B, C,* and *D* indicate combinations.

TABLE 3.2. *Primary and secondary bonds related to mechanical behaviour*[2]

Substance	Material characteristic	Covalent bonds in the chain	Dissociation energy (kcal/mol)	Groups responsible for attraction between chains	Molecular cohesion per 5 Å chain length kcal/mol
Polyethylene	Plastic	—C—C—	70–80	(CH₂)	1
Polyisobutylene	Rubber	—C—C—	70–80	(CH₂), (CH₃)	1.2
Polybutadiene	Rubber	—C=C—	70–120	(CH₂), (CH=CH)	1.1
Polyisoprene	Rubber (natural)	—C=C—	70–120	(CH₂), (CH=CCH₃)	1.3
Polystyrene	Plastic (hard)	—C—C—	70–80	(CH₂), (C₆C₅)	4.0
Polychloroprene	Rubber	—C=C—	70–120	(CH₂), (CH=CCl)	1.6
Polyvinyl chloride	Plastic	—C—C—	70–80	(CH₂), (CCHl)	2.6
Polyvinyl acetate	Plastic	—C—C—	70–80	(CH₂), (COOCH₃)	3.2
Polyvinyl aclohol	Plastic	—C—C—	70–80	(CH₂), (CHOH),	4.2
Cellulose	Fibre	—C—O—C—	80–90	(OH,) (—O—)	6.2
Cellulose acetate	Plastic	—C—O—C—	80–90	(OOCCH₃), (—O—)	4.8
Polyamide	Fibre (nylon)	—C—N—C—	70–90	(CH₂), (CONH)H	5.8
Silk fibroin	Fibre	—C—N—C—	70–90	(CHR), (CONH)H	9.8

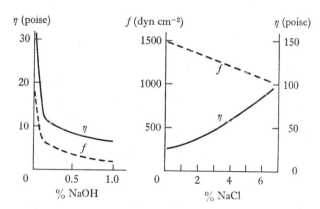

Fig. 3.7. Influence of electrolytes on viscosity and yield value of clay paste (55 % by weight). Electrolyte concentration calculated in per cent by weight of dry clay.

In dispersions with an electric double layer (clay) the rheological behaviour is dependent on the kind and concentration of the ions present, as follows from Fig. 3.7.

3.2.2 Influence of temperature

Classical physics recognized, besides the gas phase, only two thermodynamically stable phases: the crystalline solid and the liquid phase. The transition crystalline solid $\rightleftarrows$ liquid, characterized by a melting point T_m, can be defined by thermodynamic laws, which do not hold for the transition glass $\rightleftarrows$ liquid, characterized by a glass transition temperature (T_g). The difference between sudden crystallization and the gradual formation of an amorphous material by cooling is illustrated in Fig. 3.8. In this figure the change in

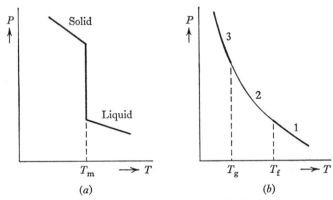

Fig. 3.8. Influence of the temperature on some physical properties of matter. (*a*) Crystalline. (*b*) Amorphous.

certain physical properties, indicated comprehensively by P, is plotted against the temperature. Such P–T curves can be recorded for properties such as density (volume), heat content, refraction, viscosity, etc. Parts 1 and 3 in Fig. 3.8*b* are both straight and joined by a curve, part 2, the *transformation interval*. Up to T_g, the temperature where the glass state begins, the material is a solid. At temperatures between T_g and T_f the material is in the glassy state and above T_f it is a liquid.

This interval makes it possible to generalize rheological behaviour, especially for polymers with their broad transformation interval; from it can be derived the conditions under which they can be worked in industrial processes.

At the low-temperature end of the transformation interval, processes such as vacuum forming or deep drawing are possible; at the high-temperature end, calendering and extrusion. Obviously, the interval can be extended to higher temperatures by going to higher molecular weights or increased cross-linking; these changes increase the viscosity and therefore shift the practical range of operation to higher temperatures.

For polymers, the range between T_f and T_g represents a series of continuous transitions. For instance on cooling a molten polymer, it can change through the visco-elastic via the rubbery or the leathery state into a solid with various degrees of crystallization. These various phase states are shown in Fig. 3.9.[4]

The decision as to whether a visco-elastic substance is 'still a liquid' or 'already a solid' is usually arbitrary because in such 'sort–hard' materials the determination of a yield value and the degree of high-elasticity (versus plasticity) is dependent on the rate of testing (see p. 21). Hence the often painful confusions in nomenclature which exist in this field.

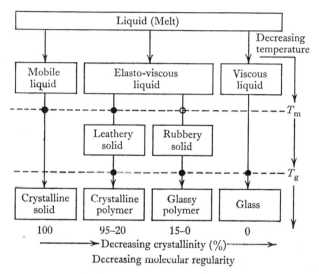

Fig. 3.9. Phase states of polymeric materials.

As an observation we point out that even in organic polymers, which at first sight seem to have a rather similar structure, the transition temperature may differ widely. See Table 3.3.

3.3 Composite solids

3.3.1 General aspects

Many solids in everyday use are not homogeneous chemical substances. Granite, concrete, wood, etc. are obvious examples. Composite solids are interesting from several points of view. In the first place, one can often use simple 'low viscosity' techniques to create strong solids; an example is

TABLE 3.3. *Glass transition temperature dependence on polymer structure*

Name	Structure	Glass transition T_g in $^\circ$ C
Polybutadiene	$-(CH_2-CH=CH-CH_2)_n-$	-80
Polyethylene	$-(CH_2-CH_2-CH_2-CH_2)_n-$	$+70$
Polyisoprene	$-(CH_2-CH=CH-CH_2)_n-$ $\underset{CH_3}{\overset{\vert}{}}$	-73

concrete, formed by mixing water and cement with sand and stones. It hardens into a strong solid by reaction between water and the cement, binding the other ingredients together.

In the second place, composites frequently show a synergistic combination of the rheological features of the two separate phases. In the simplest cases, one phase is 'reinforced' (rheologically improved) by the presence of the other. In other cases, there is a mutual enhancement of properties.

3.3.2   Reinforcement

A well-known example of reinforcement is rubber containing fine particles of carbon black. Rubber consists of long linear molecules that move relatively freely past each other. By cross-linking (vulcanization) a network is formed that preserves its shape, and reacts elastically to external forces. It can be shown (Chapter 9) that the modulus of elasticity is proportional to the number of cross-links, but the maximum elongation increases with the distance between cross-links. These facts create a problem: to obtain a very high modulus one would need to produce many more cross-links, but that would decrease the elongation to unpractical levels.

The problem is solved by the addition of filler particles (carbon black) having a very large total surface (small particle diameter). The filler surface exerts an attraction on the rubber molecules, and, in a way, 'cross-links' many of them together (multiple cross-link), thereby vastly increasing the modulus. However, the filler particle does *not* inalterably fix in space the cross-link points; the rubber molecules can slide along the surface and hence large elongations remain possible, providing relaxation of local stresses and preventing brittle fracture. Table 3.4 illustrates this point.

TABLE 3.4. *Particle surface and its effect on vulcanized rubber*
(Example: 1 cm³ of carbon black dispersed in 4 cm³ of vulcanized rubber.)

Average particle diameter	Total filler surface	Maximum elongation	Retracting force of rubber, kg/cm^2 at 300 % elongation
10^{-2} cm	3×10^2 cm²	500 %	10
10^{-4} cm	3×10^4 cm²	500 %	20
10^{-6} cm (= 100 Å)	3×10^6 cm² = 300 m²	500 %	200
No filler	No filler	500 %	10
No filler	No filler	100 %	100

3.3.3 Mutual enhancement of properties (see also § 7.3)

Composites in this class are wood and fibreglass. More recent examples are the composites of graphite fibres in resins or metals. The outstanding feature is high stiffness and tensile strength without brittleness, combined with low weight.

Small fibres (whiskers) of 'ceramics' (inorganic non-metals including boron or graphite) can be made with such perfect internal structure that their actual strength equals the theoretical magnitude calculated from primary bond strength. However, ceramics are brittle, and break by the propagation of cracks, which requires almost no energy after the first crack is formed. The composite retains the strength of such fibres, but prevents cracks from propagating through the material.

Composites with the highest strength contain aligned fibres. If such a composite is stretched parallel with the fibres, the strains in the fibre and in the matrix are virtually equal, but the matrix yields or flows in a plastic manner, so that the stress within the fibres is greater than it is in the matrix. The difference is so pronounced that in the breaking strength of the composite the contribution of the matrix can be regarded as negligible.

When the fibres are highly stressed, some that have cracks will break. But such a crack will usually be unimportant because the propagation of the crack through the brittle reinforcing material is hindered by the *softness of the matrix*. Furthermore, although the reinforcing fibres may fail, they do not all do so in one plane. For a crack to extend all the way through the material it would be necessary to pull the fibres out of the matrix one by one as they broke.

A perfect fibre of boron or silica glass has a strength of about 70000 kgf/cm² or almost 10 times that of a good nylon filament. Composites incorporating these fibres can show a strength up to 20000 kgf/cm², which equals good steel at a quarter of its specific weight.

3.4 Fluids

3.4.1 One-component systems

In simple (one-component) fluids the relationship between viscosity and potential barriers around the molecules is expressed by the activation energy E_v in equation (2.12) (p. 26).

$$\eta = A \exp (E_v/RT_{abs}). \qquad (2.12)$$

Measurements in simple liquids have shown E_v to be related to the energy of evaporation E_{vap}:

$$E_v \approx \tfrac{1}{3}E_{vap}. \qquad (3.3)$$

Both energies are related to the removal of a molecule leaving a 'hole', except that in the flow phenomena one has to visualize 'holes' in one dimension, corresponding to the formation of one-third of a real three-dimensional hole. In practice pure viscous flow as defined by equation (2.12) is rarely found, although some materials approximate to it very closely.

It is obvious that in a series of molecules with basically the same chemical constitution (homologous series) the longer molecules will produce a higher viscosity due to their larger cohesion surface. This is confirmed in Fig. 3.10 for a series of hydrocarbons. Near their boiling points, which are the (extrapolated) end points of the curves, the viscosity–temperature curves are almost straight and parallel lines, which are therefore defined mathematically by their slopes and intercepts. The slope of a given curve depends upon the class to which the compound belongs, and the intercept upon its place in the series. The difference between the intercepts of two curves measures the influence of the addition of each further group to the polymer. One may speak of a viscosity increment per group (although this increment is not always a constant).

3.4.2 Non-ionic dispersions

In a dilute solution each dissolved molecule will have an opportunity to assume its 'random coil' configuration, the size of which depends on the free energy of solution.

From a rheological viewpoint the dissolved particles have two effects:

Dispersed particles, larger than the solvent molecules, cause an increased heat dissipation in flow, resulting in a higher viscosity.

Immobilization of solvent molecules increases, as it were, the volume concentration, but it increases at the same time the volume per dispersed particle. Both effects result in a higher flow resistance.

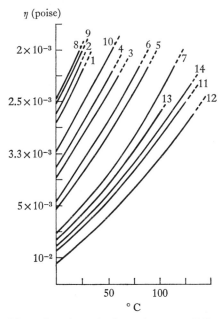

Fig. 3.10. The viscosities of various hydrocarbons at different temperatures and extrapolated to their boiling temperatures. 1. Pentane; 2. Isopentane; 3. Hexane; 4. Isohexane; 5. Heptane; 6. Isoheptane; 7. Octane; 8. Trimethylethylene; 9. Isoprene; 10. Diallyl; 11. Ethyl benzene; 12. *o*-Xylene; 13. *m*-Xylene; 14. *p*-Xylene.

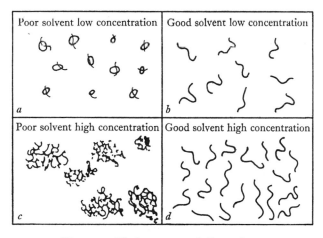

Fig. 3.11. Various sorts of dispersions.

Fig. 3.11 illustrates the different sorts of dispersions.

The viscosity of a dilute dispersion is expressed by the Einstein equation:

$$\eta_c = \eta_0(1 + KC_v) \tag{3.4}$$

where η_c is the viscosity of the dispersion, η_0 is the viscosity of the pure liquid, C_v is the volume concentration of dispersed phase, and K is a constant.

In the case of spherical particles, the Einstein equation may be valid up to 10% volume concentration, K being 2.5.

Staudinger has shown that for dilute linear-polymer solutions this equation is modified to

$$\eta_c = \eta_0(1 + KC_v M), \tag{3.5}$$

where M is the molecular weight of the polymer. According to Houwink[5] equation (3.5) is a special case of the more general equation

$$\eta_c = \eta_0(1 + KC_v M^\alpha), \tag{3.6}$$

where α varies between 0.6 and 1.6, and expresses the free flow of solvent molecules through the coiled polymer molecules. In some cases, this equation is valid up to volume concentrations of 70%. A more detailed treatment will be given in Chapter 4.

References to chapter 3

1   H. Mark, *Scientific American* **217**, 149 (1967).
2   G. J. van Amerongen, *Elastomers and plastomers* **1**, 196 (Ed. R. Houwink; Elsevier, 1949).
3   D. N. Hunter, *Inorganic polymers* (Oxford, 1963).
4   D. W. van Krevelen, *Proc. Internat. Plastics Congress*, p. 12 (Amsterdam, 1966).
5   R. Houwink, *Koll. Zeits.* **99**, 160 (1942).

4 Rheological phenomena

H. K. de Decker

4.1 Introduction

In a sense, we shall now again examine, as in Chapter 2, the structure–rheology relationship in its generalities; however, in the present chapter we start from the rheological phenomenon, and seek its interpretation through structural concepts.

At the time of the first edition of this book (1937), deformation phenomena were subdivided into ideally viscous, ideally elastic, and an ever-increasing

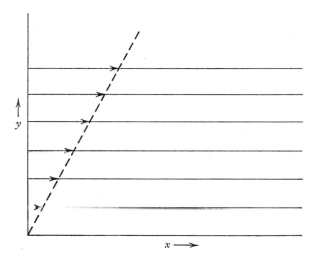

Fig. 4.1. Laminar flow in direction x.

and confusing number of exceptions and special cases, each with its own nomenclature.

Rheology now uses a new scheme recognizing that materials can have an infinite variety of shear properties, varying from hard and brittle ceramics, through metals and plastics, to simple liquids.

Based on the considerations underlying Fig. 1.1 (p. 2) we consider again in Fig. 4.1 a continuum that is subjected to some shearing deformation in the x direction. One can define the x component of strain as γ, being the x

displacement per unit of y distance, $\gamma = \mathrm{d}x/\mathrm{d}y$. For very small displacements one can write the work of deformation as:

$$\mathrm{d}W = G\gamma\,\mathrm{d}\gamma + \eta\frac{\mathrm{d}\gamma}{\mathrm{d}t}\,\mathrm{d}\gamma, \tag{4.1}$$

and the shear stress as: $\qquad \tau = G\gamma + \eta\frac{\mathrm{d}\gamma}{\mathrm{d}t}. \tag{4.2}$

These are the general equations of rheology, describing the elastic parts of the deformation by the shear modulus G, and the viscous part by the viscosity coefficient η.

For the sake of practical understanding, we start in this chapter by examining 'purely elastic' behaviour ($\mathrm{d}\gamma/\mathrm{d}t = 0$) followed by 'purely liquid' behaviour ($G = 0$), and subsequently discuss various cases where both play noticeable parts.

The key concept in understanding structure of matter in relation to rheology is the classification of elementary bonds, as given in Table 2.1, p. 15. Primary bonds provide strength and elasticity, but do not allow any 'flow' in the ordinary sense; secondary bonds, with their wide spectrum of strength and temperature dependence, are responsible for all the innumerable aspects of flow from simple Newtonian flow to paste thixotropy.

A world made of materials having only primary bonds in all directions (metals, rocks) would be solid and strong, but would not allow many things to happen, and life itself would be impossible.

4.2   Elasticity

4.2.1   Basic laws and definitions

As pointed out in Chapter 1, the 'elastic modulus' or 'Young's modulus' E or the 'shear modulus' G, express the relationship between stress and elastic strain:

$$\tau = G.\gamma \tag{4.3}$$

$$F = E.\epsilon, \tag{4.4}$$

where τ is shear stress, γ is shear, F is normal stress and ϵ is elongation.

$\epsilon = \dfrac{l-l_0}{l_0}$ where l_0 is original length and l is length under stress.

A similar relationship exists between hydrostatic pressure P and volume V (see §6.2). $\qquad P = P_0 - M_\mathrm{B}\alpha; \quad \alpha = (V-V_0)/V_0$

where P_0 and V_0 are reference values for P and V. M_B is called the 'bulk modulus' and measures the resistance against compression in all directions (hydrostatic compression). M_B has practical significance in gases and, to some extent, in other fluids (see Chapters 5 and 6). It should be noted that unilateral compression causes expansion in perpendicular directions, and the resulting deformation is shear.

The ideal case where G and E are constants, independent of the strain and the time, was first described by Hooke; equation (4.4) for this ideal case becomes Hooke's law. Examples such as the deflection of a hard steel spring or a glass rod, or the extension of an elastic band (under relatively moderate stresses) are reasonably close to the ideal case of Hooke's law of elasticity.

Deviations from the ideal case may be caused by different phenomena. First, the behaviour may be truly elastic, but there is no proportionality between stress and strain. The modulus changes at higher stress; this is usually the case for accurate measurements of elastic deformation. Secondly, the behaviour may not be truly elastic (and removal of the stress does not eliminate all the strain); in this case there is a certain amount of flow or permanent deformation, and equation (4.4) will not be adequate. If it is used nevertheless, it appears as if the modulus G (or E) is not constant and depends on time.

It is well-known that elastic deformation of materials has very definite limits, unlike flow of materials which may continue ad infinitum. Steel or glass wires cannot yield elastically beyond about 1 % of their length. Vulcanized rubber, due to a special mechanism, can elastically yield 500 % of its own length. Beyond the limit of elasticity, a mechanism for permanent deformation or flow can be activated and this new mechanism then takes over. This case is illustrated by steel, as seen in Fig. 4.2.

In other cases, there is really a blend of elastic and viscous deformation at all stresses, but at low stresses and short times of observation, the elastic part is predominant. As the force and/or the time are increased, more flow becomes apparent, and a gradual transition to viscous flow exists; this will be discussed in §4.3. Examples at room temperature are lead, gel-type asphalts, certain types of clay suspensions, paints, etc.

Under some conditions, there is no deformation mechanism available to take over beyond the elasticity limit, and the material breaks. This is a 'brittle' fracture, and no evidence of flow can be found. The pieces can be fitted together and represent the true original shape of the test specimen. Ordinary breaking of glass is an example.

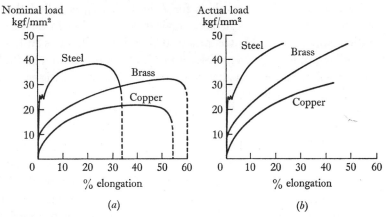

Fig. 4.2. Stress–strain curves for several metals. (*a*) Nominal values of the tensile stress, in kgf/mm², referred to the area of the original cross-section. (*b*) Actual values of the tensile stress, calculated for the actual cross-section.

4.2.2 Two mechanisms of elasticity

'Elasticity' is found if a reactive force in the material opposes deformation and wants to restore the material to its 'equilibrium' (no external forces) position and shape. It is interesting to compare the magnitudes of the 'Young's modulus' E in equation (4.4), the constant in Hooke's law, for various materials, and to compare the findings with our knowledge about structure. See Table 4.1.

TABLE 4.1. *Elasticity data*†

| | Maximum elastic deformation | | Young's modulus, E (kgf/cm²) |
Material	Yield value (kgf/cm²)	% deformation	
Glass fibre	30000	5	6×10^5
Carbon steel, annealed	3000	0.15	2×10^6
Aluminium, cold rolled	1400	0.2	7×10^5
Nylon fibre	8500	20	4×10^4
Acrylonitrile–butadiene–styrene plastic	400	3	2×10^4
Styrene–butadiene rubber, black reinforced vulcanizate	200	500	4×10

† Data are approximately for room temperature and a test taking 30 seconds to 'yield point'.

Glass consists of a three-dimensional network of atoms, mostly silicon and oxygen, all held together by strong primary bonds with their immediate neighbours. These primary (chemical valency) bonds allow a slight stretching and distortion, which is noted in the elastic data of Table 4.1.

The metals, steel and aluminium, consist of crystals, in which the atoms are arranged in an orderly fashion and are mutually attracted by strong primary bonds, as is apparent from the very high Young's modulus and the very small amount of elastic deformation.

The plastic acrylonitrile–butadiene–styrene, or ABS, consists of macro-molecules which are coiled up randomly next to and penetrating each other; at room temperature the secondary forces between neighbouring parts of macromolecules are so strong that they do not leave their original positions and respond to moderate outside forces in short times only by an 'elastic' displacement. However, the difference between the secondary forces here and the primary forces in the metals is evident from the great difference in Young's modulus and the amount of elastic deformation. In the case of Nylon, linear macromolecules are oriented in the fibre direction (see §4.2.4).

Finally, rubber deviates so much elastically that another mechanism must be responsible.

As shown by Wiegand *et al.*, the overall thermodynamics of elastic defor-mation can be treated as the compression of an ideal gas, and the total tensile force F causing elastic deformation is the sum of a free energy term dU/dl and an entropy term $T(dS/dl)$, as shown in equation (4.5):

$$F = \left(\frac{dU}{dl}\right)_T - T\left(\frac{dS}{dl}\right)_T.$$ (4.5)

In the cases of steel and glass discussed above, there is no entropy term, as has been shown experimentally; this can also be understood from the fact that the atomic lattice in these cases of elastic deformation remains as well-ordered as before, and there is no change in the 'randomness', which deter-mines probability and thereby entropy. The elasticity based on changes in potential energy only, is expressed by (4.6) and called 'ordinary elasticity', 'steel type' elasticity, or potential energy elasticity:

$$F = \left(\frac{dU}{dl}\right)_T = E.\epsilon.$$ (4.6)

'Rubber elasticity' is shown in an almost ideal way by vulcanized natural rubber at elongations below 300 %, and also to various degrees, in combina-tions with ordinary elasticity and with flow, by many rigid materials, especially by various practical rubber and plastic compounds. Rubber consists of chain-like macromolecules interconnected at certain intervals by chemical cross-links. Fig. 4.3 *a* shows three such molecules with three cross-links.

Each chain between cross-links, when in equilibrium (no external forces) will tend to the greatest disorder (highest entropy) which is calculated to be a 'random coil' shape having an average end-to-end distance x_1 related to its true stretched length L:

$$x_1 = a\sqrt{L} \tag{4.7}$$

where a is a constant.

On stretching a piece of rubber, the cross-links will be displaced, and the average end-to-end distance of the supporting chains will be increased from

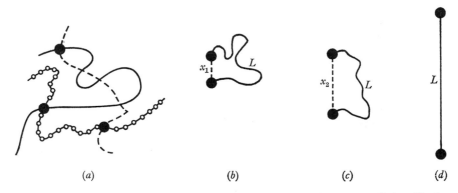

(*a*)	(*b*)	(*c*)	(*d*)

Fig. 4.3. (*a*) Three linear rubber molecules linked by three cross-links. (*b*) One molecule with two cross-links before stretching. (*c*) Fig. *b* after some stretching. (*d*) Fig. *b* fully stretched.

x_1 to x_2 as illustrated in Fig. 4.3*b* and 4.3*c*. This lowers the probability, and therefore the entropy of the configuration, according to Boltzmann's relation:

$$S = k \ln W, \tag{4.8}$$

in which k is Boltzmann's constant and W the probability. From equation (4.5), the elastic force F depends on the entropy and only the entropy in the case of ideal rubber elasticity:

$$F = -T \left(\frac{\delta S}{\delta l}\right)_T = -kT \left(\frac{\delta \ln W}{\delta l}\right)_T. \tag{4.9}$$

This can be transformed to:

$$F = \frac{\rho RT}{M_c} (\alpha - 1/\alpha^2) = kTN_c(\alpha - 1/\alpha^2), \tag{4.10}$$

where ρ is the density of the rubber, M_c the molecular weight of a supporting chain (chain between cross-links), N_c the number of such chains per cm^3, and α the extension ratio l/l_0.

This is a very basic equation; it states the fundamental law of rubber elasticity, as equation (4.6) states the fundamental law of ordinary elasticity. An improved version of equation (4.10) is:

$$F = \frac{\rho RT}{M_c}\,(1 - 2M_c/M)\,(\alpha - 1/\alpha^2). \tag{4.11}$$

To use this equation, the assumption is made that all original macromolecules are equal in length, that they have molecular weight M, and that M_c the chain length between cross-links is constant. This leaves an average 'loose end' of $\frac{1}{2}M_c$ at each end of every rubber molecule. Experimental data confirm the validity of equation (4.11) at relatively low elongation. If M is very much larger than M_c, (4.11) approaches the simpler equation (4.10).

A real rubber has a wide range of values for both M and M_c, and corrections can be derived to take this into account. Furthermore, the cross-link reaction in certain cases makes 'mistakes', resulting in connections between parts of the same molecule, forming loops that usually do not contribute to the network structure. This 'cyclization' can also be taken into account, in a way similar to the 'loose ends'.

4.2.3 Rubber elasticity in practice

It is important to note that the rubber-elastic force *is* directly related to the cross-link density, the temperature, and the elongation, as shown in equation (4.10) or (4.11) and none of the refinements change this basic fact. This force is independent of attractive forces between molecules, and operates as soon as the molecular segments are free to move through their thermal motion.

Most practical cases are complicated by the fact that the attractive forces between molecules cannot be neglected. Their influence on the elastic force can be minimized by increasing the temperature and time of the experiment. A piece of vulcanized natural rubber, having ideal rubber elasticity at room temperature in an experiment taking 30 seconds, is hard and has partial steel elasticity if tested at low temperature or in a high speed tensile test taking, for example 10^{-4} s.

It can be said that the conditions of temperature and time under which a linear high polymer readily shows viscous flow are the same conditions under which that same polymer will have rubber elasticity when cross-linked.

An interesting condition is created by incorporating hard filler particles into an 'elastic' polymer. The resulting 'reinforcement' (increased modulus and tensile strength) will be discussed in §4.4.3.

4.2.4   Crystallization

Many polymers are normally crystallized or will crystallize when stretched. In a fully crystallized and oriented fibre such as nylon, the elastic force in the fibre direction is very high (see Table 4.1). In Fig. 4.4, to move molecule *A* in its own direction the secondary forces between the atoms of *A* and the atoms of *B* have to be strained. These secondary forces occur between a large number of atoms in exactly the same position and therefore result in a strong elastic reaction.

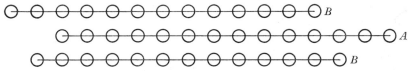

Fig. 4.4. Linear molecules in crystal lattice.

Orientation and crystallization of certain polymers is caused by stretching. The conditions under which this happens are: (1) the polymer molecules must have regularly repeating units in a linear chain such as is the case in nylon, polypropylene, and natural rubber: (2) the temperature must allow the secondary forces occasionally to be overcome by thermal motion; and (3) sufficient time should be given for the orientation process during the test.

This orientation and crystallization by stretching is used in making strong synthetic polymer fibres. The stretching is done at an elevated temperature and the subsequent cooling to room temperature prevents the crystallization from disappearing by thermal motion after the stress is removed.

A different situation prevails in rubber crystallizing under stress. Here the orientation and crystallization, and the disorientation back to random configuration, all occur at the same temperature (room temperature for natural rubber and neoprene), and the crystallization occurs each time the elongation goes beyond a certain value. At that elongation the crystalline regions formed can be regarded as 'hard filler particles' in the relatively soft rubber, and their 'reinforcing' effect is similar, as discussed on p. 93. During the crystallization the 'reinforcement' appears and disappears at a certain magnitude of deformation, dependent on temperature and time of the experiment. A summary of the types of elasticity is given graphically in Fig. 4.5.

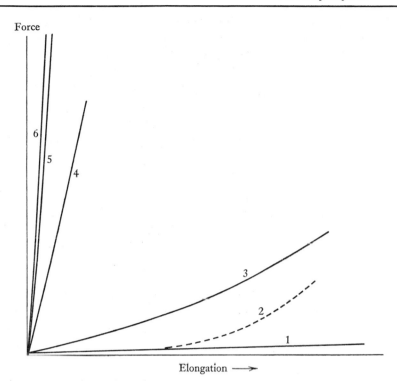

Fig. 4.5. Types of elasticity. 1. Vulcanized rubber (pure gum). 2. As 1, but crystallizing on stretching. 3. Vulcanized rubber, filler reinforced. 4. Fibre, crystalline oriented. 5. Glass. 6. Steel.

4.3 Viscosity and plasticity

4.3.1 Basic laws and definitions

Viscosity relates to the behaviour of a material under stress when the result is permanent deformation. Viscous flow is this deformation as it occurs in time. If the deformation disappears after the stress is discontinued (there is no permanent deformation) the material reverts to its original shape, and the deformation is called elastic.

'Viscosity' or 'viscosity coefficient' as a physical quantity expresses the resistance of the material against flow or permanent deformation. It is given the symbol η and defined by

$$\tau = \eta D = \eta \frac{d\gamma}{dt}, \tag{4.12}$$

where τ is shear stress, $\gamma = \dfrac{dx}{dy}$ is shear, D is shear rate and $\eta = \tau/D$. The viscosity coefficient is the shearing stress per unit of deformation rate. A convenient way to represent viscous flow is by D–τ diagrams, such as Fig. 4.8.

The definition of equation (4.12) is possible only in laminar flow as illustrated in Fig. 4.1. Laminar flow occurs only as long as a characteristic number (Re), the Reynolds number, remains below a certain limit. Above that limit turbulence occurs and equation (4.12) cannot be used. The Reynolds number is defined by $(Re) = D\rho l/\eta$, in which ρ is density, η is viscosity of the medium and D the shear rate of the motion; l is a characteristic dimension of the flow pattern.

In pure flow a shearing stress is needed to keep the material flowing, even although there is no reaction in the material to revert to its original shape. The stress is needed to overcome the tendency to stop flowing and this tendency points to an internal resistance in the material. The work done in maintaining flow is dissipated in heat. For some calculations it is useful to define viscosity as the transfer of momentum (motion in direction x of Fig. 4.1) perpendicular to its own direction.

Examples of pure flow are given by pumping hexane or light lubricating oil through a relatively wide pipe. Both liquids flow at the slightest pressure, and stop flowing when not under pressure, but the stress needed to maintain a certain rate of flow is larger for the oil than for the hexane; the oil viscosity is greater.

It is possible to understand the greater viscosity of the oil by examining its molecular structure and comparing it with the hexane. Both liquids (for the purpose of this example) have hydrocarbon molecules of the same n-paraffinic structure, or straight carbon chains with hydrogen atoms saturating all the available valencies. The hexane consists of C_6H_{14} molecules, the oil, for example, of $C_{12}H_{26}$ molcules. As explained in Chapter 2, there will be attractive forces between the molecules of a liquid, but the energy of attraction is such that the kinetic energy of the thermal motion easily overcomes it. Larger molecules will be hampered by their neighbours in many more places than small molecules. Otherwise expressed, at a given temperature T, the kinetic energy per molecule is $\frac{1}{2}mv^2 = kT$; the molecules having a larger mass m will have slower thermal motion, and are therefore more difficult to displace.

In reality, pure Newtonian flow as defined in (4.12) hardly exists in nature, although in some cases this phenomenon can be approximated to a very high degree. Flow of gases, or liquids of low molecular weight at relatively high temperatures, are examples. Many complications occur in real flow phenomena.

The more complicated patterns of flow behaviour cannot be represented by equation (4.12) with a constant viscosity factor η. Many empirical equations have been proposed to describe such flow behaviour, for instance equations such as (4.15) and (4.16), the basic form of which is

$$D = \tau^n/\eta^*, \tag{4.13}$$

where n is a constant and η^* is a 'coefficient of non-linear viscosity'. The general term used for such types of flow in older literature is 'plastic flow'. Further study of empirical equation such as (4.15) and (4.16) will not improve our understanding of rheology, unless we turn first to the molecular constitution of materials.

4.3.2   Molecular structure and kinetics of flow

A clarification and simplification of real patterns of deformation in nature is possible on the basis of molecular kinetics. In our first approach we have used simple idealized molecules in a liquid at sufficiently high temperatures (water or hexane at room temperature). The balance of attractive and repulsive forces between adjoining molecules keeps them at a constant average distance, although thermal motion makes them oscillate around this average position (see Fig. 4.6). To cause deformation, the deforming force must find molecules able to overcome the limitation of this average position; this means that the 'potential barrier' of Fig. 4.6 has to be overcome. In other words, flow will only occur if some molecules can be pushed aside to create a 'hole' for others to fall into, thereby displacing the hole and propagating the movement. This is easier at higher temperatures.

The viscosity–temperature relation can be expressed by the empirical equation (4.14):

$$\eta = A \exp(E_v/RT), \tag{4.14}$$

where A and R are constants, E_v is the activation energy representing the potential barrier for viscous flow, and T is temperature. The viscosity becomes very large if E_v is large or the temperature relatively low, or both. It follows that flow may not be noticeable in the time of measurement, but that an increase in stress or in temperature, or a longer observation time will reveal flow. The simple fact that flow is not observed under small stresses, but appears at larger stresses, has formerly been described in terms of a yield value (Fig. 4.8). Our present understanding shows the 'yield value' to be an arbitrary parameter, dependent on the material characteristics in conjunction with the conditions τ, T, t of the experiment. Below the 'yield value', the limited deformation is of the elastic class; it reverts to zero if the stress stops.

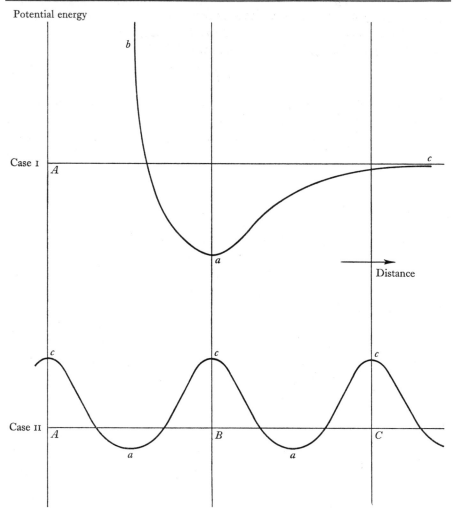

Fig. 4.6. Potential barriers between molecules. Case I: One molecule fixed at *A*. A second molecule approaching in a direct line prefers position *a*. The barrier *b* is insurmountable; the barrier *c* surmountable. Case II: Three molecules fixed at *A*, *B* and *C*. A fourth molecule 'skipping past' prefers positions *a*; positions *c* are barriers.

Measurements, in simple liquids, have shown E_v, the activation energy of viscous flow, to be about one-third of the energy of vaporization, indicating that both phenomena can be related to the mechanism of removing one molecule and leaving a 'hole'. However in the flow phenomenon one has to visualize 'holes in one dimension', requiring only one-third the energy of a real three dimensional hole.

4.3.3 Large molecules

For the more complicated flow phenomena, an understanding of larger molecules will be needed. Referring to Table 2.1 on p. 15, there are two groups of forces between elementary particles:

(1) primary or 'chemical' forces, of the order of 80 to 150 kcal/mol,

(2) secondary or 'physical' forces, of the order of 1.0 to 10 kcal/mol.

Our reasoning above, leading to an activation energy of flow of a few kcal/mol, confirms that, in simple liquids, the bonds to be overcome in flow are the secondary forces between molecules. A very large molecule, such as that of nylon or natural rubber, would require much too much energy to remove it altogether to a new spot (4000 kcal/mol for polyisoprene $(C_5H_8)_{1000}$). The fact that these materials exhibit flow under reasonable stresses at room temperature, and the magnitude of their activation energy (10 kcal), prove that only small *segments* of the macromolecule are moving with respect to one another to cause the flow of material.

These segments, or kinetic units, play a role similar to that of the actual molecules in a simple liquid. The concept to keep in mind is the flexibility of the primary bonds; it requires a high energy to *break* these bonds, but their rotation enables motion otherwise hampered by the secondary forces, as illustrated in Fig. 4.7. Various types of secondary forces exist, and with different magnitudes (see Table 2.1, p. 15).

In a high molecular weight alcohol, one can distinguish two types of secondary forces, those between $-CH_2-$ groups (about 2 kcal/mol) and those between $-OH$ groups (about 7 kcal). In the liquid state, a small stress may not cause any flow ('yield value') but as stress increases, a growing proportion of the $-OH$ bonds will be loosened. The result is a $D-\tau$ curve such as A in Fig. 4.8. There is a curvature upward, indicating gradual loosening of bonds as the stress increases. Curves such as A in Fig. 4.8 are found in many cases. In general, they indicate that the material has a series or a 'spectrum' of different secondary forces, and therefore a spectrum of activation energies. Basically, this curve should be described by a series of viscosity coefficients, each one related to one particular secondary force and its flow mechanism. Using equation (4.14) it can be said that each coefficient will have a different A and E_v, meaning a different frequency distribution and activation energy.

This complicated type of flow, curve A in Fig. 4.8, is non-Newtonian or 'plastic'. Instead of describing it with a series of viscosity coefficients, one

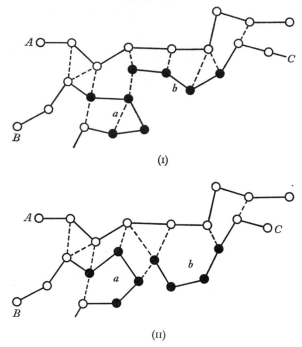

(I)

(II)

Fig. 4.7. Flow of macromolecules. Molecules A, B and C are shown (I) before and (II) after displacement of kinetic units a and b (black atoms). ○—○ primary bond between two atoms; ○- - -○ secondary bond between two atoms.

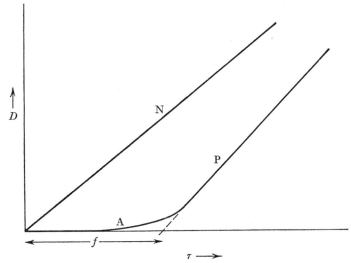

Fig. 4.8. Flow curves. N, Newtonian; P, plastic; f, yield value; D, shear rate; τ, shear stress.

usually describes it by empirical equations, such as:

$$D = \psi \tau^n = \tau^n/\eta^*, \tag{4.15}$$

or
$$D = \psi(\tau - \tau_0)^n, \tag{4.16}$$

where ψ is a 'fluidity' coefficient (a higher value meaning easier flow), and n an exponent describing the curvature of the D–τ relationship. In (4.16) the value τ_0 is a 'yield value', dependent on the conditions τ, t, and T as we have seen. As a matter of fact, the coefficients ψ and n are also dependent on temperature. Furthermore they depend on the time elapsed between different phases of the experiment.

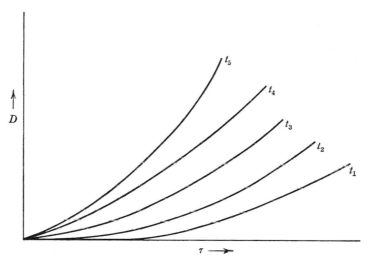

Fig. 4.9. Flow curves for different times t_1, ..., t_5 for 'thixotropic' or 'plastic' flow. D shear rate, τ shear stress. Compare Fig. 4.13.

As indicated before, there are very few instances of Newtonian flow (linear flow) in nature. Many cases have the characteristics of Fig. 4.8, curve A and equations (4.15) and (4.16) and, moreover, should really be represented in a D, τ, t diagram such as in Fig. 4.9, which is repeated for each temperature.

Curve B in Fig. 4.10 occurs when the mechanism causing flow is based on a limited number of displacements (each of which has an activation energy easily accessible for the kind of flow considered). As soon as the available mechanism is fully activated, a further increase in stress will not cause any faster flow, if there are no other mechanisms available. An example is provided by wet sand, where the flow is due to motion of sand particles in the

3

water, but too fast a motion would result in the particles touching each other and flow will be severely limited. This phenomenon (getting harder upon deformation) is usually called dilatency; it can be represented by equations (4.15) and (4.16) but in this case the exponent *n* will be less than 1.

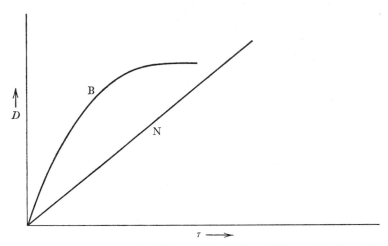

Fig. 4.10. Flow curves for 'dilatant flow' (B) and Newtonian flow (N).

4.3.4   Kinetic units and the glass transition

At various points we have referred to the kinetic units (in many cases molecules) that are responsible for flow. In the stress-free condition these units are bound together by secondary forces, and a shearing stress overcomes (some of) these secondary bonds and causes flow. It is interesting to examine more carefully the kinetic units and their secondary forces as they actually occur in various materials under flow conditions.

If a liquid is homogeneous and consists of relatively simple molecules, such as water or benzene, these molecules are the kinetic units, the forces between them are of the 'secondary bond' category, and flow will be practically Newtonian between melting point and boiling point under normal pressure.

If the molecules are quite large (mol. wt. 500 to 5000) such as paraffin wax, 'stand-oil', etc. these molecules may still be the kinetic units, but being very large, they do not easily participate in flow (the melting point of paraffins goes up with their molecular weight); they have to be given time, high temperature, and relatively high stress.

If large molecules are essentially linear, there will be an additional effect to be considered, namely orientation by flow. This orientation would tend to lower the 'viscosity', and the result would be a kind of thixotropy (see below).

The rheology of macromolecules (mol. wt. over 5000) is usually based on the motion of segments of these macromolecules (see above). These segments of about 10 or 20 atoms in the chain (plus their pendant atoms or groups) are the kinetic units in this case. Below the 'glass transition', materials are hard and relatively brittle; a stress will cause small elastic deformations or rupture. Increase in temperature will overcome the effect of the forces holding all atoms in fixed positions, and at the glass transition temperature, the kinetic units become relatively free to move past one another.

The glass transition temperature is dependent on the type of attraction between the kinetic units; in polyacrylic acid, the $—C\!\!=\!\!O\ldots H\!\!-\!\!N\!\!=$ interaction is relatively strong; in polymethyl acrylate of the same molecular weight the interactions are small. The high glass transition temperature of the acid is due in part to these strong secondary forces: 85° C for the acid compared with 3° C for the acrylate. The other factor affecting the glass transition in macromolecular materials is the flexibility of the chain: see Table 3.3 on p. 46.

Regularity in structure can cause crystallization and thereby interfere with flow behaviour without, however, changing the glass transition temperature. The magnitude of the crystallization effect on rheology depends on the heat of crystallization – the amount of energy needed to dissolve the crystal and induce flow, but also on the geometry of the molecule – the ease with which molecules can slide along without leaving the crystal. For this last point, the crystallite may be visualized as a piece of tubing, and the member molecules as wires passing through the tube; it may be relatively easy to slide one of the wires through the tube, even if it is difficult to break the tube (dissolve the crystallite). Figs. 4.11 and 4.12 illustrate crystallites in polymers.

4.3.5 Solutions and suspensions

In a very dilute solution of macromolecules, for example polybutadiene in benzene, each polymer molecule will have an opportunity to assume its 'random coil' configuration, the size of which depends on the free energy of solution. The polymer molecules have two effects on the viscosity of the benzene:

(1) Viscosity increase by any large particle substance (bulk effect);

(2) Viscosity increase by immobilization of benzene molecules inside and around the coiled-up macromolecule (swelling effect).

It is convenient to start with the bulk effect of particles suspended in a liquid. The viscosity of the suspension, η_c, is related to that of the pure liquid, η_0, and to the volume fraction, v, of the suspended phase, by Einstein's equation:

$$\eta_c = \eta_0(1 + Dv),\qquad(4.17)$$

in which D is a constant. The volume fraction can be written $v = nV_0$, if n is the number of particles per unit volume and V_0 the volume of one particle. For spherical particles in a volume concentration of 10% the Einstein

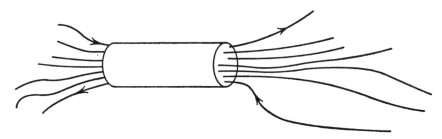

Fig. 4.11. Wires passing through a short tube illustrating crystallites and sliding possibilities for molecules.

Fig. 4.12. Crystallites in an unstretched linear polymer.

constant $D = 2.5$, and the suspension viscosity η_c is only 1.25 times the pure liquid viscosity η_0.

The fact that even dilute solutions of polymers in solvents have very high viscosities is related to the second effect mentioned above, namely the interaction (swelling effect) between polymer and solvent. Examining a linear polymer of molecular weight M, we can write an expression similar to the one first used by Staudinger:

$$\eta_c = \eta_0(1 + CKM),\qquad(4.18)$$

where C is the concentration in moles per volume.

The analogy with the Einstein equation (4.17) is obvious, K being a constant for the system, similar to D in equation (4.17). Using an unknown proportionality constant p, we can write $V_0 = pM$ where p measures the swelling effect.

It follows from experiments that equation (4.18) is valid only in dilute solutions and then only in some special cases. A different form of equation (4.18) is given by (4.19). This form allows us to define 'intrinsic viscosity' $[\eta]$ as in equation (4.20); it is the specific viscosity η_{sp} when divided by concentration and extrapolated to zero concentration.

$$\eta_{sp} = (\eta_c/\eta_0 - 1) = CKM \text{ or } \frac{\eta_{sp}}{C} = KM. \tag{4.19}$$

$$[\eta] = \lim_{(C=0)} \frac{\eta_{sp}}{C} = KM. \tag{4.20}$$

According to Houwink, this equation becomes very general, rather than special, if the dependence on molecular weight is non-linear, as shown in equation (4.21); it still refers to extreme dilution.

$$[\eta] = KM^\alpha. \tag{4.21}$$

The exponent α is found experimentally to be usually between 0.6 and 1.0, Theoretically its magnitude is given by the amount of free flow of solvent molecules through the coiled polymer molecule and, according to Kuhn, should be between 0.6 and 0.9.

The specific influence of the solvent on the space occupied by the polymer molecule can be expressed by equation (4.22):

$$[\eta]_\Phi = \Phi(\overline{S^2})^{\frac{1}{2}} M, \tag{4.22}$$

in which Φ is a constant characteristic for the solvent and S is the distance of a chain element from the centre of gravity of the coiled molecule.

Comparison of equations (4.21) and (4.22) shows that $M^{1-\alpha}$ is proportional to $(\overline{S^2})^{-\frac{1}{2}}$.

In the extreme case, a highly concentrated polymer solution approaches a homogeneous polymer swollen with solvent. Its rheology can better be understood from the point of view of the polymer itself. Transfer of macro-molecules as such will be impossible without considerable deformation and the function of the solvent has become that of a 'plasticizer', causing the attraction between sections of the polymer molecule to be less, and therefore the flow to be easier than in the pure polymer.

The function of a plasticizer can also be performed by (groups of) atoms chemically attached to the polymer chain. If such attachments lessen the average attraction between sections of the chains, a plasticizing effect is obtained which in this case is called 'internal' plasticizing.

Reverting to heterogeneous systems, a dispersion of 'hard' particles in a carrying liquid is less complicated in its rheology than a polymer solution. Einstein's equation (4.17) can be used, the total particle volume fraction being the product of particle volume and number of particles. The rheology becomes more complicated if the particles start touching each other (high concentrations and/or extreme shapes and/or mutual attraction).

An expanded form of equation (4.17) was derived by Guth for more highly concentrated suspensions where there is mutual interaction between the suspended particles:

$$\eta_c = \eta_0(1 + D_1 v + D_2 v^2 + \ldots).\tag{4.23}$$

In expressions (4.17) and (4.23) the factors D or D_1 are dependent on the outward shape of the particles. If these are spheres, $D = 2.5$ as stated before, and $D_1 = 2.5$ with $D_2 = 14.1$ in equation (4.23). For elongated shapes, the D factors are higher.

A clear example of spherical particles dispersed in a pure liquid is latex or milk. The viscosity of these dispersions is indeed related to the viscosity of water and the volume fraction of rubber or fat, in the manner predicted by (4.17) or (4.23). It follows that one can more easily handle, for pumping or stirring or gravity flow, a dispersion of a polymer in a non-solvent (viscosity between 1 and 3 × that of the medium) than a solution in a good solvent (viscosity up to 10^6 × that of solvent for 10 % concentration!).

Dispersions of 'real' solids, such as dispersions of fine glass beads or sand particles, are again similar to the latex example. An important point is the 'screening off' or 'peptizing' of the dispersed particles. If such particles have more attraction for each other than for the surrounding liquid medium, then a 'real' dispersion is impossible and coalescence or coagulation will soon follow, leading to different rheological phenomena (see below). If, however, the particles have more attraction for the liquid, they behave as a true dispersion. If substances have insufficient attraction for the surrounding liquid to form good dispersions, the trick of 'peptizing' is used. The surface of the solid particles is simply screened off by a monomolecular layer that has strong interaction with the liquid. In the case of dispersions in water, a layer of a soap is often used to cover the particles; the polar end of the soap is ionized and forms an ionic double layer preventing coagulation.

As a matter of fact, in the examples of latex and milk given above, rubber (or fat) particles are prevented from coagulating by a double layer formed by soap (or protein). In certain types of asphaltic bitumen, the asphaltene particles are well peptized and do not attract each other; the bitumen behaves as a real 'sol' or true dispersion.

One of the most interesting, and at the same time most practically important rheological systems is obtained in dispersions of solids that are not completely peptized and do not, therefore, behave as true dispersions or sols. Examples are paint, toothpaste, etc. and also some non-sol types of asphalt. Under some conditions these systems contain a skeleton of dispersed particles, touching one another at one or more points, and providing stiffness through the relatively large polar forces at the points of contact. A certain amount of shearing destroys most of this skeleton, and the system begins to flow as a true dispersion. In cases like these, a 'yield value' obviously occurs in the D–τ diagram, which is highly desirable. Also in these cases the yield value even has a clear physical definition, viz. the stress needed to destroy the skeleton. This has some similarity with the yield value of a gelatine gel (polar polymer in solvent).

At several points in our discussions we have encountered materials having a temporary structure which can be destroyed by shearing but which is repaired upon resting. The phenomenon called 'thixotropy' is based on such behaviour.

A thixotropic material is stiff and does not flow under relatively small stresses, but can be made to flow with less and less viscosity as the shearing stress is increased, whereafter the low viscosity persists even under reduced stress; but when the material is stress-free it returns to its high stiffness with time.

Thixotropic materials have a viscosity that depends entirely on the previous history (time, stress, flow). This is expressed in Fig. 4.13a. Clear examples are drilling mud, quicksand and most rubber compounds (unvulcanized). Thixotropy occurs to a lesser extent in most plastic materials, either because stress causes orientation, or because it destroys weak structural bonds.

Table 4.2 gives quantitative flow data for various materials. For a number of 'simple liquids' the flow is Newtonian and a simple viscosity number describes it. For other materials the 'Bingham' or the 'plastic' equation (4.15) or (4.16) is needed and two constants must be given.

Fig. 4.13b is a D–τ diagram showing flow patterns in different classes of materials, with a list of examples for each class.

As can be seen in Table 4.2, a sol asphalt has a very high viscosity; nevertheless it flows (slowly) under its own weight. Clay, however, can easily be deformed between the fingers (low viscosity), but keeps its shape due to its yield value θ. Clay is a modelling material, but asphalt is not.

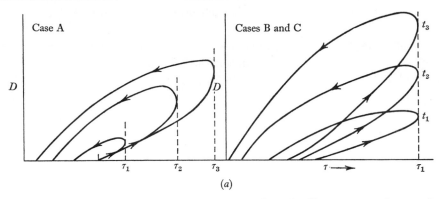

Fig. 4.13. (*a*) Flow curves for thixotropic cases. Case A: Shear stress τ increased to maximum τ_1 or τ_2 or τ_3 in same time t_0 and then decreased. Case B: Shear stress τ increased to maximum τ_1 in time periods t_1 or t_2 or t_3. Case C: Shear stress τ held constant at τ_1; shear rate (D) values shown at times t_1, t_2 and t_3.

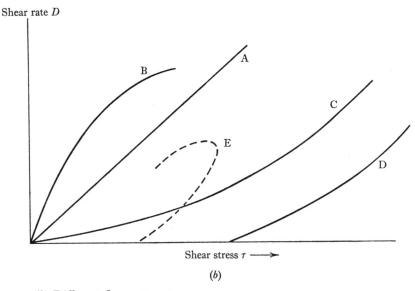

Fig. 4.13 (*b*) Different flow patterns.

    A Newtonian: water and other simple liquids; steel at 1000° C; aqueous 5 % sugar solution, 20° C.

    B Dilatant: vinyl resin pastes; beach sand; printing inks.

    C (Pseudo) Plastic: sol asphalt; most plastics and rubbers.

    D Plastic: grease; gelatine in water; moulding clay; paint; toothphaste.

    E Thixotropic: drilling mud; quicksand.

TABLE 4.2. *Viscosity data (at 20° C unless otherwise stated)*

(Viscosity in poise (dyne sec/cm²))

Newtonian flow D = τ/η

Water	0.010	Light machine oil	1.038
Pentane	0.003	Castor oil	9.86
Hexane	0.003	Asphalt (sol type)	2×10^5
Heptane	0.004	Mercury	0.016
Octane	0.007	Mercury (300° C)	0.009
Benzene	0.0065	Zinc (357° C)	0.014
Ethanol	0.012	Soda glass (575° C)	11×10^{12}
Butanol	0.029		

Bingham flow $D = (\tau - \theta)/\eta = \psi(\tau - \theta)$; θ in dyne/cm²

	η	θ
0 % Carbon black in varnish oil	49	—
10 % Carbon black in varnish oil	147	229
20 % Carbon black in varnish oil	1277	915
0 % Lithopone in linseed oil	0.4	—
32 % Lithopone in linseed oil	5	1200
45 % Bentonite clay in water	0.3	1000
55 % Kaolin clay in water	25	1500

'Plastic' flow $D = \tau^n/\eta^x = \psi\tau^n$; D in rad/s; τ in kgf/cm²

	ψ	η^x	n
Natural rubber unmasticated (70° C)	0.0036	280	3.7
Natural rubber masticated 12 min (70° C)	0.10	10	2.6
SBR rubber unmasticated (70° C)	0.83	1.2	4.05

Complete equations for deformation as related to stress contain all stress and strain components in three dimensions (such equations are given for metal elasticity in Chapter 12). For most practical purposes (and, in particular, in this book), the simple form of equation (4.12) is adequate. However, some phenomena can only be interpreted properly by the use of the complicated three-dimensional equations. Specific cases are the 'normal pressure effects' or 'Weissenberg effects' observed in flow of visco-elastic materials. These effects are illustrated by rotating a shaft in a relatively large mass of a material. If the material has a 'yield value' of stress (is visco-elastic) it will be seen 'climbing the shaft'. A simple explanation can be found by visualizing that by climbing up the shaft, the material stays outside the viscous shearing action of the remaining material. One can assume that any material will have a tendency to climb the shaft but if the fluid is purely viscous, the gravity pull will always keep it down. The extent of shaft climbing is related to the magnitude of the yield value and the speed of rotation.

The climbing of a visco-elastic material up a rod rotating in it provides a simple test for the presence of elastically recoverable strain. Even manual rotation is often sufficient.

4.3.6 Visco-elastic behaviour

So far in this chapter, we have mostly examined flow as a stationary phenomenon. After the application of a certain shearing stress, and after the material has had enough time to adjust to it, a definite flow rate may result, and it is this flow rate $D = d\gamma/dt$, that was examined in its dependence on stress (see equations (4.12, 4.15, 4.16)). In doing this, we used D as if it was independent of time, although in Figs. 4.9 and 4.13 it was pointed out that this is a simplification.

The following is a brief discussion of (1) total deformation as a function of time and stress; (2) stress decay at constant deformation (stress as a function of time); (3) deformation before reaching stationary flow, during very short times, such as occur in the 'dynamic' response under vibration or under impact.

(1) Total deformation is the sum of permanent or viscous deformation and elastic (recoverable) deformation;

$$\gamma_{\text{total}} = \gamma_{\text{visc}} + \gamma_{\text{elast}}.$$

As was shown in §4.2, the elastic deformation should be subdivided into γ_0, 'ordinary' elastic deformation (due to primary bond distortions) and γ_r, rubbery elastic deformation (due to deviation of coil molecules from their random shape). The viscous deformation γ_v can be derived from equations (4.12, 4.15, 4.16) above.

The total deformation under a given stress τ can be written as follows:

$$\gamma_{\text{total}} = \gamma_0 + \gamma_r + \gamma_v = \frac{\tau}{G_0} + \frac{\tau}{G_r} + \frac{\tau t}{\eta}. \qquad (4.24)$$

The modulus G_0, resulting from primary bond distortion, is practically independent of time, and only slightly dependent on temperature. However, the modulus G_r, resulting from the 'pull' of coiled molecules back to their random shape, is dependent on both time and temperature, because of the resistance met by coiled segments in their motions. Finally, the viscosity η is temperature dependent, as shown in equation (4.14). Combining these dependencies, one can write equation (4.24) in a more complete form as:

$$\gamma_{\text{total}} = \frac{\tau}{G_0} + \frac{\tau}{G_r} (1 - e^{-t/\beta(T)}) + \frac{\tau t}{\eta(T)}, \qquad (4.25)$$

in which $$\beta(T) = Be^{E_r/RT},\qquad(4.26)$$

and $$\eta(T) = Ae^{E_v/RT}.\qquad(4.14)$$

E_r and E_v are the activation energies of rubbery elastic deformation and of viscous flow, respectively.

The continuing deformation under constant stress is called 'creep' if it is relatively slow and concerns a 'solid' or 'plastic' material (in a liquid it would be simple flow). Creep can be observed in metals as well as in high polymers, to a degree depending on temperature and stress.

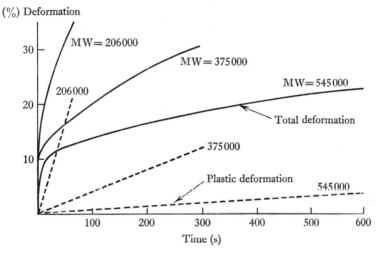

Fig. 4.14. Creep curves for polyisobutylene. — total; - - - plastic.

In polymers, the molecular weight (average and distribution) has an important influence on creep, as shown in Fig. 4.14 for polyisobutylene; the graph shows separately the contributions of the rubber-elastic and the viscous terms in equations (4.25).

(2) Stress as a time function under constant deformation is derived from Maxwell's equation

$$\frac{d\gamma}{dt} = \frac{1}{G}\frac{d\tau}{dt} + \frac{\tau}{\eta} \quad \text{or} \quad d\gamma = \frac{d\tau}{G} + \frac{\tau}{\eta}\,dt;$$

if the deformation is kept constant, $d\gamma = 0$,

and $$\frac{d\tau}{\tau} = -\frac{G}{\eta}\,dt$$

or $$\tau_t = \tau_0 \exp\left(-tG/\eta\right).\qquad(4.27)$$

The ratio $\eta/G = \lambda$ may be called the relaxation time. It determines the rate of stress decay; if λ is very small, the stress will rapidly approach zero. In true materials, such as polymers, the expression (4.27) is not satisfactory and must be replaced by

$$\tau_t = \tau_{01} \exp\left(-t/\lambda_1\right) + \tau_{02} \exp\left(-t/\lambda_2\right), \qquad (4.27a)$$

indicating that the material has a series of relaxation times (relaxation spectrum) and that part of its reaction will be very slow, and part rapid.

It will be noted that both the total deformation (4.25) and the stress decay (4.27) contain a term having a coefficient $\exp\left(-t/x\right)$; in (4.25) the x is a retardation time, in (4.27) it is a relaxation time. In both cases x indicates the rapidity of the response of the material to shear, and both deformation and stress will approach stationary states through an asymptotic function. If x is very small, the response is so fast that the asymptotic growth of deformation (or decay of stress) cannot be noticed and stationary states are reached 'instantaneously'. True liquids and true solids would have $x = 0$ (ideal Newtonian or ideal elastic behaviour). High molecular weight materials at temperatures slightly above their glass transition temperature have spectra of x containing very large values, and their responses are very slow.

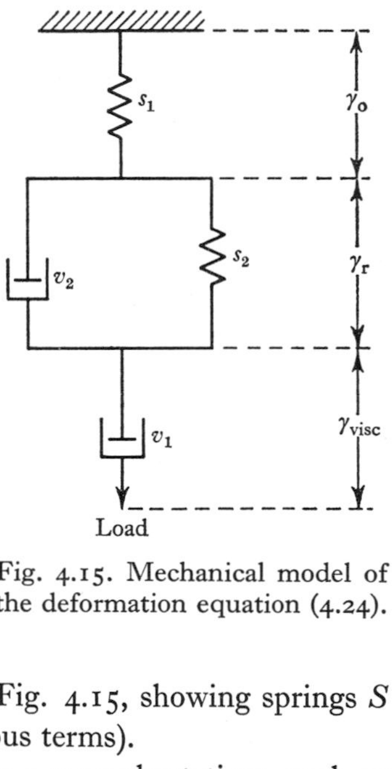

Fig. 4.15. Mechanical model of the deformation equation (4.24).

The well-known 'Maxwell models' and their modifications can be used to illustrate the visco-elastic behaviour of materials (see Fig. 4.15, showing springs S for elastic terms and 'dashpots' V for the viscous terms).

(3) If a deforming stress occurs only during a very short time, such as happens in impact, or changes rapidly in magnitude and direction, as in vibration, there is usually not enough time to reach any kind of visible flow phenomena. Nevertheless, the viscosity or plasticity of the material (internal friction between molecules) is a very important factor in the response of the material to this kind of stress.

In impact, a 'plastic' material will return less elastic energy, appear more 'dead', if the relaxation times (or retardation times) are larger. This means less resilience for more 'viscous' materials, having values of $\lambda = \eta/G$ that are high.

In vibration, a material returning less elastic energy will therefore show more energy loss and this leads to rapid 'damping' of a free vibration and a high damping factor k. It also leads to a retardation in the response of the material to a periodic force; this retardation can be expressed as a phase difference between force and deformation.

The stress needed to keep a test piece in harmonic vibration can be derived as follows.

The shear in harmonic motion is $\gamma = \gamma_0 \sin \omega t$; the stress needed is

$$\tau = G\gamma + \eta \frac{d\gamma}{dt},$$

or $\quad \tau = G\gamma_0 \sin \omega t + \eta \gamma_0 \omega \cos \omega t = G\gamma_0 \left(\sin \omega t + \frac{\eta}{G} \omega \cos \omega t \right).$

Thus, the stress has one component in phase with the shear (sinus term) and determined by the elastic modulus G, and another component which is 90° out of phase (cosinus term) and the relative magnitude of which follows from a damping factor $k = (\eta/G)\omega = \lambda\omega$. The resulting total stress runs ahead of the shear by a damping factor k, which is again proportional to the relaxation time $\lambda = \eta/G$, and to the frequency of the vibration. Fig. 4.16 illustrates this vibration.

All materials are situated between two extremes. The extremes represent, respectively, the pure elastic state (solid metal or glass) and the pure liquid state (water at room temperature). In a purely elastic material, upon impact, all energy is returned elastically; in vibration, the deformation and stress are simultaneous without phase difference and damping is zero.

In a purely liquid material upon impact, all energy disappears into thermal energy, and in vibration, the phase difference between stress and strain is maximal (90° or $\frac{1}{2}\pi$) and damping is extreme.

Many practical materials at room temperature show some evidence of relaxation and damping at the same time as elasticity. Obvious examples are wood, rubber, silk, nylon, PVC, asphalt, etc.

At a given frequency of deformation, liquids show damping roughly proportional to their viscosity, because their G values are all equally small. In other materials such as thermoplastic polymers, both η and G have variable and finite values, so that two of the three variables, λ, η, and G, are needed to characterize their damping.

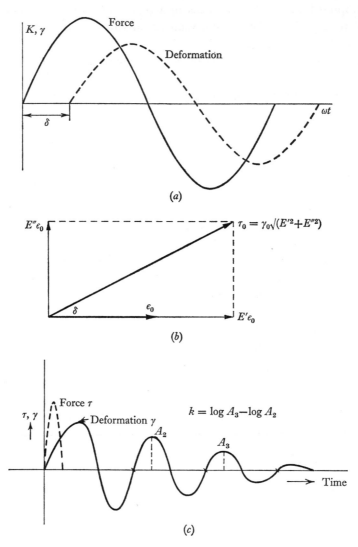

(a)

(b)

(c)

Fig. 4.16. (*a*) Periodic force and deformation as a function of time. (*b*) Vector diagram relating stress τ, deformation γ, moduli E, and loss angle δ. (*c*) Damping of free vibration.

4.3.7 Yield value and plasticity of crystals

It has been shown in §4.2.1 that elastic deformation reaches a limit beyond which other mechanisms take over. We also know that there is really always some plastic deformation, even if very small, under any condition of time and temperature. Nevertheless, in many rigid materials, metals in particular, the total deformation is so very predominantly elastic up to a certain stress, and then becomes so predominantly plastic, that it is useful to agree on the definition of a transition point.

This transition point (which is not a material property!) can be defined by selecting a very small amount of plastic deformation which has to be reached and surpassed at that point. Thus, an 'elasticity limit' or 'yield value' is conveniently defined as in equation (4.28):

$$\gamma_Y = \gamma_E + \gamma_{PM}. \tag{4.28}$$

The yield value γ_Y is the sum of the elastic deformation γ_E and a predetermined amount of plastic deformation, γ_{PM}. In this definition, the time factors of the experiment have to be defined also. The magnitude of γ_{PM} is 0.2 % in some practical metallurgical definitions.

The nature of the plastic deformation mechanism that takes over beyond the yield value in crystalline materials is different from that in amorphous materials, notwithstanding the fact that a certain similarity may be observed in the macroscopic appearance of flow in the two types of material.

The flow or plastic deformation of amorphous substances is basically a process in which the constitutent atoms, molecules, or 'units' change places 'individually'. In crystals, a large number of them, held together by a certain elastic coupling, move simultaneously in such a manner that a very definite change of shape is brought about, which may be described in general as due to a process of slip or shear along definite crystallographic planes. Such a process can take place unhindered only in an ordered arrangement of particles, such as exists in an undeformed crystal lattice.

Each normal crystalline material, even a pure metal crystal, has 'flaws', or places where the regular structure has an imperfection. These flaw may be of many types (impurities, cavities, etc.). They may cause a 'dislocation', that is a 'jump', in the regular lattice pattern, as illustrated in Fig. 4.17. This dislocation then, as soon as it is formed, will be able to run through the entire length of the crystal plane, thereby effecting a shift of one atom distance, as shown in Fig. 4.17. The direction in which the dislocation runs is such that the resulting deformation complies with the external stress. It is easy to see, and can be shown mathematically, that the force necessary is considerably less than the force needed to shift an entire crystal plane based on total energy of attraction. It is equally obvious that the temperature motion only enters in determining the force needed to create dislocations; the principal mechanism, the running of the dislocation, is based on a zipper-like action not involving much energy.

In fact, as the movement of a dislocation leaves a perfect lattice in its wake, the crystal, statistically speaking, remains unaltered in the course of the deformation process. This situation, however, is wholly altered if, as has been assumed by Taylor, the distance over which a dislocation can be propagated

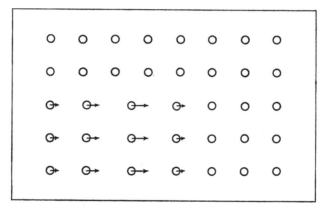

Fig. 4.17. Metal crystal with dislocation. In the next deformation step, atoms move to the right (arrows) bringing the dislocation one step back to the left.

is not given by the dimensions of the crystal, but is limited by the same system of faults or flaws which, as was suggested above, may play a part in the formation of the dislocations. In fact, the supposition presents itself that, upon reaching some irregularity in the crystal lattice, be it foreign atoms or an unfavourably situated flaw, instead of its being propagated in a regular manner, the dislocation will cause a local distortion of the lattice, which will be the seat of a system of stresses. As the deformed state of the crystals is thus characterized by the presence of an ever-increasing number of arrested dislocations, an internal field of stress arises, which may be expected to influence the further development of the glide process.

It can now be understood how a metal can undergo strain hardening, or annealing, and how it can be reinforced by alloying with other metals. Strain hardening, as seen above, is caused by the plastic deformation itself, which leads to 'opposing' stresses around lattice flaws; these are also sometimes described as lattice distortion or fractioning; their result is that further gliding through the moving of dislocations is stopped. This explains the upward slope of the true stress–strain curve for plastic flow in metals, as shown in Fig. 4.2. Another result is that 'strain hardened' metal maintains purely elastic behaviour up to a higher yield value than before hardening.

If a strain hardened (cold rolled sheet, cold drawn wire) metal is exposed to a sufficiently high temperature for a sufficient length of time, it softens and becomes similar to the metal in its original state. The thermal motion at the higher temperature is able to restore atoms to their less stressed positions, ideally going back to the perfect lattice with its minimum of flaws, and the elasticity limit (yield value) and flow behaviour are restored to their original situation. This heat treatment which causes return to the non-hardened state, is called 'annealing'.

The utility of mixtures of metals (alloys) can be based on different effects. In the field of rheology, the utility of alloys is based on the hardening effect of one atom species inserted in the crystal lattice of another species. An example is brass, an alloy of zinc and copper, which is considerably harder than either of the pure metals. Another is carbon steel, essentially iron with 0.2–1 % carbon: the carbon is soluble in the iron near the melting temperature and insoluble at room temperature but after rapid cooling ('chilling or quenching'), a number of C atoms remains in the iron lattice, causing the steel to be much harder and stiffer than pure iron. Fig. 4.2, p. 54, gives the stress–strain curves for these alloys.

After this survey of crystal plasticity in its relation to crystal elasticity, it is important to remember the distinction between 'three-dimensional' crystals (metals, diamond, salts, oxides) and other crystalline materials, having primary bonds in two dimensions (mica, graphite) or in one dimension (asbestos, silk). The discussion in this subsection was related only to three-dimensional crystals. The other types are dominated in their rheology by the fact that in one or two directions there are only secondary forces to hold the material together. That means that visco-elastic deformations along these directions are relatively easy, and leave the primary structure in the other directions virtually intact. Graphite is easy to deform by a sliding motion of its crystal platelets along one another. Asbestos or silk are relatively easy to deform by sliding the crystal forming 'threads' along one another. As we have seen above in §4.3.4 this may still require considerable force, but in none of these cases is there displacement of atoms away from the positions fixed by their primary bonds, as is the case in metals and similar crystals.

A further complication arises in high polymers of regular linear structure by the fact that their crystallinity may not be perfect; the material may consist of crystallites and amorphous material, each polymer molecule participating in one or more crystalline and one or more amorphous regions. This type of material (see Fig. 4.12) is closer to a plastic or rubber containing a reinforcing filler, the crystallites playing the part of the filler (see §4.2.4 above).

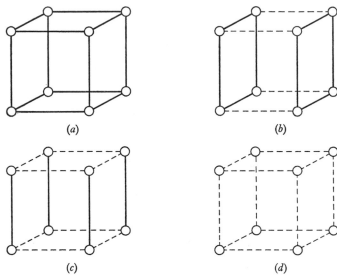

Fig. 4.18. Primary bonds ◯—◯ and secondary bonds ◯- - -◯. Possible arrangements in three dimensional crystals.

Fig. 4.18 shows the possible configurations of primary and secondary bonds in three dimensions. It should be concluded from this figure that certain 'molecular' crystal lattices (ice, many organic chemicals in crystalline form) will show behaviour similar to that of metals (*d* is similar to *a*). However, the replacement of all primary bonds by secondary bonds in all three dimensions (the molecules being regarded as stiff units, like the atoms or ions in other crystals) should result in all stresses, for elastic as well as for plastic deformation, being smaller than for metals. This is indeed the case.

It is interesting to note a relatively simple inorganic compound, phosphorus pentoxide, which was shown to exist in 3 of the 4 possible configurations of Fig. 4.18. Depending on its thermal history, it forms a completely stable atomic network (—P—O—P—O---) in three dimensions,

(Fig. 4.18*a*) or a fibre structure (Fig. 4.18*c*), or a purely molecular lattice (Fig. 4.18*d*) having P_4O_{10} molecules as the units in a lattice held together by secondary forces. The latter configuration, incidentally, is the only one that shows the well-known hygroscopic behaviour of this compound; the weak secondary forces allow water molecules to enter and react rapidly. In contrast, configuration (*a*) can actually be immersed in water and remain there for days without visible reaction.

4.4 Strength

4.4.1 Basic concepts

From the standpoint of rheology, strength (sometimes called ultimate strength) is the maximum force that a material can resist without rupture. Inherent in the concept of rupture is the non-reversibility of the process; the material separates into pieces that will not reconstitute the original coherent material. Whereas, by definition, elastic deformation is reversible and plastic flow can conceivably be reversed (in the case of water flowing through a pipe this is very simple), the phenomenon of rupture is irreversible. Bonds between parts of the material suffer a kind of total breakdown in rupture that is not analogous to the straining of bonds in elastic deformation, or the partial breakdown and re-arrangement of bonds in plastic flow.

Rupture appears to us as a macroscopic separation of two pieces of material previously bound by a great number of bonds between atoms or molecules (primary or secondary bonds). It would be an easy simplification to assume that the breaking process overcomes all these elementary bonds at once and therefore needs a force roughly equivalent to the sum total of all these bonds. Nevertheless, this simplification is a fallacy, for various reasons that will be discussed next. It is also evident from experimental data that this simple assumption may be in error by a factor of 1000 or 10000, as illustrated by Table 2.2 on p. 22.

We shall now first correct our molecular (structural) interpretation of processes leading to rupture, and subsequently, introduce a better concept, namely the concept of 'maximum capability' of the material (maximum capability for uptake of energy and/or deformation). This will lead to an understanding of strength and rupture.

Macroscopic observation of many breaking phenomena, and microscopic study of most, reveals that there is always a failure point or a failure region where the rupture starts, rather than a sudden simultaneous separation of an entire break surface.

Assume that, for one reason or another, the external force succeeds in separating a group of atoms (molecules or kinetic units should be visualized in other cases, but for our purpose the discussion of atoms as in a glass or metal is appropriate and can be transferred to these other cases without effort) from their opposite numbers in the network, as shown in Fig. 4.19, where group A is separated from A'. If the same external force persists on the sample, then the separation of A and A' will result in considerable stress concentration in the regions B_1 and B_2. This concentration is due to the

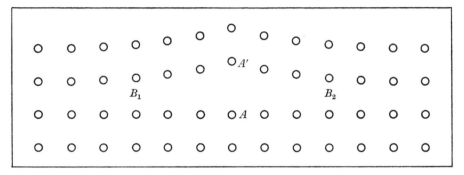

Fig. 4.19. Small separation in the crystal lattice.

'notch factor'; it operates at the tip of any discontinuities in matter where it increases stresses to many times their overall value. Calculations according to applied mechanics show that the concentration next to a simple elliptical hole in a material is given by equation (4.29):

$$F_{\max} = F\left(1 + \frac{2a}{b}\right). \tag{4.29}$$

where F is the external stress perpendicular to the long axis a of the ellipse, b is the other axis and $F_{\max}$ is the maximum stress at the edge.

If a sharp notch is described as an ellipse, this leads to an axis ratio of $a:b$ of 100 or more, and the maximum stress becomes 200 or more times the external stress. It follows that stress concentration in the regions B of Fig. 4.19 may lead to propagation of the rupture or separation, even if the external stress is several orders of magnitude below the stress needed to separate two atoms. In this concept, the material rips apart from a given starting point, and does not need the force expected for a simultaneous neat separation of all the atoms involved.

As a matter of fact, materials always contain 'flaws' in their atomic structure, or *Lockerstellen* as the German literature calls them. These flaws can be jumps in the regular succession of atoms in a crystal lattice, or small (a few atom distances) areas of lower particle density in an amorphous resin or glass. These flaws can conveniently be called 'vacancies'. Also, at the exterior surface of the sample, there will be many particles that are in a non-ideal position and thereby represent flaws in the form of a surface 'notch' of atomic dimensions.

Furthermore, any foreign matter or impurity, present by accident or on purpose, necessarily creates a small area of imperfection in the material, a flaw typified by distortion of the regular structure and stress concentration around it. These foreign inclusions can either be different atoms in a crystal lattice (see earlier) or small crystals of a foreign nature dispersed between regular crystals or amorphous structure.

It has even been shown that a (pure) crystalline material may have an 'interstitial' extra atom, the lattice around it being distorted to create a place between regular lattice points, thus causing the opposite of a vacancy.

To demonstrate the influence of all these flaws on the surrounding material, one can either use the notch theory as shown above, or one can follow the more complete and refined method of calculating potential energy curves for atoms surrounding a flaw and compare them with 'normal' atoms. In both cases, the result is that atoms surrounding a flaw will need considerably less stress than the other atoms to be permanently dislocated. Thus we can say that atoms surrounding flaws will be obvious candidates for participation in a separation area as soon as breaking stress is reached. Recent research on extremely pure single crystals virtually free of flaws, confirms the expectation that these will be many times stronger than practical materials.

It should be pointed out here, that a similar mechanism is responsible, in metals, for plastic deformation and for ultimate break, namely flaw + stress → dislocation → running or ripping action, either to permanent deformation or to rupture. In other rigid materials, such as glass, there is no plastic deformation between elasticity and rupture, because the running of a dislocation depends on crystal structures, and these are not available in glass. In the more 'plastic' rigid materials, the plastic deformation mechanism may actually *repair* pre-existing or emerging flaws, thereby improving the breaking stress; however, this is a time dependent phenomenon, and the breaking stress will be considerably lower in a fast experiment (impact or brittle failure) because no time is given for the flow and its repair mechanism. A simple mechanism causes the breaking of rubbery elastic materials. We have seen that chain molecule sections between cross-link points are in a random coil position when unstressed, and that these are pulled from the random coil to a more elongated position by stressing the material. This elongation process has its natural limits, namely where each chain section is fully stretched to its own length. It can be calculated that the actual elongation at break corresponds roughly with the ratio between end-to-end distances for fully stretched and fully coiled positions (Fig. 4.3 b and d) of a typical vulcanized rubber. The tensile strength, in turn, is simply the force needed to reach this ultimate elongation. This will be shown in an example later in this section (p. 96).

4.4.2   Maximum capability for deformation

From the above discussions it can be seen that materials have limited capability to deform, and that this limitation determines the 'strength' or breaking stress. The capability to deform can be examined from the point of view of actual deformation (change of shape), or from that of energy uptake. The latter point of view will be discussed first.

If a force causes deformation of a sample of material, it performs work. This work remains as free energy (purely elastic deformation as in steel, or decreased entropy as in rubber) or is dissipated in thermal motion (heat) through flow phenomena. As long as a mechanism for conversion of deformation energy into heat remains available, and as long as this disposes of all additional energy beyond the maximum amount that can be stored elastically, the material will not break. In that case it has unlimited capability for deformation. Such unlimited capability for deformation (or for energy uptake) does not mean that the breaking strength is infinite; it only means that forces up to a certain magnitude can be exercised for any length of time without leading to rupture.

The magnitude of the forces that are allowable, as well as the capability for energy uptake, are dependent on time and temperature. We have seen how a plastic such as PVC can either flow or break brittle, depending on the temperature of the material and the rapidity of stress increase. At any time, the work W carried out on a piece of material is used partially to increase the free energy E, and partially to be dissipated in heat H, as expressed in equation (4.30):

$$W(t) = E(t) + H(t). \tag{4.30}$$

The symbol (t) denotes that each of the quantities is time dependent. If breaking occurs, $W(t)$ is the work of fracture. In the simplest case, there is no heat dissipation, so that $H(t) = 0$, and $W(t) = E(t)$ and the maximum work one can perform on the test-piece before rupture equals the maximum free energy stored elastically which, as we have seen, is simply related to the magnitude of elastic deformation the material can undergo. At the point of breaking there is then a sudden irreversible transformation of this maximum free energy into heat. Glass and other brittle materials illustrate this mechanism.

If there is dissipation during deformation, as expressed in equation (4.30), the picture is more complicated. In the first place, any increment of work done, ΔW, can be dissipated as heat (by a 'creep' mechanism):

$$\Delta W = \Delta E + \Delta H. \qquad (4.30a)$$

In addition, if no more work would be performed, there would be a gradual dissipation of free energy as heat (by a 'relaxation' mechanism):

$$0 = \Delta'E + \Delta'H, \qquad (4.30b)$$

which decreases the stored free energy by $\Delta'E$, and automatically increases the capacity for more work without breaking, by the amount $\Delta'E$. To obtain rupture in this picture, the work W must be performed at such a rate $\Delta W/\Delta t$, that $\Delta E/\Delta t \geqslant \Delta'E/\Delta t$. Again, as in the simplest case, when breaking occurs there is a sudden irreversible transformation of the stored free energy into heat. There will be no rupture as long as the rate $\Delta W/\Delta t$ is slow enough to allow ΔE to remain below $\Delta'E$. The only possibility of rupture under these conditions consists of 'creep failure', caused by gradual decomposition of the material structure.

It has been shown for rubbery materials that the tearing energy can be related directly to the work performed at the tip of a cut or tear. As we have seen before, breaking strength is directly related to tearing strength, because breaking is really always the propagation of a tear from an initial flaw. It was shown by Mullins that the tearing energy (which is proportional to the deformation work W used in equation (4.30)), is dependent on temperature and time in the way given by the Leaderman–Tobolsky–Ferry transformation, which means dependence on the relaxation times in the material.

It is essential to note that the energy approach to breaking gives us an understanding of the mechanism, and establishes the principle that maximum capability for storing elastic (free) energy determines the breaking strength, but fails to offer an estimate of the magnitude of this capability. Such an estimate is only possible through the relationship between elastic energy and elastic deformation. By studying the structure of a material, an idea of its maximum allowable elastic deformation can be obtained. As a general consideration, we must remember that the allowable elastic deformation may appear large or small for the same material, depending on the shape of the test-piece. A sharp notch or cut will cause stress concentration at its tip, and the material may locally reach its maximum elastic deformation while showing very little overall deformation. From the allowable elastic deformation, the corresponding allowable stress can be calculated, and this constitutes the breaking stress – as we shall illustrate for rubber.

For *metals* we know that deviations from elastic behaviour occur at small deformations (e.g. o.1 %) and we have reasoned that a flow-by-dislocation mechanism is started at flaws when that deformation is reached. That same flow mechanism leads to strain hardening by lattice distortion which increases the allowable elastic elongation. If the stress is increased, the process is repeated up to the point where the lattice distortion is so considerable that the next stress increase leads to break. If stress increases occur very slowly or at sufficiently high temperature, the situation is modified by annealing, or the disappearance of blocking mechanisms by diffusion, leading to greater plastic compliance at the same or lower stress.

For *glass* and three-dimensional *resins*, the amorphous structure provides a wide range of deviations from the 'ideal' angles and lengths of valency bonds between atoms (or, even worse, between multi-atom units in resins). Situations with greatly strained angles and lengths occur with statistical frequency; external stress causes some of these situations to be corrected, and others to be aggravated; the latter initiate the rupture process which must obviously occur at the elongation needed for a *few* extreme situations to be beyond repair, not at the elongation which the *average* configuration would be able to sustain.

This explains the fact that glass has very little capacity for deformation, and that break occurs while the greater majority of the bonds has hardly been distorted, leading to a breaking strength corresponding to this relatively small distortion (deformation).

The absolute level of breaking strength in metals and glass is similar; the fact that metals have a plastic flow mechanism interposed between elastic deformation and rupture does not greatly change the fact that the breaking is, in both cases, the rupture of primary bonds by a ripping action starting at a discontinuity.

The measure of a material's ability to retain its strength in the *presence of cracks* is determined by the work of fracture of the material ($W(t)$ in equation (4.30)), which is to say the energy required to break it. Glass has a very small work of fracture and well-designed strong steel a very high one, the difference being in the term $H(t)$ in equation (4.30). The inherently strong materials such as silicon carbide, boron and graphite all behave somewhat like glass; their work of fracture is small and so they are quite vulnerable to the presence of cracks.

Metals are normally used to bear large stresses because they can accommodate themselves to cracks. With a metal the engineer obtains high strength without having to take extreme care to eliminate all but the tiniest cracks. Polymeric materials such as polyethylene are rather like metals in resistance to cracks, although they will not take as much stress as metals.

The basic chemical reason why metals and polymers are so much more resistant to cracks than ceramics is that the interatomic forces in metals and the intermolecular forces in polymers do not depend on a particular directional alignment to achieve strength. Moreover, the chemical bonds of metals and polymers are unsaturated, that is, the atoms or molecules of such materials are readily capable of forming new bonds. Ceramics, on the other hand, have highly oriented forces and saturated bonds. In metals and polymers, atoms or molecules always slide over one another at the leading edge of a crack. The crack therefore cannot penetrate the internal structure of a metal or polymer as easily as it can the structure of a ceramic; a much larger stress is needed to make a crack run through a metal or a polymer and divide it into two pieces.

The familiar forms of ceramics are represented by glass, chalk and sand; by abrasive powders such as corundum and carborundum, and by gems such as diamond and ruby. It is because ceramic crystals are so vulnerable to cracks that ceramic materials are often used in the form of a powder, to which they are easily reduced. As a powder they scratch steel because in fact they are harder than steel.

In sum, if a ceramic material is unscratched, it can be very strong. If it is flawed in any way, however, it breaks easily. The lack of resistance to cracks is called brittleness. Ceramics are usually fragile because ceramic crystals are almost always marred by cracks or surface irregularities. Even if they are not, such imperfections can be introduced all too readily.

To improve the strength of glass, the obvious method is to reduce the number of flaws both internally and on the surface; this appears particularly effective in thin glass fibres.

Uniform cooling in the entire cross-section during glass formation from the melt, promotes a flawless structure; this is much easier to accomplish if the cross-section is small, as is the case in fibres.

For *rubber*, the breaking process has been discussed on p. 86.

For *rigid plastics*, the reasoning is as for glass, but secondary forces replace the primary bonds of the glass case. As the secondary bonds are highly sensitive to temperature, one observes that the tensile strength of a plastic usually diminishes at higher temperature; the tensile strength of PVC drops from 700 to 200 kgf/cm^2 if the temperature changes from 0 to 80° C.

For *liquids* there is no rupture phenomenon except under very extreme conditions of speed, when even the shortest retardation times cannot follow the stress increase; 'atomization' is the result.

4.4.3   Reinforcement

Reinforcement is sometimes defined in its broadest sense, as any change in a material causing any of its desirable mechanical properties to be increased. Thus, reinforcement with respect to hardness, modulus, impact resistance,

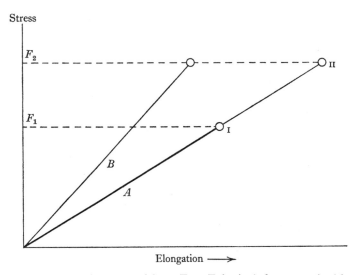

Fig. 4.20. Tensile strength increased from F_1 to F_2 by 'reinforcement'; either modulus increased from A to B, or compliance (elongation at break) increased from I to II.

abrasion resistance and tensile strength is sometimes mentioned. For the purpose of this book, we prefer the more direct definition which links reinforcement with 'force' or 'strength'. We assume the word reinforcement as meaning a change in a material causing increased resistance to deformation, including rupture. Increases in elastic modulus or in breaking stress are the most significant examples.

A higher elastic modulus will result if the retractive force in a material is increased without changing its range of elastic compliance. A higher breaking strength will result from either this higher modulus or a longer range of elastic compliance, or both. Fig. 4.20 illustrates this.

(a) Metals

Any structure that blocks the running of dislocations, and any feature that prevents the forming of dislocations, will increase the range of elastic deformation and therefore usually the breaking strength. We have seen how this works out in the examples of brass and carbon steel, and also in the case of 'strain hardening'.

(b) Glass and thermosetting resins

The means available for reinforcement are elimination of flaws and substitution of atoms with strong bonds and a large diameter (allowing more deviations from the ideal valencies without causing undue stress). The elimination of flaws is well-known in the production of fibreglass, which has higher breaking strength than regular cast, rolled or blown glass.

(c) Rigid thermoplastics

This class of materials offers good illustrations of two mechanisms of reinforcement. The first is that of 'blocking of flow' where the flow would lower the elastic resistance of the material; the second is that of 'redistribution of stress' around flaws.

'Blocking of flow' is effected by fillers, small particles of a hard material mixed with the plastic. The plastic under stress, being amorphous and held together by secondary forces between macromolecules, suffers some plastic flow, even if the stress is small and much of the strain is elastic (see above §§4.2.1 and 4.3.6). At greater stress, the plastic flow becomes more pronounced, and as more flow occurs, flaws develop leading to ultimate failure. Now the presence of rigid small particles (assumed spherical for the purpose of calculations) will increase the viscosity coefficient η according to the Einstein–Guth–Gold equation (4.23), to a higher value η_R:

$$\eta_R = \eta(1 + D_1 V + D_2 V^2), \tag{4.31}$$

where V is the total filler volume fraction. This means that total deformation γ under stress τ is now reduced:

$$\gamma = \gamma_{\text{elas}} + \gamma_{\text{visc}}; \ \gamma_{\text{visc}} = \tau t / \eta_R. \tag{4.32}$$

If we examine the case of relatively large stress and short times, the overall 'rigidity' or elasticity modulus can be written as $E = \tau/\gamma$, and it is increased by the filler because γ is lowered through its viscous term.

It is obvious that the filler particles must be well-dispersed, and completely surrounded (wetted) by the plastic, otherwise they would engender flaws and the overall effect would be early break, and the opposite of reinforcement. The larger the filler particle and the more its surface deviates from a smooth sphere, the greater is the chance that crevices and flaws will be created near the filler surface. Therefore, good reinforcing fillers are of very small particle size and have a surface affinity for the plastic they reinforce. Polyethylene reinforced with carbon black is a good example. We shall see later how the smallness of the filler particles is even more vitally important in rubber reinforcement.

The other reinforcing mechanism in thermoplastics, which we call here 'redistribution of stress', is derived from the incorporation of 'yielding elements' into the material. If a stress concentration around a flaw is sufficient to initiate rupture, but there is close at hand a part of the structure than can deform rapidly enough under the stress, no rupture occurs. The local deformation has redistributed the stress. Such local deformation, yielding rapidly to a stress concentration, is supplied by rubber molecules embedded in the thermoplastic. The rubber molecules, provided their secondary attraction to the plastic molecules does not impede their motion too much, are relatively mobile, their kinetic units having relaxation times of fractions of seconds. They provide local fast mobility yielding to stress concentrations at flaws. This mechanism is responsible for the impact resistance of special types of polystyrene. Polystyrene itself at room temperature is quite brittle, but the presence of a few per cent by volume of rubber (such as SBR or BR), if well-dispersed, has a dramatic effect on impact resistance, bringing it from about 0.2 ft lbf/in to about 2. The types of rubber chosen are obviously not attracted to polystyrene molecules by secondary forces much greater than in the homogeneous rubber. Effective dispersion is greatly promoted by a 'grafting' procedure resulting in each rubber molecule bearing a number of polystyrene side chains, thereby giving it an anchorage in a relatively wide area around the rubber molecule. This also prevents demixing which would occur over longer periods of time.

In ABS plastics, this principle is part of the plastic structure: a rubbery polybutadiene or SBR chain carries side chains of acrylonitrile/styrene copolymer that are in the rigid plastic condition. The rubber chain molecules provide the low relaxation times to prevent brittle fracture: the impact resistance of ABS can go to 5 or 7 ft lbf/in.

For the 'redistribution of stress' it is not necessary to add a mobile polymer chain; a liquid plasticizer performs the same function and improves the impact resistance by short relaxation times for flow, as shown by plasticized PVC compared to rigid PVC. However, the recovery of the material after removal of the stress is considerably less in the case of a liquid plasticizer, because no elastic memory whatsoever is contributed by a true liquid. Moreover, the tensile strength is greatly reduced by liquid plasticizers, but only slightly by rubbery polymer chains.

(d) Reinforced concrete and glass laminate

Two categories of materials show a remarkable reinforcement, which is a combination of improvements in breaking strength (in elongation and bending) and in impact strength. In both categories, a medium is reinforced by embedding into it a network of wires having a vast capacity for elastic retraction (yield point at very high stress); the wires are steel in the case of reinforced concrete, and they are glass in the case of laminated plastics (polyester, etc.).

It is essential for both these cases that there is very strong adhesion between the wires and the surrounding medium. The mechanism then is based on the wires being able to redistribute stress concentrations, thereby eliminating them from the many flaws present in the material, and compelling almost the entire bulk of the material to carry the stress.

(e) Reinforced rubber

Fillers or crystallites can both reinforce vulcanized rubber, as explained in §§4.2.3 and 4.2.4 above. In either case, there are three functions that the 'hard particle' has in the rubber matrix. We suggest here a combined theory of rubber reinforcement, incorporating all three functions, namely the increased number of cross-links (Kuhn–Flory), the filler volume effect (Einstein–Guth–Gold), and the sliding nature of filler-to-rubber adsorption points (Chapter 9).

For a systematic approach, we use the three quantitative principles underlying these functions, calling them (a), (b) and (c).

(a) Elastic stress (modulus) is proportional to the number of chain sections trying to return to random coil positions, or to half the *number of cross-links* connecting these chains (equation (4.10) p. 56). Cross-links can be either chemical bonds (e.g. sulphur linkages) or physical attachments (e.g. participation in crystallites or adsorption on filler particle surface). See Fig. 4.21. The chemical cross-links are unalterably fixed in the network structure; the physical cross-links can slide or even disappear in one place and reform in another. The physical cross-link arrangement is continually changing during deformation, but their total number does not vary much (the filler particle surface is always completely covered as long as no vacuoles are formed; hence the average number of adsorption points is constant).

We assume that the stress, in accordance with equation (4.10), is proportional to the total number of cross-links, so that *all chemical plus all physical links* must be added up to arrive at the elastic stress.

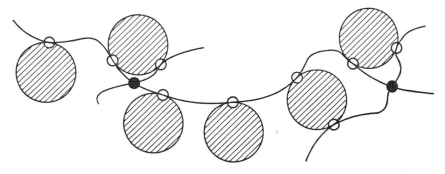

Fig. 4.21. Rubber reinforcement by fixed and sliding cross-links. — Rubber molecule; ● fixed (chemical) cross-link; ○ sliding cross-link; ⊘ reinforcing black filler particle.

(*b*) Any deformation of a medium is restricted by non-deformable particles (filler) embedded in the medium. This restriction is expressed as an additional force opposing the deformation. It occurs in viscous as well as elastic deformation and is best expressed as an *effect of total filler volume* in accordance with the Einstein–Guth–Gold equation (4.31).

(*c*) Maximum elongation of a chain network can be found geometrically when a chain segment is fully stretched from its most probable coil form to a straight line (see Fig. 4.3, p. 56); it can not be elongated further, and break occurs. The maximum distance L between cross-links is related to the equilibrium distance x by equation (4.7) on p. 56. To a first approximation we can use the *average chain length between chemical (fixed) cross-links to determine maximum elongation.* Stretching is not prevented by physical cross-links, because these can slide and rearrange.

These three principles are used to derive tensile data of elastomers in three steps illustrated in Fig. 4.22 (*a*).

Step 1. *Elastic stress* (modulus) is derived from principles (*a*) and (*b*).

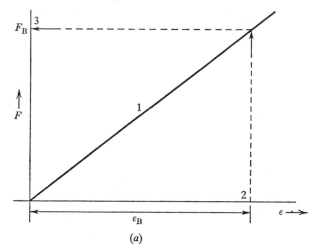

(a)

Fig. 4.22. (a) Derivation of breaking stress F_B for elastomers. Step 1. Establish curve $F = f(\epsilon)$. Step 2. Calculate breaking elongation ϵ_B. Step 3. Find F_B from steps 1 and 2.

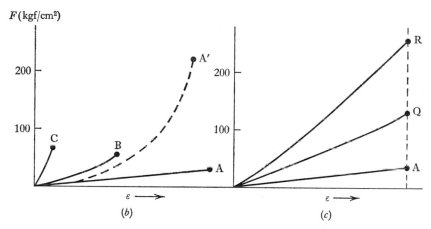

(b)　　　　　(c)

Fig. 4.22. (b) Stress F verses strain ϵ for pure gum SBR vulcanizate. (c) For reinforced SBR vulcanizate.

This establishes a curve for stress F as a function of elongation ϵ as given in equation (4.33) and Fig. 4.22a; the curve contains no indication of an end point.

$$F \approx (N_x + N_y)(1 + D_1 V_F + D_2 V_F^2)(1 + \epsilon), \tag{4.33}$$

where N_x is the number of chemical links, N_y is the number of physical links, D_1 and D_2 are constants, and V_F the total volume of filler particles.

In a non-crystallizing elastomer containing a filler, N_y is proportional to the surface of the filler particles. Thus the stress F depends on both the total surface and the total volume of the filler; this has been confirmed by measurement.

Step 2. *Maximum elongation* is derived from principle (*c*) using equation (4.7). This establishes an *end point* for the curve of stress versus elongation, as shown in Fig. 4.22*a* and equation (4.34).

$$\epsilon_B = E\sqrt{M_c} - 1 \qquad (4.34)$$

where E is a constant containing known molecular data, and M_c the chain length (average) between chemical cross-links. It is clear that increasing the number of chemical links, while augmenting the stress, also diminishes the breaking elongation, and this is experimentally correct.

Step 3. *Tensile strength* (breaking stress) is found by substituting maximum elongation ϵ_B from equation (4.34) into the *F–ε* curve of equation (4.33). See Fig. 4.22*a*.

Table 4.3 gives data calculated for a common case of black reinforced SBR, a non-crystallizing elastomer. Fig. 4.21 illustrates the true dimensions of a reinforced rubber structure.

TABLE 4.3. *Quantitative illustration of reinforcement mechanism (see Fig. 4.22)*

A rubber vulcanizate consists of: rubber, SBR 1500, 100 parts; filler, HAF black, 50 parts, and is vulcanized with conventional other ingredients to obtain the following properties: modulus (300 % elongation), 85 kgf/cm²; tensile strength, 250 kgf/cm²; elongation at break, 600 %.

Assumed average molecular weight (before vulcanizing) for the rubber: 200000.

From the kinetic theory of rubber elasticity (corrected for filler bulk effect) we calculate the following approximate numbers based on the above data.

(*a*) Per original rubber molecule:
Number of chemical cross-links (fixed) 15.
Number of adsorptive cross-links (sliding) 40.

(*b*) Per black particle of 300 Å diameter:
Number of adsorptive cross-links (sliding) 10000.

(*c*) This means one adsorptive contact between a rubber molecule and a black particle every 28 square Å of black surface, which is very reasonable.

The relationship between structural and rheological data is satisfactory, taking into account that many refinements and corrections will be needed to obtain a fully quantitative correspondence.

Fig. 4.22*b* shows the influence of chemical cross-linking on a pure gum SBR (no fillers). Going from A to B to C, the modulus grows by increased cross-linking, but simultaneously the maximum elongation diminishes, so that the resulting tensile strength is hardly improved by more cross-links. If the elastomer were crystallizing, elongation would orient molecules and create crystallites which act as reinforcing filler particles; the same number of chemical cross-links leading to the end-point A in SBR would lead to A' in natural rubber.

Fig. 4.22*c* shows the combined effect of filler and cross-links in SBR. Chemical links only lead to A, as in Fig. 4.22*b*. Combining these with the reinforcing filler, and counting *all* cross-links, would lead to Q and finally allowing for the effect of filler volume leads to R, the experimental value.

4.5 Time and temperature effects in rheology

At various points in our discussion of rheology in this chapter, the influence of time and temperature of the experiment has been mentioned. In §4.3.6 a complete survey was given of visco-elastic behaviour, indicating the dependence of stress–strain relationships on time, and defining relaxation times and retardation times in the material.

It will be useful, for a clear understanding of rheological responses of materials, to summarize briefly the effects of time and temperature.

4.5.1 Time effects

The time element enters into all rheological responses of matter because it takes time for a sample to comply with a stress. The basis of elementary rheology rests on two idealized cases: relaxation time at infinity (elastic) and retardation time at zero (flow).

In all intermediate cases there are finite values for these characteristic times (the name relaxation time is conveniently used for both types).

The material's response to stress is then characterized by an exponential function exp (λ) of the relaxation time λ, as shown in equations (4.25) and (4.27). Moreover, we have seen that a given material should really not be characterized by one relaxation time, but by a series (spectrum) of relaxation times, where each of the individual values in the series corresponds to one particular elementary process (molecular motion) in the material.

The greatest importance of relaxation time spectra appears under the following conditions:

(1) Blends of materials having very different spectra: liquids with thermo-plastics, elastomers and plastomers, etc.

A hard plastic (polymer below its glass transition temperature) can be made to resist brittle fracture under shock type load by adding to it a liquid plasticizer or a rubber (polymer above its glass transition temperature). But no great differences are caused by blending when the materials are tested under slow loading conditions such as in a regular tensile test.

(2) Very slow or long-duration experiments, such as creep of metals and plastics: these reveal deformation mechanisms that are not discovered in normal-duration testing; for example, a zinc casting can show deformation (creep) and even break under a constant stress well below the elasticity limit (at room temperature), provided a long enough time is given (several months or years). The same is true for steel at higher temperature, but well below its melting point. Examples with plastics and fibres at room temperature are well known (flat spotting of nylon-cord tyres after long standing in cold weather, etc.). In all these cases the assumedly infinite relaxation time of elastic bodies shows components that are finite, and measurable, provided sufficient time is given.

(3) Very fast experiments, in which the force is increased or decreased in a very short time (milliseconds). In contrast with the long experiments mentioned under (2), here are conditions where very short relaxation times become apparent, in cases where they could have been assumed to be equal to zero. The study of impact deformation, and of response to vibrating forces, reveals these short times – or their absence. A material that flows readily if given time under a moderate stress may behave as hard and brittle if the same stress is imposed very suddenly. Most thermoplastics are examples of this.

4.5.2 Temperature effects

The temperature concept enters into the understanding of all rheological phenomena because the thermal motion of atoms and molecules is the basic mechanism enabling materials to deform.

The greatest importance of the temperature dependence appears under the following conditions:

(1) Transition zones. A temperature of 'second order transition' (which is often a 'glass transition temperature') indicates the point where the thermal energy needed for the kinetic units of a material to become relatively free to move past each other is reached. Below that transition temperature they are in fixed positions, and their response to stress is elastic; above that temperature an increasing amount of flow is possible as the temperature goes up.

Transition zones can also indicate first order transitions from one structure to another (one crystal structure to another or one amorphous density of packing to another) but these are rheologically less important than the second order transition from glasslike (hard) to plastic (soft or rubbery).

The first order transition between crystal and melt is of course highly

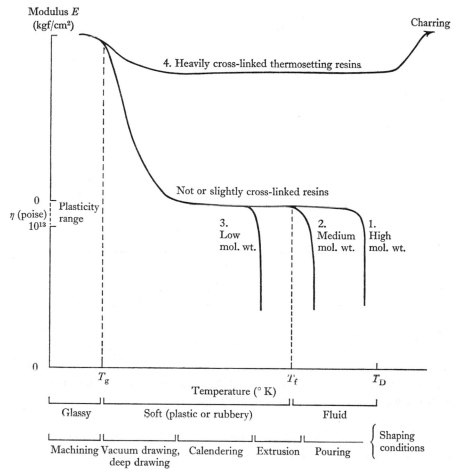

Fig. 4.23. General form of the rheology–temperature curve.

important for rheology. It delineates vast changes in rheological behaviour, such as occur when steel or ice melt.

The practical importance of transition temperatures is very great. Metals are cast from the liquid state or forged from the malleable state. Glass is cast or blown from the semi-liquid state. Plastics and rubbers are formed from the semi-liquid state.

The transition between 'solid' and 'liquid' behaviour in polymers can span a wide temperature interval. The essential changes can be represented by a curve giving modulus or viscosity against temperature.

In Fig. 4.23 such curves have been drawn for four cases. No. 2 refers to a

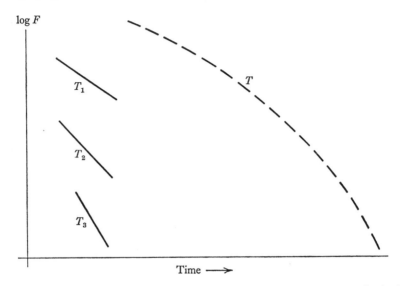

Fig. 4.24. Stress relaxation measured at higher temperatures T_1, T_2, T_3 and calculated at room temperature T. Stress F and $T < T_1 < T_2 < T_3$.

slightly cross-linked polymer of medium molecular weight worked out in some more detail and shows T_g and T_f for that particular polymer. T_D is the temperature above which chemical decomposition occurs. This temperature is independent of the molecular weight, being a chemical characteristic connected with the weakest primary bonds in the molecules.

The influence of the molecular weight demonstrates itself in the horizontal shift of curve 1 towards curve 3. Cross-linking (curve 4) results in reduction of the plasticity range, flow of the molecules being impeded by the intermolecular bridges.

Fibres are often 'spun' from the liquid state, and then drawn to orient the crystallization while at the transition temperature.

(2) Temperature cycles are often used to bring a material into a desired rheological state. Thus, annealing (for moderately long periods at elevated temperature) relieves a material from internal stresses and may result in it returning to a shape which it had before deformation. This phenomenon is sometimes called 'elastic memory', it is easily explained when remembering

that thermal motion of kinetic units is needed to restore a sample to its original shape after a deforming stress is removed.

Conversely, quenching by sudden cooling from a high temperature is used to preserve a state obtained, or to freeze in tensions.

(3) The relationship between temperature and time of the experiment has often been stressed. 'Annealing' to relieve internal stress can be accomplished either by a long waiting period at normal temperature, or by a shorter treatment at higher temperature. The fact that relaxation times are intrinsically temperature dependent explains this and many other phenomena.

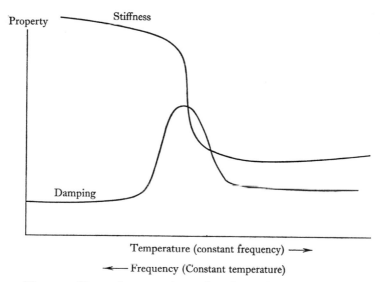

Fig. 4.25. Dynamic properties as functions of temperature and frequency of experiment.

According to the Williams–Landell–Ferry relationship, one can do the following. The relaxation of a material can be plotted as stress against time (at a given temperature T_1). The plot for a higher temperature T_2 shows a decay to much lower values at the same times. These various plots can be combined to a single plot (see Fig. 4.24), showing stress decay over very long times at the standard temperature T, by transforming to the lower temperature the data obtained at high temperatures.

Also, the temperature–time relation is clear in dynamic experiments (see Fig. 4.25). At the second order transition, which indicates the 'liberation' of kinetic units, relaxation times appear that are commensurate with the frequency of the stress cycle in the experiment. This then causes a large damping factor and phase difference between stress and strain. In warming up a material and testing with a given frequecy ω_1, a transition temperature T_1 will be found where the damping characteristics go through extremes and where the dynamic stiffness changes suddenly. At temperatures above T_1, the picture of a 'rubbery' substance emerges, with moderate damping and low stiffness. However, had we tested with a much higher frequency ω_2, then we would have to heat to a higher temperature T_2 before the corresponding relaxation times were short enough to match ω_2, and the transition therefore depends both on temperature T and frequency ω (or time factor t).

References to chapter 4

First and second reports on viscosity and plasticity (Amsterdam, 1938–39).

R. Houwink, *Elastomers and plastomers* **1** (New York & Amsterdam, 1950).

F. R. Eirich, *Rheology*, **1–4** (New York, 1956–67).

Rubber Reviews for 1965 (Division of Rubber Chemistry of the American Chemical Society).

J. W. S. Hearle & R. H. Peters, *Fibre structure* (Manchester, 1963).

G. E. Dieter, *Mechanical metallurgy* (New York, 1961).

A. Kelly, 'The nature of composite materials', *Sci. Amer.* **217**, 160 (1967).

J. R. van Wazer *et al. Viscosity and flow measurement* (New York, 1963).

J. P. Pfeiffer, *The properties of asphaltic bitumen* (New York & Amsterdam, 1950).

H. K. de Decker, *Phosphorous pentoxide crystal structures and physical chemistry* (Amsterdam, 1941).

A. Schuringa & H. K. de Decker, *Measuring rheological properties* (in Dutch) (The Hague, 1946).

H. K. de Decker, *Strength of Rubber* (Division of Rubber Chemistry of the American Chemical Society, September 1956).

H. K. de Decker & D. J. Sabatine, 'Dynamic response of elastomer blends', *Rubber Age* **99**, no. 4, 73 (1967).

5 Gases

R. Willem de Decker

5.1 Introduction

In gases the molecules are not packed tightly together, as they are in a liquid. Consequently, the density changes substantially as the pressure or the temperature is changed. This characteristic of gases causes a number of interesting phenomena which do not occur in liquids or solids.

The following discussion is mostly centred on one gas 'air', but the general principles are applicable to all gases.

5.2 Basic properties of gases

The molecules constituting the gas move randomly, keeping an average distance that is large compared to their size. The *absolute temperature* T is proportional to the mean kinetic energy of the molecules, and is related to pressure and density of the gas by the 'equation of state'. For an ideal gas (molecules having negligible volume and mutual attraction) the equation of state is that of Boyle–Gay Lussac:

$$p = \rho RT/M, \tag{5.1}$$

in which p is pressure, ρ density, R the universal gas constant, and M molecular weight.

Sound in a gas originates as a local deviation in pressure. Such a deviation will move radially outward from its origin, forming a pressure wave or *sound wave* that consists of a succession of reversible and adiabatic compressions and expansions moving with a velocity known as the *speed of sound*.

Based on Newton's second law and the principle of continuity, it can be shown that the speed of sound a is related to the changes in pressure p and density ρ in the wave, as follows

$$a^2 = dp/d\rho. \tag{5.2}$$

This equation can be transformed, using the thermodynamic and state equations, into

$$a^2 = \gamma p/\rho, \quad \text{or} \quad a = \sqrt{\left(\gamma \frac{RT}{M}\right)}, \tag{5.3}$$

in which γ is the ratio of the specific heats. The equation shows that the speed of sound in a given gas (γ and M constant) depends on the temperature, but comparing different gases at the same temperature, it will mainly depend on the molecular weight. At sea level the speed of sound in air is about 335 m/s (1100 ft/s), in hydrogen about 1280 m/s (4200 ft/s), and in Freon about 92 m/s (300 ft/s).

Gases have hydrostatic *compressibility* for which we can define a *modulus of bulk elasticity* as follows

$$dp = -E\,dV/V, \tag{5.4}$$

in which dp is pressure increase and dV is the increase in molar volume V. By using the definition of density, $\rho = M/V$, it can be seen that

$$dV/V = -d\rho/\rho,$$

and therefore the modulus can be written (using also equation (5.2))

$$E = \rho\frac{dp}{d\rho} = \rho a^2, \quad \text{or} \quad a = \sqrt{(E/\rho)}. \tag{5.5}$$

Thus, the speed of sound in a gas is directly related to the bulk elasticity modulus of that gas. The modulus is 1.03 kgf/cm² (14.7 lbf/in²) for air at sea level; for comparison, the bulk modulus of water is 19250 kgf/cm² (280000 lbf in/²), or about 2×10^4 times that of air.

Viscosity of a gas is expressed in the usual fashion by a viscosity coefficient η.

$$\tau = \eta\frac{dv}{dy}, \tag{5.6}$$

where τ is shear stress, dv/dy change of the velocity in direction x along the y axis. The viscosity coefficient for air at 0° C is 1.7×10^{-4} poise, or about 100 times lower than that for water at 0° C.

Viscosity is caused by the transfer of momentum from one flowing layer to the next. In a gas all molecules are completely mobile and this allows them to transfer more momentum per second when their thermal motion is greater. Therefore, gas viscosity *increases* with temperature (cf. the *negative* (exponential) law for liquids in Chapter 6).

In addition, it is clear that density has a major impact on the viscosity. An interesting equation links gas viscosity with the density ρ, the velocity $\bar{v}$ and the mean free path λ of the molecules:

$$\eta = \tfrac{1}{3}\rho\bar{v}\lambda. \tag{5.7}$$

The actual change of η with temperature and pressure depends on the equation of state.

Air viscosity is 1.72 at 20° C but 2.29 at 100° C, both at 1 atmosphere, and is about 1.82 at 20° C and 20 atmospheres (viscosities in 10^{-4} poise).

To eliminate the influence of density on viscosity in a gas, the '*kinematic viscosity*' v is often used: $v = \eta/\rho$.

The velocity of a flowing medium has a direct influence on the *hydrostatic pressure* it exerts. This is of major importance in gases, and is quantitatively expressed by *Bernoulli's equation*, derived from the principle of conservation of total energy:

$$\tfrac{1}{2}\rho v^2 + p = \text{constant.} \tag{5.8}$$

5.3 Elasticity (compressibility)

Compressibility should be taken into account in any description of the flow of gases. However, this complicates matters seriously. It can be shown that for flow with a speed that is low compared to the speed of sound, a gas can be treated as incompressible ($E = \infty$). Experiment has shown that this is a good assumption when $v^2/a^2 \leqslant 0.1$ (in air, $v \leqslant 100$ m/s (330 ft/s)).

As the speed of the gas flow (v) past an object is increased the effects of compressibility will become more and more pronounced. When moving, the gas molecules bounce against the surface of the object and are deflected from their intended course. This deflection takes the form of a pressure pulse which radiates outward from the body in a more or less spherical pattern and moves with the speed of sound (a). Then at any given time, the pressure distribution around the body will be a summation of all the previously generated pressure pulses, as shown in Fig. 5.1. The Mach number (Ma) $= v/a$ indicates the relative velocity of flow. When the speed of the object equals the speed of sound, which in turn depends on the bulk modulus, as seen in equation (5.5), the pressure pulses propagate forward with the same speed as the object and a substantial (theoretically infinite) pressure rise is created in front of the object (Fig. 5.2). This pressure rise is abrupt and is called the *shock-wave*. In fact it is so abrupt that the pressure rise is usually treated as a discontinuity in the flow.

The assumption of a discontinuity is quite accurate since both theory and experiment have shown the thickness of a shock-wave in air to be of the order of 10^{-5}–10^{-4} cm, approximately the same as the mean free path between the molecules.

If the speed of the object exceeds a then it leaves a conical shock-wave behind it which travels at the speed of sound (supersonic airplanes).

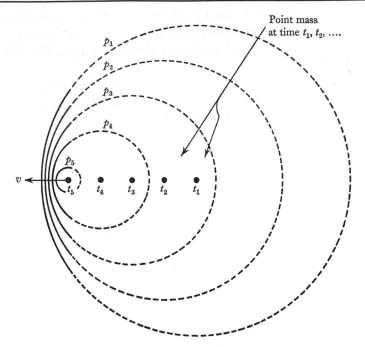

Fig. 5.1. Pressure distribution at a speed below that of sound.
$v < a$; $(Ma) < 1$ (subsonic speeds).

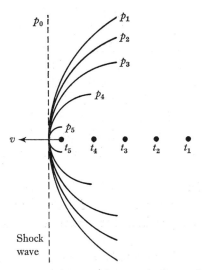

Fig. 5.2. Pressure distribution at the speed of sound (shock wave).
$v = a$; $(Ma) = 1$ (transonic speed).

5.4   Viscosity

The effects of viscosity are usually studied in the absence of compressibility effects (i.e. at speeds below $(Ma)^2 = 0.1$) since this allows a relatively simple analysis of many basic phenomena.

A gas flowing past an object is influenced by a viscous force (if the boundary layer is laminar) and an inertial force (turbulent boundary layer). The viscous force is proportional to $\rho v^2 L^2$ (if L is a typical dimension of the object), but the inertial force to $\eta v L$. The preponderance of one force over the other is indicated by the Reynolds number

$$(Re) = \frac{\rho v^2 L^2}{\eta v L} = \frac{\rho v L}{\eta}. \tag{5.9}$$

If two situations have the same Reynolds number, their flow patterns are similar. Thus, for a gas flow with constant η and ρ past two bodies, one twice as large as the other, the flow patterns are the same only if the speed past the larger body is half the speed past the smaller body.

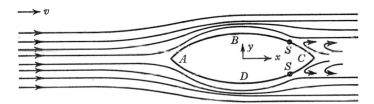

Fig. 5.3. Flow about a body with convex surfaces.

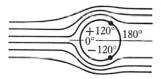

Fig. 5.4. Flow about a cylindrical body.

5.4.1   Boundary layer separation

Observation shows that in flow past convex surfaces, the flow often separates from the surface. This is referred to as boundary layer separation. To explain this consider the flow about a body with convex surfaces (Fig. 5.3). The gas flow is accelerating over the regions AB and AD and decelerating over the regions BC and DC. In an ideal, non-viscous flow, the total energy would be preserved during the acceleration and deceleration phase and the flow would be at the same speed at point C as it was at point A. However, in a viscous fluid, energy is dissipated in the boundary layer. As a consequence, at a certain point (point S in the figure) the molecules will come to their original velocity and beyond that point, will experience an acceleration in the opposite direction. The only way the flow can accommodate this is by separating from the surface at point S. At the same time a region of 'reverse flow' exists from C to S as shown in the sketch. For instance, on a body of circular cross-section, the flow will separate at the $\pm 120°$ station as shown in Fig. 5.4.

References to chapter 5

A. H. Shapiro, *The dynamics and thermodynamics of compressible fluid flow* (New York, 1953 (volumes 1 & 2).
L. Prandtl & O. G. Tietjens, *Applied hydro and aeromechanics* (New York, 1957).
H. Schlichting, *Boundary layer theory* (New York, 1960).

6 Simple liquids

H. K. de Decker

6.1 Definition of ideal simple liquid

Materials can show 'liquid behaviour' (that is, viscous flow) to various extents under many conditions. Chapter 4 summarizes the types of viscous or visco-elastic behaviour that are typical of liquids and shows how deviations from pure viscosity are related to deviations from a simple liquid structure, that is from the ideal picture given below. Chapters 8, 9, and 11 contain more detailed accounts of the various 'fluid' states observed in plastics, rubber, and glass, The present chapter will review the simplest cases of viscous flow.

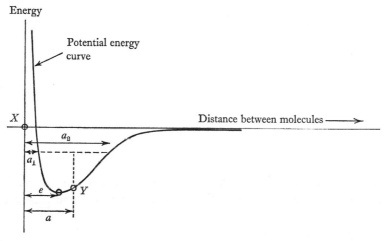

Fig. 6.1. Attraction between two molecules in a liquid. Molecule X at the origin; molecule Y at an average distance a, with oscillations between a_1 and a_2.

The pure, simple liquid that forms the ideal starting point of this review consists of spherical molecules. Their mutual attraction, symmetrical on all sides, prevents them from moving around independently and the temperature of the experiment is low enough to prevent evaporation and keep the liquid in condensed form. The *average* distance between two molecules is small enough to keep them well inside each other's attraction area (see Fig. 6.1). At any particular instant, there is a short-range ordering given by this average distance. The real structure at each instant (drawn in Fig. 6.2) shows only small deviations from the ideal (crystalline) order (dotted in Fig. 6.2).

A distribution function g(r) can be defined, giving the statistical density of molecules at a distance *r* from any given molecule. In a simple liquid, a maximum in this function will occur at the distance *a* in Fig. 6.1 or AP in Fig. 6.2. Beyond this first pronounced maximum, further maxima of decreasing clearness may be detected, but generally the distribution becomes practically uniform in a statistical sense beyond a distance of about 3*a*.

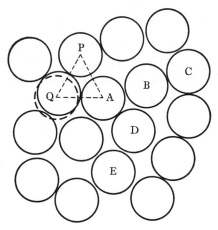

Fig. 6.2. Planar representation of momentary position of molecules in a liquid. Short-range order is illustrated by ABC or BDE; ideal (average) position is illustrated by APQ.

This distribution was fully confirmed by X-ray diffraction in simple liquids, and can be demonstrated in two dimensions with beads scattered on a flat surface.

The ideal rheological behaviour belonging to this ideal liquid is *Newtonian flow*, caused by molecules changing places with the help of their thermal motion; there are no elastic components in this deformation. However, this same liquid can show two types of elastic reactions which will be discussed first.

6.2   Static elasticity (compressibility)

The place-changing motions of molecules in a simple liquid are so frequent and rapid that such a liquid will quickly fill the space given to it, and therefore has no shape of its own.

The simplest type of elasticity displayed by fluids (both gases and liquids) is the elasticity of compression. Once a liquid fills a given space, a reduction of this space causes compression of the fluid and an elastic reaction is felt. It should be remembered that this compression elasticity is *hydrostatic* in nature; the elastic reaction force exerts equal pressure on every part of the confining wall surface.

This kind of elastic behaviour is called compressibility. For a liquid, it involves very slight shifts in the average distance between molecules, just as in the case of 'steel elasticity' for solids. Compression reduces this average distance and the thermal motion seeks to restore the original equilibrium distance, causing a very strong force. A 'bulk modulus' M_B can be defined as the reciprocal of the compressibility coefficient, thus

$$M_B = (p_2 - p_1)v_1/(v_1 - v_2), \tag{6.1}$$

in which p and v are the pressures and volumes at two stages of compression 1 and 2. It will be seen from Table 6.1 that the bulk modulus of liquid mercury in compression is of the same order of magnitude as the Young's modulus of solid lead in extension.

TABLE 6.1. *Bulk modulus of liquids and Young's modulus of solids*

	Bulk modulus	Young's modulus
	(in kgf/cm² at 20° C)	
Water	2.0×10^4	—
Alcohol	1.2×10^4	
Mercury	2.7×10^5	—
Lead	8.5×10^4	1.7×10^5
Aluminium	7×10^5	7.3×10^5
Carbon steel	2×10^6	2.0×10^6
Air	1.05	—

Water and alcohol have lower bulk moduli than the metals because the non-ideal shape and polarity of their molecules leads to structures not having densest packing, which are gradually destroyed by increased compression.

The isothermal compressibility of fluids for three temperatures is represented in the *p–v* diagram of Fig. 6.3. The isotherm for T_1 shows the liquid compressibility in its sector LA. If the volume is increased beyond A, a two-phase system forms by evaporation; no compressibility force exists until, at point B, only vapour remains, and sector BG shows vapour compressibility.

The isotherm at T_c is a continuous compressibility curve for the 'critical' fluid, changing from vapour to liquid at the critical point C. The isotherm at T_2 is a typical gas isotherm, subject to the law of Boyle ($pv =$ constant) or the less idealized equations of state derived by van der Waals and others.

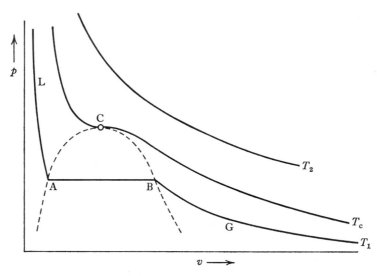

Fig. 6.3. *p–v* diagram showing three isotherms. *p*, pressure, *v*, molar volume (reciprocal of density).

6.3   Dynamic elasticity (impact)

Another type of elasticity of liquids appears when the liquid is deformed at very high flow rates. In ordinary flow, molecules move so rapidly in a 'displacement' mechanism due to their Brownian motion, that any displacement of part of the liquid (through an external force) is rapidly transmitted to neighbouring parts of the liquid, hence flow (see §6.4.1 below). But if a part of the sample is moved more rapidly than the displacement mechanism, then the disturbing force meets molecules opposing it. In a qualitative way, it can be seen that above a certain limiting velocity related to the displacement mechanism, strong elastic phenomena will appear in liquids. A more quantitative treatment of the limiting velocity has been given in Chapter 5 for the more simple case of gases.

6.4 Viscosity

6.4.1 Simplest case

The ideal liquid of Figs. 6.1 and 6.2 will, under moderate shear rates, show Newtonian flow (see Chapter 4, p. 59) according to the equation

$$\tau = \eta D = \eta \frac{d\gamma}{dt}, \qquad (6.2)$$

where τ shearing stress, γ shear, t time, η viscosity coefficient. The viscosity coefficient η is a constant dependent on temperature only, and there are no elastic components in the deformation.

The mechanism whereby this ideal viscous flow occurs was derived by Eyring. To cause flow, the deforming force must find molecules able to overcome the limitation of their average position. This means that a 'potential

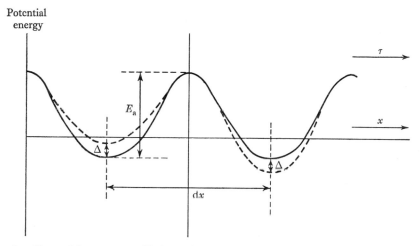

Fig. 6.4. Potential energy profile in path of moving molecule. E_a activation energy for flow, Δ energy difference caused by stress, τ shearing stress, x distance.

barrier' must be overcome by some molecules which can then be pushed aside to create a 'hole' for others to fall into, thereby displacing the hole and propagating the movement (displacement mechanism).

According to Eyring, this is a monomolecular reaction, the reaction consisting of one molecule passing from one side of a potential barrier to the other. The shearing stress τ in direction x will do an amount of work $(\tau \, dy \, dz) \, dx$ in displacing an elementary particle $dx \, dy \, dz$ over the distance dx. It will therefore become easier for the particle to overcome the existing potential barrier E_a by going in the direction x. This is illustrated in Fig. 6.4. The

energy barrier to be overcome now is only $E = E_a - \frac{1}{2}\tau\,dx\,dy\,dz$ instead of E_a
Based on these principles, Eyring derived an expression for η equivalent to the empirical equation:

$$\eta = A \exp(E_v/RT), \tag{6.3}$$

in which η is defined by (6.2), A and R are constants, T is the absolute temperature, and E_v is the activation energy representing the potential barrier for viscous flow.

At higher temperature a simple liquid has an increased specific volume V_s. This represents an increase with temperature of the average distance between molecules (distance a in Fig. 6.1) and may be used to explain that the 'shift' or 'displacement' mechanism becomes easier at higher temperature, and the viscosity lower. Actually, the temperature dependence of simple liquid viscosity is relatively minor if observed while V_s is kept constant. Such an experiment with a liquid requires, of course, very high pressures and is therefore mostly of theoretical importance.

The viscosity of some relatively simple liquids can be represented, within 10 % of observed values, by the equation

$$\eta = A'/(V_s - B), \tag{6.3a}$$

in which $V_s = f(p, T)$ and A' and B are constants characteristic of the liquid (B relates to the total volume of the molecules). Carbon disulphide and carbon dioxide are examples. In the latter case, the approximation of equation (6.3a) is even valid for a gas as well as a liquid, in a temperature range from $-15°$ C to $40°$ C and at specific volumes of 1.0 to 100 cm^3/g. Examples of liquids closely approaching ideal flow behaviour are molten metals or liquified rare gases.

6.4.2   Shape, size, and interaction of molecules

Comparing simple liquids having molecules of different shape and size, the viscosity will generally be higher as the molecules are larger or less compact in shape. Differences between the normal hydrocarbons illustrate this. Hexane (C_6H_8) has a viscosity of 0.003 poise at room temperature, against 1.0 poise for light machine oil (average formula $C_{20}H_{22}$). Even higher members of the series are solid at room temperature. An extreme example is polyethylene, a 'high polymer' of linear structure that may have as many as 100000 carbon atoms in one molecule.

Above their melting point, such polymers exhibit flow with very high viscosities, for example 10000 poise at 150° C. However, flow in this case is due to the motion of 'kinetic units' (as discussed in Chapter 4) and not to displacement of the entire molecule as such. Neighbouring sections 'pull back' on a kinetic unit that is moving away, thus creating an elastic component. The entire flow phenomenon is visco-elastic and cannot be described with the simple Newtonian equation.

Liquids having strong interaction between molecules show a higher viscosity than those with weak interaction. The comparison between water and ethanol on the one hand, and hexane on the other (Table 6.2 and Fig. 6.5) illustrates this. The polar attraction between water molecules is relatively strong, and liquid water has order of a longer range than the short-range order existing in an ideal liquid (Fig. 6.2). It has been shown that actual associations exist between the dipole molecules of liquid water and that only 0.2% of all water molecules are moving as single units. Still, the associations are on average not larger than 2 or 3 water molecules.

TABLE 6.2. *Viscosity in poise*

Temp. (° C)	Mercury	Water	Ethanol	Hexane
0	0.0170	0.0179	0.0177	0.0040
20	0.0157	0.0100	0.0119	0.0032
40	—	0.0066	0.0083	0.0026
60	—	0.0047	0.0059	0.0022
80	—	0.0036	—	—
100	0.0122	0.0028	—	—

The temperature variation of viscosity derived according to Eyring (equation (6.3)), accounts for the viscosity change due to faster thermal motion. In cases of strong polar attraction between molecules, such attraction results in a high activation energy; a higher potential barrier has to be overcome to move a molecule than in the case of simple van der Waals' attraction. Consequently, the temperature viscosity curve is steeper for liquids with polar attractions between molecules. Comparison of the data for water and ethanol with those for mercury and hexane illustrate this point (see Table 6.2 and Fig. 6.5).

Cooling down to the crystallization temperature brings all molecules into a pattern of long-range order where they are fixed, although the thermal motion causes oscillations around the ideal locations. The 'displacement' mechanism required for flow is not provided by the thermal motion, and the external force has to provide all the effort. For water the viscosity coefficient goes from 0.018 poise (liquid water) to 4×10^6 poise (ice).

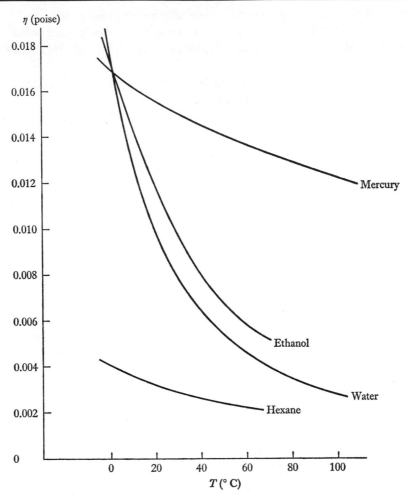

Fig. 6.5. Viscosity–temperature relationship for simple liquids.

6.4.3 Liquid crystals

Among the real liquids having molecules of not too extreme size or shape, there are some interesting cases of long-range order within the liquid state. The following conditions must be fulfilled for this kind of order to occur:

molecules relatively large;

strong interaction forces between molecules;

strong orientation of these forces in a few directions only;

geometry favourable for orientation.

This long-range order can follow one of several patterns, called 'liquid crystals'. The patterns have received special names:

nematic (linear arrangements; orientation of molecules in one direction);

smectic (planar arrangements perpendicular to orientation);

cholesteric (planar arrangements parallel to orientation).

Which of these patterns is likely to be formed is roughly predictable by qualitative guide lines based on the lateral and terminal interaction forces between molecules.

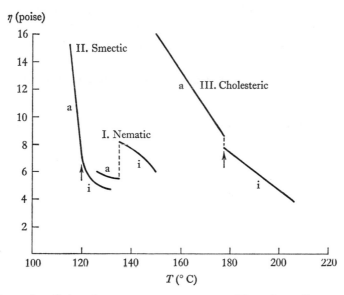

Fig. 6.6. Examples of viscosity–temperature curves: I. Nematic, *p*-diacetoxy stilbene chloride. II. Smectic, ethyl-*p*-azoxy benzoate. III. Cholesteric, cholesteryl benzoate. On each curve, a, liquid crystals; i, isotropy.

Obviously, liquid crystals are a transition between solid and liquid forms. About one in every 200 organic compounds exhibits this behaviour. In each case the specific pattern is formed inside a certain temperature range only. An example is anisylidene-*p*-aminophenyl acetate with a nematic range from 82° to 110° C; another is cholesteryl nonanoate with a cholesteric range from 78° to 91° C.

The rheology of liquid crystals is that of true liquids, but complicated by the incidence of long-range order. These molecules tend to orient themselves not only under the influence of mutual attraction, but *also* under that of flow itself. The viscosity becomes dependent on the flow rate (compare Chapters 3 and 4 for these non-Newtonian phenomena).

Fig. 6.6 illustrates some cases. For a nematic liquid, when cooling down to the transition temperature, long strings of head-to-tail oriented molecules are formed. These strings orient themselves in the direction of flow and actually *decrease* the viscosity at the transition. For the other two types of liquid crystals, the molecules are oriented in the direction of flow, but association takes place perpendicular to it. Hence a pronounced increase in viscosity at the transition; for cholesteric materials this increase is even discontinuous.

Orientation in these liquids can also be caused by a magnetic field. If this places the nematic molecules perpendicular to the flow direction, a very strong increase in viscosity is noted, for example, three times as high as the nematic viscosity (curve ɪa in Fig. 6.6), and even above that for the regular isotropic liquid (curve ɪi in Fig. 6.6).

Liquid crystals have recently gained renewed interest through their optical and electrical properties. In these properties they behave as solids with special sensitivities due to the fact that their structure is suspended in a liquid medium.

Reference to chapter 6

F. R. Eirich, *Rheology* (New York, 1965–67).
First and second report on viscosity and plasticity (Amsterdam, 1938 & 1939).

7 Dispersions

H. K. de Decker (§ 7.1), M. van den Tempel (§ 7.2) and H. K. de Decker and B. B. Boonstra (§ 7.3 and 7.4)

7.1 Solutions (by H. K. de Decker)

7.1.1 General

Solutions are blends of two or more components, in which each component is dispersed on a molecular scale between molecules of the other components.

Usually the component with the larger concentration is called the solvent; the other components are solutes. However, in liquid solutions of solid materials (e.g. sugar in water) the solvent (water) may have a lower concentration than the solute. Conversely, some solids may contain substantial proportions of a liquid without losing their solid character (PVC plastic with liquid plasticizer).

Solutions are either solid or liquid. Solid solutions can be crystalline (metals, ceramics) or amorphous (glasses, plastics, rubbers). Only a brief account of truly liquid solutions remains to be given here.

If the molecules of a solute B are associated in groups, the blend of these groups with solvent A is not a solution in the strict sense, although its rheological, optical and other properties may be those of a real solution. This is an example of a multiphase system, where B retains its own identity in small particles dispersed in A. Examples of associated solutes are soap or gelatine in water (Chapter 17).

7.1.2 Elasticity

Usually the elastic characteristics of liquid solutions do not differ significantly from those of pure liquids (see Chapter 6). Some solutions of very large molecules having polar groups, such as gelatine, form an exception to this rule. These molecules form a skeleton network by mutual association, thereby immobilizing a relatively large volume of solvent (e.g. 90% water, 10% gelatine). Such solutions may show solid elasticity under small stresses. However, these network structures break down under higher stress, and a 'yield value' separates the elastic region from the viscous flow in the stress/ shear rate diagram. These effects will be discussed under viscous flow.

7.1.3 Viscosity

In the ideal case a blend of two liquids A and B shows a viscosity derived from the component viscosities and the blend ratio. Thus, lubricating oils are blends of many hydrocarbons, and the viscosity of a particular oil can be adjusted by adding a calculated proportion of a lower or higher viscosity component.

If the blend viscosity is properly expressed in terms of the component ratios and their properties, most deviations from ideal behaviour can be accounted for[1] by considering the molar heats of association of each of the components and the molar heat of mixing q_m. If the molar ratios of A and B are m and $(1-m)$, the viscosity of the mixture η_m can be derived from the viscosities of the components as follows:

$$\log \eta_m = m \log \eta_A + (1-m) \log \eta_B - \mathrm{f}(q_m, T). \qquad (7.1)$$

Furthermore, the molar ratios should be calculated from the 'physical' molecular weights, or in other words, the weight of the associations of molecules that actually occur in the blend, rather than the molecular weight of single molecules.

The term $\mathrm{f}(q_m, T)$ in (7.1) can be written in different forms. In cases where q_m is practically zero, such as 'ideal' pairs of liquids, the term disappears. A good example is a mixture of hexane with decane, in which there is no association of either component. However, mixtures of benzyl benzoate with toluene, or ethyl ether with carbon disulphide show no term $\mathrm{f}(q_m, T)$, but do show evidence of association of one or both components.

If the term $\mathrm{f}(q_m, T)$ has a significant value, the blend viscosity will deviate from the ideal and may be either higher or lower than the sum of the component viscosities, for example acetic acid and water or ethanol and water show an increased viscosity. At low temperatures and around the 50/50 blend the viscosity values are several times those of the components. It is well-known that such mixtures form molecular complexes in solution which may not correspond to known solid compounds. It is not difficult to understand the increased viscosity of the mixture. Any disruption of the structure of a pure liquid component A, by the molecules of B, will tend to impede flow, and association between A and B would cause an even greater impediment.

Examples of blend viscosities lower than those of the components are aniline plus *o*-nitrophenol, different kinds of electrolytes with water, or *p*-nitrotoluene in alcohol. The reduction in viscosity only occurs in a liquid, such as water or alcohol, which has a well-ordered structure. The addition of a component B disturbs and breaks down this structure. A further observation clarifies this point even better. At lower temperatures, the greatest depression of viscosity occurs because the original structure is more perfect; at higher temperature the structure has already been disrupted somewhat by thermal motion. Electrolyte ions will tighten the structure of water and increase the viscosity if they are small hydrated ions; they will loosen the structure and lower the viscosity if they are large and barely hydrated. Figs. 7.1 and 7.2[1] illustrate the cases of blend viscosities.

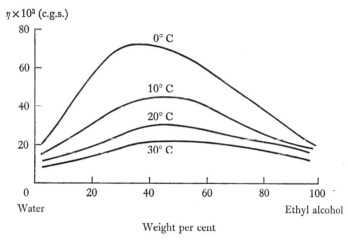

Fig. 7.1. Viscosity of ethanol–water mixtures. Increased viscosity by association.

If component B consists of very large molecules, a further principle becomes important, namely the disturbance of flow by a large particle. The viscosity of the 'solution' η_c can then be expressed as that of the 'solvent A', η_0, by the Einstein equation:

$$\eta_c = \eta_0(1 + Dv), \tag{7.2}$$

in which D is a constant (normally 2.5) and v the volume fraction of B assumed to be present as spherical particles.

For high molecular weight polymers in solvent the largest effect on viscosity is not the size of the polymer molecules B, large as they are, but the degree of uncoiling they undergo in the solvent A (see Chapter 4). The following cases should be distinguished.

(1) Good solvation. The viscosity of the solution is high because the B molecules uncoil thus allowing maximum solvation by A. A good example is 'rubber cement', a solution of rubber in benzene or toluene, which is extremely viscous at rubber concentrations of 5 to 10 %.

(2) Poor solvation. *B* molecules are tightly coiled and contribute much less to the viscosity of the solution.

(3) Case (1) combined with polarity. A continuous network of *B* is formed and the viscosity may become so high that an elastic yield value exists (gelatine in water).

(4) Case (2) combined with polarity. The lyophobic parts of *B* will associate in small clusters (micelles), and turn the lyophilic part towards *A* (soap in water).

The following comments can be added concerning polymer solutions.

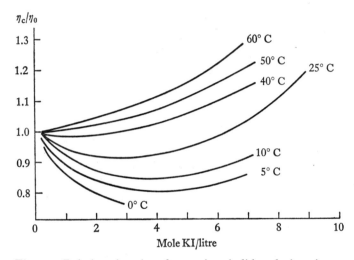

Fig. 7.2. Relative viscosity of potassium iodide solutions in water. Lower viscosity at low temperature.

Concentration. The viscosity depends on both concentration and molecular weight so that it is convenient to express the relationship as a power series in both variables.

Structure of polymer. Deviations from the strictly linear structure cause the factors K and α (see p. 69) to change, even in infinitely dilute solutions.

Interaction between polymer and solvent. This effect can be understood most easily if expressed as cohesive energy density (C.E.D.). If the C.E.D. of polymer and solvent are equal, the polymer will distribute at random in the solvent forming undisturbed random coils, and the solution will have the highest possible viscosity. If the C.E.D. of the polymer is considerably higher than for the solvent, the polymer molecules will coil up as narrowly as possible, minimizing the number of contacts with solvent molecules. This is so for a poor solvent and the solution viscosity is low. This effect modifies the factor K in equation (4.18) or (4.21), and is proportional to what is sometimes called the 'swelling index', the free energy of swelling or of solvation. A solvent allowing a polymer the complete development of undisturbed random coils is sometimes called a 'θ-solvent'.

Flow rate. The intrinsic viscosity (and more clearly any viscosity at higher concentration) decreases with increasing flow rate, due to the orientation of macromolecules and hence the flow is non-Newtonian (see Fig. 4.9).

7.2 Fluid dispersions (by M. van den Tempel)

7.2.1 Viscosity of dispersions

The relation between the rheological properties of a dispersion, and its composition and structure, is a very complicated one, because several different mechanisms may contribute to the stress when a deformation is applied. It is instructive to start from the simplest possible system, in which the only mechanism that must be considered is the influence of the dispersed particles on the flow of liquid surrounding these particles. By making the system gradually more complicated it is possible to investigate the effect of each separate mechanism on the rheological properties. In particular, it becomes possible to investigate which mechanism is responsible for the pronounced elastic or plastic properties that are frequently present in dispersions.

The usual starting point is the (much abused) equation derived by Einstein[2] for the viscosity of a very dilute dispersion of rigid spheres in a Newtonian liquid of viscosity η_0:

$$\eta = \eta_0(1 + 2.5\, C_v). \tag{7.3}$$

According to Einstein's analysis, the suspension behaves as a Newtonian liquid, and the size of the dispersed spheres does not affect the viscosity. The rigid spheres embedded in the uniformly sheared liquid rotate with an angular velocity equal to one-half of the shear rate. The liquid motion is thereby disturbed, in the sense that near the surface of the sphere a rotation is superimposed over the uniform shear. This rotational motion dies out at increasing distance from the sphere in proportion with (distance)$^{-4}$. In the derivation of equation (7.3) it has been assumed that each of the spheres moves in a field of flow that is not disturbed by the presence of other spheres. This means that the validity of equation (7.3) requires $C_v \to 0$; measurable deviations from the behaviour prescribed by this equation may already be expected at volume concentrations exceeding about 1%.

The attempts that have been made to verify equation (7.3) by experiment are, in reality, investigations determining to what extent the conditions under which the equation is valid can be satisfied. The careful measurements of S. G. Mason[3] should especially be mentioned.

It will now be our task to investigate how the rheological properties of the system are changed if the restrictions imposed in deriving equation (7.3) are dropped one by one. The restriction that the suspending liquid be a Newtonian fluid does not appear to be a very severe one, although this subject has not been thoroughly investigated. If the suspending medium is a Hookean solid, all relations derived as extensions of equation (7.3) remain valid in principle, if moduli are substituted for viscosities. A similar procedure is often used for non-Newtonian (visco-elastic) fluids as the suspending medium although it is recognized that this cannot be correct in general. In the following sections, the materials constituting the phases will always be considered incompressible, although the effect of compressibility has been studied in some detail.[4]

7.2.2　Deformable spheres

A general method for investigating the rheological behaviour of a (dilute) dispersion of deformable spheres was proposed by Fröhlich and Sack.[5] They studied the properties of a dilute suspension of elastic (Hookean) spheres in a Newtonian liquid, but their method can easily be extended to systems in which any of the phases may be either liquid or solid, and even to very concentrated systems.[4] It is shown in the original paper that the suspension, if subjected to shear, will show visco-elastic behaviour because the deformation of the spheres requires a certain time due to the adhering liquid. The result of their analysis is

$$\tau + \lambda_1 \frac{d\tau}{dt} = 2\eta \left(D + \lambda_2 \frac{dD}{dt} \right), \tag{7.4}$$

where η is given by equation (7.3) and

$$\lambda_1 = \frac{3\eta_0}{2E} \left(1 + \tfrac{5}{3} C_v^3 \right),$$

$$\lambda_2 = \frac{3\eta_0}{2E} \left(1 - \tfrac{5}{2} C_v^3 \right). \tag{7.5}$$

These equations are valid if $C_v \ll 1$ and the deformation of the particles is so small that they remain essentially spherical. It follows from these equations that the effect of deformation of the spheres can only be observed so long as a steady state has not been established; the steady-state viscosity is still completely described by equation (7.3). In dispersions of moderately rigid particles ($E > 10^7$ dyn/cm²) in a low-viscosity fluid ($\eta_0 < 10$ poise) the steady state is established in a time of the order of microseconds. The characteristic time λ_1 is equivalent to a Maxwell relaxation time, since equation (7.4) reduces to the well-known equation for a Maxwell body if the retardation term containing λ_2 is neglected.

The effect of emulsified droplets of an immiscible fluid can be investigated by the same method, when it is realized that surface tension causes emulsified droplets to behave like elastic spheres. If the system consisted of two very pure immiscible fluids, without a trace of surface-active emulsifying material, the macroscopic rheological behaviour would be affected by flow occurring in the interior of the dispersed particles. Oldroyd[6] found equation (7.4) to be valid for this case, but with different values for the rheological parameters η, λ_1 and λ_2. In particular, the steady-state viscosity η is no longer given by equation (7.3) but by an expression (7.6) that had already been derived by G. I. Taylor:[7]

$$\eta = \eta_0 \left(1 + \frac{\eta_0 + \frac{5}{2}\eta_i}{\eta_0 + \eta_i} C_v \right). \tag{7.6}$$

Here, η_i is the viscosity of the liquid in the droplets. Moreover, the most important modification is the substitution of the quotient of interfacial tension σ, and droplet radius a, for the elastic modulus E, of the spheres in equation (7.5). Since the interfacial tension in most systems of interest is of the order of 10 dyn/cm², this means that time effects due to drop deformation may just become perceptible in the millisecond range if the droplets are fairly large and the surrounding fluid has a sufficiently high viscosity. It also follows that, under the conditions used in deriving equation (7.4), the steady-state viscosity of an emulsion is not affected by interfacial tension or droplet deformation.

The deformation of the droplets in a field of uniform shear gives rise to ellipsoidal particles with length L and width B:[7]

$$\frac{L-B}{L+B} = \frac{2a\eta_0 D}{\sigma} \frac{\eta_0 + \frac{19}{16}\eta_i}{\eta_0 + \eta_i}. \tag{7.7}$$

Even under extreme conditions this quantity will not exceed 0.01 in aqueous dispersions of sufficiently small drops (e.g. latex). The deformation may become appreciable for rather large drops emulsified in a viscous liquid.

Measurements of the internal liquid motion in emulsified droplets have shown[3] that it is practically impossible to produce a system of sufficient purity to satisfy the conditions used in deriving equation (7.6). Surface-active materials are usually present at least in trace amounts, and this is sufficient to make droplets behave like solid (elastic) particles without notice-able internal circulation. This is because the tangential stress exerted by the flowing external liquid on the surface of the particle can be compensated by a gradient in interfacial tension as soon as surface-active materials are present.[8] At the interface between two very pure liquids, however, such a gradient cannot exist and the tangential stress must be compensated by a stress due to motion of the internal liquid.

7.2.3   Non-spherical particles

When a system containing dispersed non-spherical particles is subjected to shear, the particles will rotate and the average orientation will change to cause some degree of alignment. The main result is that the various types of particle interaction forces will become noticeable at even lower concentrations than in a dispersion of spheres. Although the behaviour of an isolated elongated particle in a sheared liquid has been investigated in great detail, we shall not discuss the results of these investigations because it is generally found that the rheological properties of such systems are mainly determined by particle interaction. This, of course, is no longer correct in the extreme case where the particles are flexible macromolecules; a suspension of such particles is not a 'multiphase system' and a study of its rheological behaviour should be developed along quite different lines (see p. 48).

7.2.4   Concentrated dispersions

A very large number of more or less empirical relations has been proposed between the viscosity and the concentration of a fluid dispersion, and several attempts have been made to extend the hydrodynamic theory of Einstein to more concentrated systems. The results of the extended theories are not useful at concentrations beyond a few per cent. Comparison of the empirical equations with experiment is difficult, because other types of interaction cannot be excluded and their effect becomes easily predominant in more concentrated systems. Since any interaction will always result in an increased viscosity, the tendency is to take the lowest viscosity ever measured (at a given volume fraction of dispersed material) as indicating the behaviour of a system in which only hydrodynamic interaction occurs. A typical and useful example of an empirical equation obtained in this way[9] is shown in Fig. 7.3.

Another method to obtain essentially the same result makes use of the fact that most other types of particle interaction give rise to flocculation or aggregation of the particles. This has an enormous (often predominating) effect on the rheological properties, but the effect decreases rapidly with increasing rate of deformation. It is sometimes possible to carry out viscosity measurements at sufficiently high shear rates, such that further increase of the shear rate has no noticeable effect on the measured viscosity. The liquid has become Newtonian at these shear rates, and the 'limiting' viscosity is attributed to hydrodynamic interaction because all the aggregates are permanently disrupted. It should be realized that the motion of each individual particle is still affected by the action of non-hydrodynamic interaction forces, and therefore the rheological properties must deviate to an unknown extent from those of a system in which such forces are absent.

Fig. 7.3 also shows the effect of concentration on viscosity as predicted by an extension of the theory of Fröhlich and Sack to higher concentrations.[4] The behaviour depicted in Fig. 7.3 is shown by a wide variety of systems, ranging from aqueous dispersions of globules of rubber or bitumen to dispersions of mineral fillers in asphalt.

It is evident that, in more concentrated dispersions, the field of flow in the neighbourhood of a particle depends on the average interparticle distance. In a system of given volume concentration the average distance between particles increases with increasing particle size, and also when particles of widely differing sizes are present. It is, therefore, not surprising that the viscosity of more concentrated dispersions increases with decreasing particle size, if all other conditions remain unchanged. Also a dispersion of equal-sized particles has a somewhat higher viscosity than a similar dispersion containing particles with a wide distribution of sizes. Thus, in several types of synthetic rubber latex the globules are smaller than about 1000 Å, and concentrated (60%) latex of acceptable flow properties could only be made after methods had been developed for increasing the average globule size to the micron range.

7.2.5 Systems of interacting particles

The effect of (non-hydrodynamic) particle interaction far outweighs all other effects on the rheological properties of dispersed systems. This becomes evident when it is considered that actual viscosities, at not too high shear rates, may easily be several orders of magnitude higher than those in Fig. 7.3. In particular, the pronounced elastic or plastic properties of many dispersed systems must be attributed to particle interaction forces resulting in the formation of aggregates. In most systems, various types of repulsive and attractive forces are operating simultaneously. Since the technological

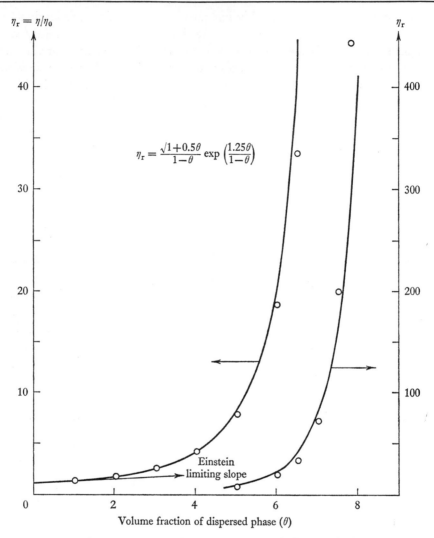

Fig. 7.3. Relation between volume-fraction of dispersed phase and relative viscosity of well-stabilized dispersions (○ Mooney, ref. 9; Van der Poel, ref. 4).

importance of many dispersed systems derives from their typical 'plastic' properties, it is evident that close control of these properties requires a detailed understanding of the various types of interaction forces, and of their effect on the mechanical properties of the system.

Fig. 7.3 shows that appreciable viscosity changes will result from small variations in the volume concentration, particularly in the more concentrated system. It is, therefore, not surprising that attempts have been made to explain high viscosity values in terms of particle 'swelling', either due to penetration of continuous phase liquid in the interior of the particle or to the formation of a shell of dispersing agent, emulsifier, or immobilized liquid around the particle (solvation, hydration). An experimental method for distinguishing between particle swelling and particle interaction, as a cause of high viscosity, may be found in the observation of non-Newtonian behaviour. Particle swelling is not expected to result in pronounced visco-elastic (plastic) properties. Based on this criterion the typical rheological behaviour of dispersed systems discussed in §2.4 appears to be due to particle interaction in the large majority of cases.

7.2.6   Electroviscous effects

Although the elastic and plastic properties of dispersions are due to particle aggregation resulting from a net attractive force between dispersed particles, it is of interest to study first the effect of purely repulsive forces on the mechanical properties of a dispersion. Such repulsive forces will not give rise to aggregation, and as far as is known at present they will not result in appreciable deviations from Newtonian behaviour.

The presence of a diffuse electrical double layer may affect the viscosity even if the average distance between the particles is larger than twice the double layer thickness, that is if there is no electrostatic interaction. This is called the *primary electroviscous effect*. The existence of this effect is due to the deformation of the diffuse electrical double layer if the liquid flows around the particle. It can easily be visualized that the double layer around a spherical particle will lose its spherical symmetry when the dispersion is subjected to shear; the double layer tends to restore its original structure and this gives rise to additional energy losses, and hence to an increased viscosity. The magnitude of this effect depends on the extension of the diffuse double layer (i.e. on the ionic concentrations) and on the potential difference between particle and surrounding fluid. The latter quantity is usually expressed in terms of a ζ-potential, which is a rather ill-defined quantity measuring the potential drop in the 'outer' part of the diffuse double layer. The effect becomes appreciable in (dilute) aqueous dispersions at very low salt concentrations[10, 11]. The addition of even small amounts of electrolyte reduces the extension of the diffuse double layer to such an extent that the primary electroviscous effect is no longer noticeable.

Whereas the primary electroviscous effect is connected with the behaviour of single, isolated particles, the electrostatic interaction of two or more charged particles gives rise to the *secondary electroviscous effect*. If the dispersed particles are identical the electrostatic interaction is necessarily repulsive; it is interesting to observe that a purely repulsive interaction must give rise to an *increased* viscosity, just as any other type of interaction such as pure attraction or a combination of repulsion and attraction. In the case of the secondary electroviscous effect the increased viscosity can be understood from the apparent increase of the particle size (and hence of the volume fraction of dispersed material) because the particles cannot approach one another as closely as in the absence of electrostatic repulsion. The magnitude of the effect depends mainly on the extension of the diffuse double layer, and it is only appreciable in aqueous dispersions at very low ionic concentrations. No deviations from Newtonian behaviour have been observed that may be attributed to this effect.

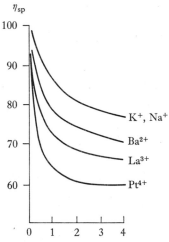

Fig. 7.4. Influence of various electrolytes on the viscosity of agar-agar sol (electro-viscous effect).

The *tertiary electroviscous effect* can be observed in solutions of macromolecules carrying charged groups along the chain. The electrostatic interaction between these groups depends on the ionic concentrations in the solution, and because the shape of the macromolecules is determined by the interaction between its elements it is evident that the viscosity of the solution may be varied by the addition of electrolytes. The phenomenon is illustrated in Fig. 7.4.

7.2.7   Plastic dispersed systems

As was discussed in §2.4.4, the normal state of a dispersed system is one in which the particles are aggregated, because the net force acting between the particles will usually be attractive over at least one range of interparticle distances. The typical 'plastic' behaviour of such a system will appear if the attractive force is sufficiently strong to produce a coherent, irregular network of interlinked chains. The chains consist of more or less linear aggregates of particles. The network should extend through the entire volume occupied by the material, and the continuous liquid phase fills the pores of the network.

This is the structure responsible for the rheological properties of materials like moulding clay, where electrostatic forces predominate and also of materials like butter and margarine, or grease, where van der Waals forces between the dispersed solid particles are more important.

Although the irregular network model of an aggregated dispersion resembles that of a rubber-like polymer, the properties are quite different. Firstly, the flexibility of the chains is very much less than in a polymer system, because particle chains contain a relatively small number of links between junction points. Therefore, the configurational entropy does not contribute significantly to the free energy and hence to the elastic forces. Secondly, deformation of a network of particle chains proceeds by rupture of bonds, followed by rearrangement of the structure and reformation of bonds in a stress-free arrangement. It is evident that such a deformation mechanism must result in elastic, solid-like properties only at very small deformations, that is when the majority of the interparticle bonds has only been stretched but no appreciable breaking of bonds and particle re-arrangement has occurred. On the other hand, the behaviour of the material is more liquid-like in larger deformations because the system is easily deformed when a sufficient number of bonds is broken, and when the deformation is discontinued the material has no tendency to revert to its original shape (before the deformation) because new bonds will immediately be formed in the final configuration.

Reformation of broken bonds between particles occurs spontaneously if they come sufficiently close together. The rate of reformation of bonds in a flowing dispersion depends mainly on the rate of removal of continuous phase liquid from between the approaching particles. This rate may vary over an enormous range in any single dispersion, because it depends not only on the nature of the continuous fluid, but also on: (1) particle size, shape and orientation, (2) the magnitude of the interaction forces as a function of interparticle distance, and (3) the extent of binding of the moving particles to other parts of the aggregates.

In particular the last factor accounts for the observation that the aggregation process may continue for very long periods: the rigidity of plastic dispersions (moulding clay, putty, grease), is often found to increase slowly during storage over periods of up to many months. On the other hand, a sufficient number of bonds to produce a coherent network may already be formed in a time less than one second (see Fig. 7.5).

7.2.8   Steady-state viscosity

It is relatively easy to acquire a rough idea about the time-scale in which interparticle bonds are formed after distortion and re-arrangement of the structure. In a steady-state viscosity measurement at sufficiently high shear rate, the particles will move past one another at such a high velocity that no time is available for the squeezing out of liquid between the approaching surfaces, and no effective bonds will be formed. This behaviour remains unchanged at still higher shear rates, and this means that the viscosity of the dispersion is independent of the shear rate in this range, that is the dispersion behaves as a Newtonian fluid. At lower shear rates, small and short-lived aggregates are formed and this must result in an increased viscosity. The shear rate at which deviations from Newtonian behaviour become just perceptible is often of the order of 10 to 100 s^{-1}, and this means that the minimum time required for the formation of bonds in such systems is about 0.1 to 0.01 s.

The increased viscosity of such slightly aggregated dispersions can be explained as a result of the immobilization of liquid between the particles of an aggregate.[9] Reference to Fig. 7.3 shows that a dispersion containing a volume fraction of dispersed particles of, say, 50% would have a relative viscosity of about 8 at shear rates sufficiently high to prevent the formation of aggregates. At somewhat lower shear rates the volume fraction of dispersed particles together with immobilized liquid might increase to, say, 0.55 and the relative viscosity would rise to 12.5. This provides a basis for a quantitative analysis of non-Newtonian flow in dispersions[12] which, however, can be used only for the initial stages of the aggregation process. This is because the use of Fig. 7.3 implies that the behaviour of the system is still mainly determined by hydrodynamic interaction, that is, the extent to which the liquid motion is disturbed by the presence of particles and aggregates. As soon as aggregation has proceeded to a stage where a more or less continuous network is present throughout the entire volume, a considerable part of the stress is supported by the bonds in the network chains and the hydrodynamic forces become less important, particularly at the low shear rates where a continuous network may exist.

Steady-state viscosity measurements have been made at much lower shear rates than can be attained in most commercial instruments. Two important results have emerged from such measurements. Firstly, the viscosity was found to increase with decreasing shear rate down to the lowest shear rate values that could be investigated (about 10^{-6} s^{-1}). A typical result is shown in

Fig. 7.5 from which it appears that the viscosity does not become constant at very low shear rates as it does in most polymeric systems. This means that some of the bonds contributing to the strength of the network require a time of at least 10^6 s for their formation, which is not unexpected on the basis of the discussion of the preceding paragraph.

Secondly, measurements at these low (constant) shear rates can be used to obtain information about the system *before* a steady state has set in. The shear stress, measured as a function of the deformation (which is simply proportional to the time) usually shows a pronounced maximum before it

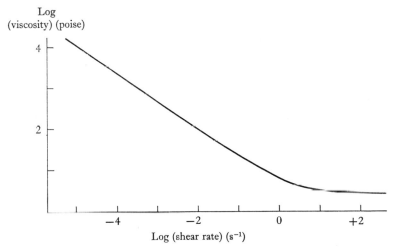

Fig. 7.5. Steady-state viscosity of dispersed system, measured between coaxial cylinders down to very low shear rates. The curve may be displaced both in the horizontal and the vertical direction by varying the composition of the dispersion, but the shape remains the same.

drops off again to the steady-state value that is used for computing the viscosity.[13] The typical curve shown in Fig. 7.6 is equivalent to a stress–strain curve of an engineering material; in particular, the maximum of the stress represents the 'strength' of the material and it is closely related to the property called 'yield value'.

The deformation at which the maximum in the stress occurs is often of order of magnitude one, and the shape of the stress–shear curve before the maximum resembles that of an elastic solid. The behaviour is, however, quite different since only a small fraction of the strain is recovered when the stress is suddenly removed. This means that already a considerable amount of structural breakdown and irreversible re-arrangement occurs at small deformations, but nevertheless the material can support an increased stress if the deformation is further increased. This peculiar behaviour is due to the rapid reformation of bonds in sufficient number and strength to compensate for the loss of strength due to breaking of bonds. This process continues until the maximum possible stress is attained; then the recovery of strength due to reformation of bonds drops to a lower level which ultimately determines the steady-state viscosity. The reason for the change in bond reformation around the region of maximum stress is not well understood; the problem here is not

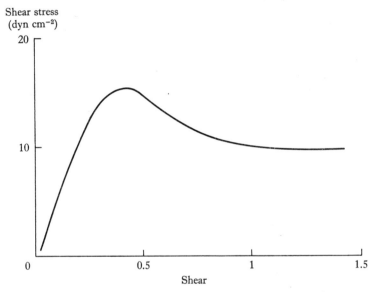

Fig. 7.6. Stress–shear curve of a 10 % emulsion of water in oil, measured at a shear rate of 0.01 s^{-1}.

to explain the decreasing strength of the structure beyond the maximum, but the surprisingly high strength just before the maximum where a large amount of structural breakdown has already occurred.

7.2.9 Yield value

The yield value of a plastic material may be defined as the maximum stress that can be supported by the material before it loses so much of its strength that its behaviour becomes more liquid-like. This is a rather vague description; nevertheless, the concept of a yield value has been found useful in many practical applications because the transition from solid-like to liquid-like behaviour occurs rather abruptly in many materials. Measurements of a yield value are often carried out under conditions where the material (e.g. asphalt, grease) is subjected to non-uniform stress (e.g. in a penetrometer); results of such measurements cannot be translated into material properties (i.e. properties independent of the instrument used in their measurement). More sophisticated techniques, such as the slow deformation of the material between concentric cylinders described in the preceding paragraph, result in a maximum stress value that has been shown to be correlated with the yield value determined by any other method, and which has the advantage that it is a material property that one can hope to interpret in terms of structure and composition.

As might be expected, the maximum stress (or the yield value) depends on the rate of deformation (or rate of loading) used in its measurement (see p. 21). As a rule, the strength increases with increasing rate of deformation, and this is the reason why impact testing of plastic materials does not result in appreciable deformations even though the stress may be in excess of the 'yield value' for a very short time. It is interesting to observe that a 'yield value of the deformation' has met with much less interest in the literature, although it undoubtedly gives just as much information about the behaviour of the material as the usual yield value of the stress. The 'yield strain' should be measured in shear, and it is simply the deformation attained in a given time under the influence of a shearing stress equal to the yield value. Its usefulness derives from its relation with rather ill-defined properties like plasticity, brittleness, shortness or toughness, which are more dependent on deformations than on stresses.

7.2.10   Small deformations at low stress values

The behaviour of plastic dispersed systems under conditions where the stress and the deformation remain below their yield values is of some practical importance, because it is often required that the materials are virtually solid under the influence of a small force, such as gravity. The interpretation of rheological properties in terms of structure and composition is also less complicated at very small than at large deformations. The obvious methods for investigating the behaviour of plastic materials in very small deformations are creep (or stress relaxation) and dynamic (vibration) measurements. The use of vibration techniques for dispersed systems has met with less success than in the case of polymeric systems, mainly because no interesting relaxation phenomena could be detected in the range of the time-scale accessible by the usual techniques. This is not unexpected in view of our present views about the structure of these materials, from which it becomes understandable that the most interesting relaxation process (breaking and reformation of inter-particle bonds) occurs in a time between about 0.01 s and several weeks. The rapid processes can, however, only be observed after the structure has been disrupted nearly completely in a large and rapid deformation, and not in a creep or dynamic experiment at very small deformations.

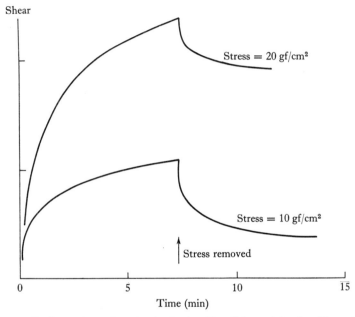

Fig. 7.7. Deformation of a plastic fat (25% solid particles in oil) under a constant shear stress (creep experiment).

The behaviour of plastic dispersed systems in creep has been studied extensively, both in simple shear and in unilateral compression. If the deformation and the stress remain well below their yield values, the materials show typical visco-elastic behaviour (Fig. 7.7). Analysis of the creep curves in terms of the theory of linear visco-elastic behaviour has, however, been much less successful than in the case of polymeric systems. The reason is that the behaviour of plastic systems is usually far from linear, in the sense that two experiments carried out on identical samples with different values of the stress result in deformations increasing more than proportional to the stresses. This means that the rigidity of the material decreases when the stress is increased (and, also, the deformation because we compare the deformations after equal times of loading). This is not surprising when it is realized that structural breakdown does not begin suddenly at the yield value, but contributes already to the deformability at very small deformations. The deviation from linear behaviour can be measured, and the results of such measurements have been used in an analysis of creep properties in terms of a network structure formed by interacting particles.[14] For systems containing dispersed solid particles in oil the 'rapid' elastic modulus and the shape of the creep curves were in agreement with predictions from theory if it was assumed that only van der Waals' forces were operating between the particles.

Measurements in unilateral compression, or in any other way than in simple shear, are more difficult to interpret because the stresses are not uniform in the entire sample. An unknown part of the stress is then used for transport of liquid through the pores of the particle network, and not for deformation of the whole sample. This results in some regions of the sample becoming more 'wet' and other regions becoming more 'dry', a phenomenon readily observed when walking on wet sand or compressing cylinders of clay. Only under very special conditions has it been found possible to estimate which part of the external force is used for deformation, and which part for liquid motion.[15] Because the liquid motion is determined by the average pore size in the network which, in turn, is determined by the total surface area of the dispersed particles, this method[15] can be applied for determining the average particle size in dispersions where more direct methods are difficult to use.

7.2.11 Thixotropy

If a plastic dispersed system is subjected to a large and rapid deformation, and the deformation is then suddenly discontinued, the material should become more rigid due to the reformation of the aggregate network. If the network formation occurs so fast that it is virtually complete in the time required for a viscosity measurement, the material is called non-Newtonian or pseudo-plastic. This is the behaviour that may be expected in a dispersion containing a high concentration of very small particles, with strong attractive forces acting between them (e.g. clay). In other materials (e.g. paints, ink) the strength of the network increases during many minutes, or even longer times, and such materials are called thixotropic. These terms were introduced at a time when the basic phenomena responsible for this behaviour were not well understood, and it is evident that their use should be restricted to cases where a crude and qualitative description of the behaviour is considered sufficient.

7.3 Solid dispersions (by H. K. de Decker and B. B. Boonstra)

7.3.1 Definition and scope

Most solids around us are not homogeneous chemical individuals, but *composite* solids. Granite rock and cement concrete are the most obvious examples.

In nature, composite solids often result from precipitation or solidification of various substances from one and the same liquid or fluid phase. Molten lava contains various chemical compounds and when it cools down they all crystallize in more or less intricate patterns, hanging together through van der Waals' forces, resulting in primary rocks. Saps circulating through a young tree contain chemicals which form lignin and cellulose in meaningful patterns, predetermined by biological causes. The two components of the wood so formed cohere through various types of secondary forces.

In man-made materials, the components can be brought together by a number of means. In creating metal alloys we may operate from a common liquid which solidifies into more than one chemically distinct solid phase, or we may use a solid solution from which several crystalline individuals separate with time.

Modern techniques make more and more use of powdered metals to make articles of intricate shape such as gears,[16] by shaping under compression and sintering. Mixing of metal powders with finely divided secondary metals or ceramic materials results in composites with very desirable properties. Aluminium flakes are always covered by a strong oxide layer. In melting and compressing, this aluminium oxide is dispersed in the matrix metal and the resulting composite shows a much higher resistance to creep at high temperatures than the cast and annealed metal alone. Similar effects are obtained with alumina dispersed in copper. Hard but brittle materials such as tungsten carbide can be embedded in a softer metal (cobalt) by this technique and the resulting material (called a cermet, from *ceramic* and *metal*) shows outstanding performance cutting steel at high speed. Many of these cermets find application in missiles and space craft.

However, in many other cases we first produce one solid phase in the form of particles (grains, dust, fibres) and then embed these in another, usually softer, continuous solid formed from the liquid state. In concrete, the gravel and sand particles are embedded in and held together by thin layers of cement hydrate formed from a solution of cement in water. In reinforced plastics or rubbers, very small particles (300 Å diameter for instance) or very small pieces of fibre are mixed into the polymer while it is in the 'plastic' or semi-liquid state. After the polymer has been hardened by cooling or cross-linking, the extraneous particles are bound to the matrix mostly by secondary forces.

In this book, we are interested in composite solids because they demonstrate two rheological principles. One is the convenience of using simple low-viscosity moulding and shaping techniques to produce strong materials of 'infinite' viscosity and high elasticity. The other is the principle of 'reinforcement', discussed in Chapter 4, whereby two solids can show a synergistic combination of the mechanical properties of the two separate phases. Such combinations may be of low modulus (flexibility) with high strength, or very hard with low brittleness, etc.

7.3.2 The principle of easy shaping

Table 7.1 gives examples and shows, in each case, the nature of the dispersed solids, the nature of the vehicle or carrier, and the mechanism by which the vehicle is transformed into a binding solid (setting).

TABLE 7.1. *Principles of easy shaping*

Material	Dispersed solids	Vehicle or carrier	Setting mechanism
Concrete	Gravel, sand, cement	Water	Cement hydrate formation
Road asphalt	Gravel, sand, clay	Asphaltic bitumen at 149° C	Bitumen solidifies at room temperature
China clay	Dehydrated clay particles	Water	Hydrate formation and water evaporation
Spray paint	Pigment particles	Binder dissolved in solvent	Solvent evaporation
Aluminium cermet	Aluminium oxide	Aluminium	Solidification after melting or sintering
High-speed cutting material	Tungsten carbide	Cobalt	As above

To keep the viscosity of the material low during its shaping or form-giving stage, two requirements must be met: (1) low viscosity of the vehicle itself and (2) the lowest possible contribution of the dispersed solids to the viscosity of the dispersion.

Whereas (1) is obvious, it will be worthwhile to show how (2) can be achieved. For the simple case of spherical particles, it can be calculated that the densest possible packing of particles of equal diameter contains about 26% void space. Therefore, a dispersion of spherical particles of equal size would have to contain more than 26 volume per cent of a liquid vehicle to be sufficiently fluid. However, the strength of the vehicle after solification, and its bonding with the dispersed solid, are sometimes not sufficient to form a strong composite if more than 26% by volume of it is present. Moreover, the vehicle

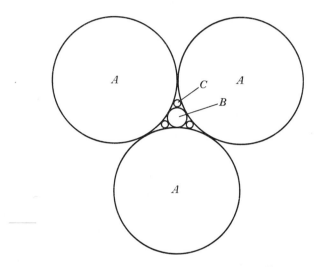

Fig. 7.8. Voids between solid particles *A* filled with smaller particles *B* and *C*.

may be more expensive than the fillers. In these cases, the possibility of filling the voids between particles of size *A* with smaller particles of size *B*, and filling the voids between *B* with still smaller particles *C*, etc, presents itself as a logical solution (Fig. 7.8). This is one of the basic features of a good concrete or a road asphalt. The particle size distribution is such that a minimum of carrier is needed (often only 5–10%), which still provides good liquidity before setting, and forms only a thin layer of cement between the particles.

The amount of vehicle can be kept so small that it only provides a minute lubricating layer between the particles enabling them to behave as a free-flowing particulate solid. In the composite solid obtained in these cases, air space is found between the particles but this does not impair the load carrying capacity of the composite. Hard top road asphalt is an example of this.

7.3.3 The principles of composite materials and reinforcement [17, 19]

A composite (solid) material in the sense of modern materials technology, is more than a combination of several solids; it has special properties, and is in some respects better than any of the components.

There is much variety in the composition as well as in the properties of composites. Without any attempt at completeness or at a systematic analysis, the categories shown in Fig. 7.9 indicate the range of this class of materials.

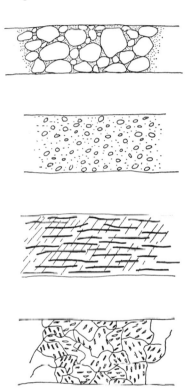

Type I. Space mostly filled by hard particles; particles touching each other; may be large. Compressive strength as for hard particles, much weaker in tension. Examples: concrete, asphalt, some cermets (tungsten carbide in cobalt).

Type II. Space largely filled by matrix; particles small and isolated, bound mostly by secondary forces to matrix. Provides stiffness as well as resistance to brittleness. Examples: rubber with carbon black, plastics with grafted rubber.

Type III. Space divided equally between matrix and filler; filler particles are elongated (fibres) and usually oriented. Provides the high tensile strength of the fibres without their brittleness. Examples: fibreglass (in polyester or epoxy resins), ceramic whisker composites in metals and cermets.

Type IV. Space largely filled by matrix, particles small and isolated; both components are metals. Provides stiffness by blocking glide mechanism. Examples: steel, duraluminium, dispersion hardened metals containing oxides or other bonded hard materials.

Fig. 7.9. Survey of composites.

Type I shows the bonding of solid particles by a substitute such as cement to form concrete. The weakness of concrete is in tensile strength, and this can be overcome by yet another method of 'composition', namely by inserting steel rods or wires in the fluid mass before setting. The steel structure in the reinforced concrete distributes the tensile stresses in such a way that local concentration will not be large enough for brittle fracture (see below).

Type II might be called the 'reinforced polymer' type. The essential feature is that embedded particles provide centres which act as multiple cross-links in one respect (causing stiffness in rubber) but also act as redistribution centres for stresses (causing impact strength in plastics, and allowing high elongation in rubber). This is more fully discussed in Chapter 4.

Type III has obtained more prominence in recent years. Some important examples are found in nature. Wood is a composite of cellulose (high tensile strength and low stiffness) with lignin (high stiffness but low tensile strength). Bone is a composite of collagen (a soft but strong protein) with apatite (hard but brittle).

A synthetic composite that has been known for some time is fibre glass (glass fibres embedded in resin). Very recent developments have led to composite of graphite or boron fibres in resins or metals. These are particularly important where extreme stiffness and tensile strength need to be combined with minimum weight.

Type IV is treated in Chapter 12.

Fig. 7.10 shows the stiffness and strength of some materials, both per unit weight. It is obvious that the winners are materials such as boron, graphite, sapphire, etc. Metals and plastics are poor by comparison. Nevertheless, the latter groups of materials do much better in a practical sense because they are not brittle.

Brittleness, as indicated in Chapter 4, is the property of breaking easily by propagation of cracks. As we have seen, any small crack or surface irregularity, or even an internal flaw, can lead to breakage under relatively small stresses if there is no mechanism preventing the propagation of cracks. 'Brittle' materials have a very low work of fracture under conventional conditions of stress rate and temperature, and this is typical of 'ceramics' or inorganic non-metal materials.

Brittleness is caused by the short range of intermolecular forces. If material at the tip of a crack can flow under high stress, there will be no brittleness, because the high local stress needed for brittle fracture cannot develop; it is dissipated through flow. Thus, plastic flow resists the propagation of cracks, and prevents brittleness. The work of fracture depends on the amount of plastic deformation needed for breakage. This is illustrated in Fig. 7.11. It will be clear that a rapid (impact) stressing of the material will more often lead to brittle fracture than slow stressing: there is no time for plastic deformation to extend from α to β in Fig. 7.11. A lower temperature has a similar effect, as it reduces the degree of plastic flow. It has become possible to use the extraordinary tensile strength of ceramics without the disadvantage of

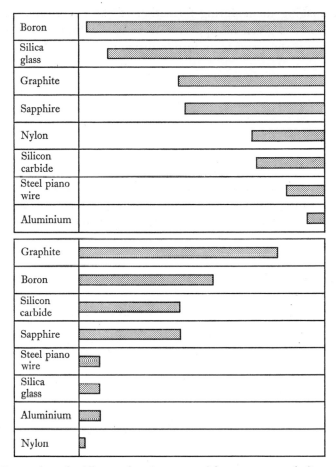

Fig. 7.10. Strength and stiffness of various materials are compared. Strength (top) is represented as greatest free-hanging length; in the case of boron it is 189.4 miles. Stiffness (lower) is represented on an arbitrary scale indicating relative stiffness per unit weight. (From Scientific American, ref. 17.)

brittleness by forming them into fibres or 'whiskers' which are single crystals in the form of elongated threads. To avoid brittle fracture the oriented fibres are embedded in a plastic matrix (metal or high polymer). The following reasons can be given for their high tensile strength and lack of brittleness:

(1) The fibres are small and have an almost ideal structure; therefore hardly any cracks will be present in the fibres and little opportunity for brittle fracture exists.

(2) The matrix is such that it adheres strongly and uniformly to the fibres

without damaging them during formation of the composite. The external stress therefore, is given no opportunity to concentrate at any point in an individual fibre; the stress is uniformly distributed.

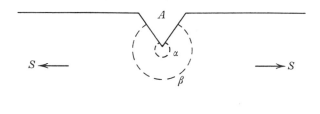

Fig. 7.11. Material having a crack at *A*, subjected to tension stress *S*. Plastic deformation limited to very small area with boundary α (brittle fracture) or larger area with boundary β (no brittle fracture).

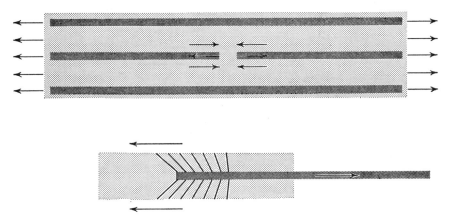

Fig. 7.12. A broken fibre in a fibre-reinforced composite causes little damage. The reason is depicted schematically at the top the broken fibre and two unbroken ones in a matrix. When the central fibre breaks, with the material stressed as shown by the solid arrows, the two pieces of fibre attempt to pull away from each other but are prevented from doing so by the shear forces (arrows) in the adhering matrix. Forces at work on a broken end of a fibre are represented in more detail in the illustration at bottom. (From Scientific American, ref. 17.)

(3) In the very few cases where a fibre breaks inside the composite, it will not lead to failure of the material because plastic deformation of the matrix prevents the crack from propagating. Fig. 7.12 illustrates this point.
Composites reinforced with whiskers have been made in various laboratories and have demonstrated some remarkable properties.

The largest tonnage of composite material of type III now being manufactured is in the form of fibre glass. Glass, because of the manner in which silica melts, is easily drawn from the molten state into thin, high strength fibres. Fibres so prepared can readily be put into a matrix of unsaturated polyester resin. The resin is originally in liquid form, and the chemical reaction to make it set around the fibres (curing) can be promoted at low temperature and low pressure. This means that the glass fibre is not grossly damaged in the formation of the composite. The fact that the fibres remain largely intact is one of the chief advantages of glass-reinforced plastics. Large pieces of such material can easily be built up by the application of successive layers of resin and glass.

When it is not important to minimize weight, tungsten (sp. gr. = 19.4) fibres, which retain high strength up to more than 1500° C, can be used in a composite. Tungsten filaments have been introduced into metal matrices for use at temperatures of 1000° C and higher. The metals cobalt and nickel make ideal matrices because they do not oxidize readily at high temperatures.

Fibres of tungsten, silica coated with carbon, graphite and boron are introduced into matrices by electroplating or chemical deposition. Another technique makes the matrix and the reinforcing fibre in one operation by the controlled melting and cooling of certain metal alloys. This is a refinement of the principle of steel-making, where various crystals of non-carbon compounds reinforce the soft iron matrix.

Composites of type III are ideally suited for carrying large tensile loads in one direction but may not be as effective as other materials under compression or under shear occurring at an angle to the fibres. Careful design can overcome such deficiencies. A case in point is a large hollow sphere used as a buoyancy tank in a deep-diving submarine. A United-states Navy research group made a suitable structure from sections of glass-reinforced plastic. In it, the fibres run *radially*, so that they support one another against buckling when the vessel is under pressure.[18]

Fig. 7.13 illustrates 'materials systems', which are the various ways of combining solid materials to meet engineering problems.

7.4 Gas dispersions. Two–phase systems involving gas (by H. K. de Decker and B. B. Boonstra)

7.4.1 Scope

There are two distinct classes of dispersions (two-phase systems) in which gas is one of the phases. One class has a gas as the continuous phase; examples

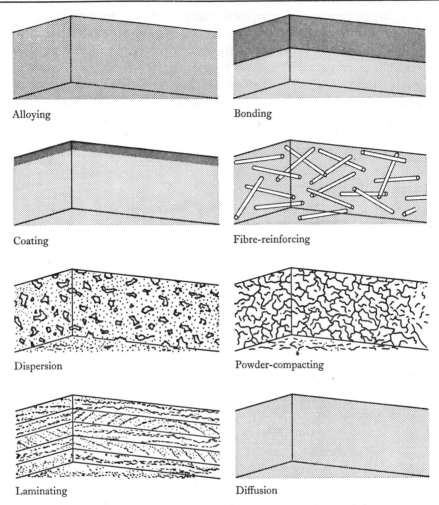

Alloying

Bonding

Coating

Fibre-reinforcing

Dispersion

Powder-compacting

Laminating

Diffusion

Fig. 7.13. Materials systems. Composite types of long standing at left; more recent
ones at right. (From Scientific American, ref. 17.)

are fog, smoke, 'spray', etc. In the other class the non-gaseous phase is
continuous; examples of this are lather, sponge, solid and liquid 'foams', etc.

The rheology of these materials is best treated as being essentially the
rheology of the continuous phase; complications caused by the dispersed phase
are then relatively easy to understand, usually as a 'filler' effect. Table 7.2
surveys dispersions in both classes.

TABLE 7.2. *General features of gas dispersions*

Class	Continuous phase	Discontinuous phase	Examples
I	Gas	Liquid	Fog, mist, clouds
	Gas	Highly viscous liquid or plastic	Smoke Paint spray Cosmetic spray
	Gas	Solid	Clouds of dust, ice, carbon black Fluid bed catalyst or heating media
II	Liquid	Gas	Soap lather Beer foam
	Highly viscous liquid	Gas	Whipped cream
	Soft plastic	Gas	Expanded plastic 'leather'
	Rubber	Gas	Sponge rubber shoe soles Foam rubber† mattresses
	Hard plastic	Gas	Styrofoam insulating blocks
	Solid	Gas	Expanded cement building blocks Sponge metal† filters Porous steel

† In open cell foam rubber and in sintered metal filters both the gas and the solid phase are continuous.

7.4.2 Continuous gas phase

In most cases of this class, the concentration of the dispersed phase is relatively small. One cubic metre of a rain cloud may contain 10 cm³ of condensed water subdivided into 10^8 droplets, so that one droplet weighs 10^{-7} g on average, corresponding to a diameter of 0.06 mm. The basic properties of the air remain unchanged. The elastic compressibility is that of air itself. The volume occupied by the droplets does not interfere with the elasticity of the air.

The viscosity of the dispersion in laminar flow is the gas viscosity corrected for the volume of the filler, in accordance with Einstein's equation (4.23) as expanded by Guth and Gold:

$$\eta = \eta_0(1 + 2.5C_v + 14C_v^2),$$

where η_0 is the viscosity of the pure gas, and C_v is the volume fraction of the filler.

At 10% of filler, which is a high concentration for this sort of dispersion, the viscosity is 1.36 times the pure gas viscosity. In the very unlikely case that 50% of the space would be occupied by the dispersed phase, the viscosity would amount to five times the gas viscosity, which still is a very low value indeed. The fact that the viscosity of dispersions in gas are so extremely low is the main basis for their practical applications. It not only allows easy mixing of the materials by the simple expedient of dispersing them in air or another gas, but transportation of such a dispersion is also extremely easy and the reason for their application as sprays.

Two complicating factors to be dealt with before the practical use of gas dispersions can be understood are settling and coalescence.

(a) Settling

According to Stokes' law:

$$V = \frac{gd^2C}{18\eta},$$ (7.8)

where V is the velocity of settling, d the diameter of particle, C the density difference and η gas viscosity. The most practical way to prevent or to retard settling, is to stir the gas medium or to reduce the diameter of the suspended particles. In the latter case, the system may appear stable over reasonable periods of observation.

Fog is a good example. Fog particles (water in air at 20° C) settle at a rate of 1 m/s if their diameter is 0.2 mm, but only at 1 cm/s if the diameter is 0.02 mm (20 μm). Most dispersions of class 1 contain particles having diameters of 1–100 μm.

(b) Coalescence

The other complicating factor is coalescence. As we have seen in other suspensions, liquid particles will tend to combine during mutual collisions. An obvious means to prevent this is a 'protective coating', usually a surface-active ingredient that lowers the surface tension and often provides a repulsive double layer on the particle surface.

A modern type of transportation for powdery materials is conveyance by pipeline. This method is used increasingly for the long distance transport of cement, grain, pigments such as carbon black, etc.

Another important technical application of solid dispersions is the fluidized bed catalyst. The catalyst is subdivided into very small solid spheres to provide a large reactive surface. These spheres are then suspended in a stream of the reacting gas mixture they must catalyse (e.g. cracking or hydrogenation). The essential point is that the gas stream moves upward with a velocity just large enough to compensate for the gravity settling rate of the particles. As a result, the solid dispersion is maintained at a 'stationary level', while the gas 'flows through' it. Similar techniques are used to treat pigment particles with gaseous chemical agents (e.g. carbon black with air). A different application is the use of a fluidized bed as a heating medium to increase the heat transfer coefficient of hot air or steam.

7.4.3 Continuous solid or liquid

Here again, the basic properties of the continuous phase are usually not greatly changed by the presence of the dispersed phase, in this case the gas bubbles, except in compression. The elastic modulus of extension of a

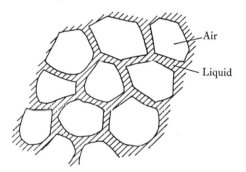

Fig. 7.14. Magnified cross section of a 'stable' liquid foam; outward pressure in air bubbles counteracts inward pressure resulting from liquid surface tension.

solid material will be reduced in proportion to the relative area of gas bubbles and non-gaseous material in a cross-section. The ultimate tensile strength is influenced similarly by the gas phase, but in addition, the bubbles may also cause local inhomogeneities, resulting in a 'notch' effect, and thereby greatly reducing the tensile strength below that calculated from a simple dilution of the solid material.

In a *liquid* foam, the gas bubbles usually occupy a volume larger than that of the liquid. If the liquid contains a sufficient amount of surface active ingredient, it is possible to reduce the continuous liquid phase to a network of thin membrane separating gas bubbles (see Fig. 7.14). In this case (soap lather, etc.) the foam acquires properties entirely different from the continuous (liquid) phase, and may show a semi-permanent structure. When the surface tension of the liquid is sufficiently low to allow it to remain spread out in thin lamellae, drainage of liquid between the bubbles can reach a point where the capillary force in the thin lamellae is sufficient to counterbalance the gravity force and the liquid will 'stay in place'. At this point of equilibrium, one can see the liquid lamellae as encased between two mono-molecular layers of surfactant (e.g. a water lamella between two layers of soap molecules). The interfacial tension between water and the soap layer (the polar side is towards the water) has a positive value; it can be shown that, to a good approximation, the vertical height h of a stable lamella is given by

$$h = \frac{2\theta}{dg},\qquad(7.9)$$

where d is the thickness of the lamella, g is the specific gravity of the liquid and θ the surface tension of the liquid–gas interface. Soap bubbles which will produce colour rings and patches by light interference have a thickness of between one and ten times the average light wavelength, say 10000 Å (10^{-4} cm) and their diameter may be 10 cm. These dimensions are comparable to a sheet of newspaper 0.1 mm thick, covering the walls of a small room (4×4 m). From equation (7.9), the interfacial tension of such soap bubbles is of the order $\theta = (hdg)/2 = (10 \times 10^{-3} \times 1)/2 = 5 \times 10^{-3}$ gf/cm. This is the interfacial tension between soap solution and air.

Foams with (semi-) stable lamellae exhibit the bulk elasticity of a gas in a container.

Gas dispersions in solid (or plastic or rubbery) materials, are usually generated from a liquid foam formed by beating (air), by blowing a stream of gas into the liquid, or by chemically generating a gas inside a liquid or plastic material. The liquid phase is then allowed to harden by cooling down a liquid plastic such as polystyrene or the chemical setting of Portland cement, etc. Bubbles in a discontinuous gas phase dispersion generally have diameters of between 100 and 1000 μm.

In a consideration of viscosity, only those cases where flow of a liquid or plastic phase occurs are relevant. Where air bubbles act as filler particles in the sense of the Einstein equation (4.23), they increase the viscosity, as much as fivefold if the volume of the bubbles is 50% of the total volume. A foam with a very high percentage of gas volume shows special properties, masking the ordinary viscosity of the pure liquid.

An interesting case occurs when both gas and solid phase are continuous. This can be achieved by imperfect packing of solid particles, as in the case of open cell sponge rubber (from latex) or sintered metal filters (see Fig. 7.15).

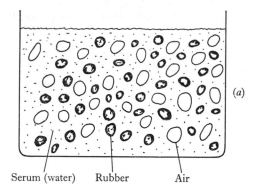

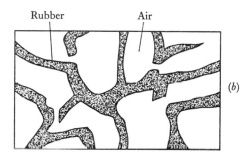

Fig. 7.15. Two-dimensional diagram showing formation of foam rubber. (*a*) Rubber latex with air bubbles. (*b*) Water has been evaporated, the rubber coalesced and vulcanized, to give two continuous phases; contraction not shown.

References to chapter 7

1 *Second report on viscosity and plasticity* (Amsterdam, 1938).
2 A. Einstein, *Ann. Phys. Leipzig* **19**, 289 (1906); **34**, 591 (1911).
3 S. G. Mason *et al.* see e.g. *J. Colloid Sci.* **14**, 13 (1959) which contains references to earlier papers.
4 C. van der Poel, *Rheol. Acta* **1**, 198 (1958).
5 H. Fröhlich & R. Sack, *Proc. Roy. Soc. London* **A185**, 415 (1946).
6 J. G. Oldroyd, *Proc. Roy. Soc. London* **A218**, 122 (1953).
7 G. I. Taylor, *Proc. Roy. Soc. London* **A138**, 41 (1932).
8 V. C. Levich, *Physico-chemical hydrodynamics* (New Jersey, 1962).
9 M. Mooney, *J. Colloid Sci.* **1**, 195 (1946).
10 G. J. Harmsen, J. van Schooten & M. van der Waarden, *J. Colloid Sci.* **10**, 315 (1955).
11 F. S. Chan, J. Blachford & D. A. J. Goring, J. Colloid Interf. Sci. **22**, 378 (1966).
12 M. van den Tempel, *Emulsion rheology* (Pergamon Press, 1963).
13 V. P. Pavlov & G. V. Vinogradov, *Colloid J. U.S.S.R.* **28**, 346 (1966).

14   M. van den Tempel, *J. Colloid Sci.* **16**, 284 (1961).
15   E. M. de Jager, M. van den Tempel & P. de Bruyne, *Proc. Roy. Dutch Acad. Sci.* **B66**, 18 (1963).
16   L. Holliday, *Composite materials* (Elsevier, 1966).
17   A. Kelly, 'The nature of composite materials', *Sci. American* **217**, 160 (1967).
18   D. R. Elliott (to UniRoyal Inc.), U.S. Pat. 3490, 638 (20 Jan. 1970).
19   P. M. Sinclair, 'Composites', *Industrial Research* (October 1969).

8 Thermoplastic materials with limited elasticity (thermoplastics)

R. D. Andrews (§ 8.1), R. Houwink (§ 8.2) and H. K. de Decker (§ 8.3)

8.1 Linear polymers (by R. D. Andrews)

8.1.1 Introduction

The characteristic feature of thermoplastics is that they are formable (plastic) and can be fabricated into useful shapes by heating, after which they re-solidify when cooled to room temperature. This behaviour can be distinguished from that of the thermosetting polymers which form a permanent chemically cross-linked network when heated, and once this has taken place no further softening or melting can be produced by heating. Materials of the thermoplastic type have been used for centuries and include such materials as pitch, asphalt, sealing wax, beeswax, paraffin and colophonium. Modified natural materials have also found application in this area: such as casein plastics (made from milk) and, most important, the cellulosic plastics made from the natural cellulose in wood or cotton fibres. The nitration of cellulose produced the first large-volume commercial plastic, celluloid, and both cellulose nitrate and acetate have been used as the base for photographic film.

The majority of thermoplastics used today are synthetic polymers with a linear chain structure. These will be treated first in this chapter. The other thermoplastic materials, such as asphalts and natural resins, are different in structure and rheological properties; these latter will be discussed in §§ 8.2 and 8.3.

Linear polymers can be either amorphous or crystalline. The amorphous polymers are glassy and transparent at room temperature. Examples of these would be polystyrene and polymethyl methacrylate (Plexiglas, Lucite, Perspex). Because of the tendency of these materials towards brittleness, copolymers have been developed which are more impact resistant, such as styrene/acrylonitrile copolymer, polystyrene/rubber blends or graft copolymers, and the ABS polymers (acrylonitrile/butadiene/styrene). The polycarbonate polymers are also a new class of glassy materials with superior impact strength. Crystalline thermoplastics include such familiar polymers as polyethylene, polypropylene and nylon. These crystalline polymers are generally turbid or milky in appearance, and thick specimens are opaque. New and interesting materials in this category include the polyacetals (poly-formaldehyde) which have improved strength and dimensional stability; they are sometimes referred to as engineering or structural plastics, which can in some cases replace metals.

(a) Transitions

The thermoplastics show characteristic solid-state transitions, the most important of which are the glass transition in the case of amorphous polymers, and the melting point in the case of crystalline polymers. Above the glass transition, amorphous thermoplastics are rubbery materials and at still higher temperatures, visco-elastic liquids. The fact that they are considered to be

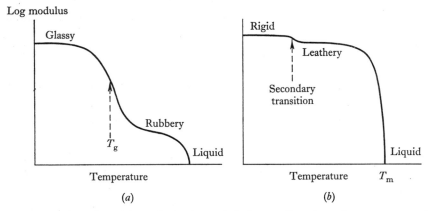

Fig. 8.1. Rigidity (logarithm of elastic modulus) as a function of temperature. (a) Amorphous polymers, showing glass transition temperature T_g. (b) Crystalline polymers, showing a secondary transition as well as the melting point T_m.

thermoplastics rather than elastomers is therefore due simply to the fact that their glass transition temperature lies above room temperature. If the glass transition is above room temperature but too close to room temperature, however, the material is not suitable for practical use, since it softens too easily on relatively mild heating.

The temperature dependence of the rigidity in both amorphous and crystal-line thermoplastics is indicated schematically by the sketches in Fig. 8.1. The states of the material in different temperature ranges are indicated in these figures (compare also Fig. 4.25). The amorphous thermoplastics generally show a drastic softening at the glass transition to a rubbery state and then a very gradual change to a liquid state or melt without any marked transition point. The crystalline polymers generally show a more abrupt transition from a solid to a relatively fluid liquid at the melting point.

These polymers also generally show so-called secondary transitions at lower temperatures; and crystalline polymers, since they are not completely crystalline, will show amorphous-phase or glassy transitions in addition to their melting point. In such a case the melting point occurs at the highest temperature. The glass transition is also generally considered to be the highest temperature transition in the amorphous polymers, but there has been some discussion recently of the possibility of higher temperature transitions in the liquid state. If such transitions exist, they are clearly less important than the major glass transition or the melting point. Secondary transitions below the major transition can be of importance, however, and there is a widespread opinion that the secondary transitions are related in an important way to the toughness or impact strength of the polymer.

(b) Response to stresses

The response of thermoplastics to mechanical stresses depends not only on the temperature range in which the stress is applied, but also on the magnitude of the stress. At very low stresses, linear visco-elastic behaviour can be observed. At somewhat higher stresses or strains, this visco-elastic behaviour becomes non-linear. If stress is increased still further, plastic yield or fracture will take place. If plastic yield takes place before fracture, the material is said to be 'ductile'. The elongation at break in the case of ductile polymers will be of the order of perhaps 200 % for amorphous glassy polymers, and may be as high as 1000–1500 % in the case of crystalline polymers.

If the stress–strain curve shows a 'yield peak', the sample characteristically forms a localized neck which then propagates over the length of the sample while the extension increases at an essentially constant load. This propagation load value is frequently referred to as the 'drawing stress'; this term comes from the textile industry where such drawing is carried out as a continuous process on a textile fibre or yarn, in order to enhance its tensile strength. When neck propagation has spread over the entire specimen, the 'drawn' material extends elastically up to the fracture point. This sequence of behaviour is illustrated in Fig. 8.2. Although neck formation is a very commonly observed feature of plastic yield in solid polymers, it is not a necessary characteristic, and in many cases yield takes place homogeneously in the specimen, without the localization associated with the necking process.

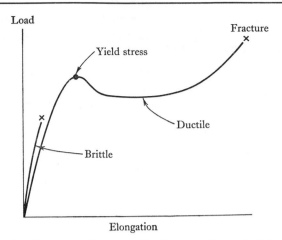

Fig. 8.2. Stress–strain curves (load versus elongation) for a
brittle and ductile material.

8.1.2   Elasticity

Since thermoplastics are relatively rigid materials at room temperature, their
elastic behaviour in this temperature region is very similar to that observed in
other rigid materials such as silicate glasses, inorganic crystals and metals,
although thermoplastics have a much lower resistance to deformation (see
Table 4.1). From a thermodynamic standpoint, this elasticity involves
primarily internal energy changes in the material due to the displacement of
atoms and molecular groups in potential energy troughs. This behaviour
contrasts strongly with the type of elasticity observed in elastomers where the
stresses and work of deformation are primarily associated with entropy
changes resulting from the change in shape of the molecules as their end-to-end
distances are extended (see Chapter 9).

Pure energy and pure entropy elasticity are obviously two extreme idealized
cases, and very few materials can be found which exemplify either of these
perfectly. A perfect crystal lattice which would show pure energy elasticity is
almost never found; most crystals have defects of one sort or another, and
metals and polymers are usually polycrystalline in nature. Crystalline
polymers may also have a very high content of amorphous material (as much
as 50 %). For this reason, they are frequently referred to as 'semi-crystalline'.
The amorphous regions of these polymers will show some degree of entropy
elasticity. In the same way, very few perfect elastomers are found, natural
rubber providing one of the closest approaches to this ideal state. Most

elastomers show internal energy changes on deformation, due to such effects as breaking or re-arrangement of internal secondary bonding patterns between the molecular chains, and also more extreme effects such as crystallization.

8.1.3 Birefringence

A very useful experimental technique in elucidating the nature of the elastic behaviour of polymers is the measurement of birefringence or double refraction. The birefringence phenomenon in amorphous, glassy polymers has been applied in a practical way in the case of engineering photo-elasticity, where it is used to study stress distributions in mechanical parts of complicated shape. In order to do this a model is machined from transparent plastic and stresses are then applied to this model in the same way that they will be in the practical situation. By placing the stressed model between crossed polarizers, a pattern of dark lines or 'fringes' is observed which are analogous to contour lines of stress levels in the sample. This technique has found important application in cases where an exact mathematical analysis of the stress pattern would prove extremely difficult or tedious, or where a direct experimental verification of the mathematical analysis is desired.

An example of the application of birefringence measurements to the study of the elasticity of a common thermoplastic can be shown for the case of polystyrene (Fig. 8.3). Birefringence (n) is defined as the refractive index of the material in the stress direction minus the value in the perpendicular direction; that is, $n = (n_1 - n_2)$. The birefringence can be measured directly by optical compensators of various sorts. It is found in general that for rigid materials the birefringence produced by an applied stress is directly proportional to the stress applied (σ). The case of a direct proportionality between birefringence and stress obviously represents a case of linear photo-elasticity, completely analogous to the case of linear elasticity where strain and stress are directly proportional. When this linear relationship holds, the effect can be most easily defined by a stress-optical coefficient, which is the ratio of birefringence to applied stress. That is,

$$\text{stress-optical coefficient } C = \frac{\Delta n}{\sigma}. \tag{8.1}$$

When stress is expressed in dyn/cm², a value of the stress-optical coefficient of 10^{-13} cm²/dyn is defined as one Brewster. For glass C has a value between 1 and 10 Brewster; for celluloid and phenol-formaldehyde resin it lies between 10 and 100 Brewster.[1]

Polystyrene in the glassy state shows a stress-optical coefficient of the order of 10 to 15 Brewster. The value decreases slightly with increasing temperature in the glassy state. As temperature increases into the glass transition region, the stress-optical coefficient shows a drastic change from small positive to very large negative values, of the order of − 5000 Brewster. This large negative value is characteristic of polystyrene in the rubbery state. This behaviour is shown schematically in Fig. 8.3.

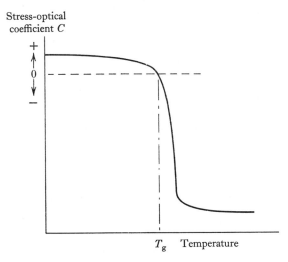

Fig. 8.3. Stress-optical coefficient C versus temperature for polystyrene; T_g glass transition temperature.

In order to understand this effect, it is necessary to realize that the birefringence in polystyrene is produced primarily by orientation of the benzene ring side groups. The benzene ring has a strong polarizability anisotropy, that is, a difference in polarizability in the plane and perpendicular to the plane. When this polarizability anisotropy (benzene ring) becomes oriented, due to the application of mechanical stress, the bulk material will consequently have a different average refractive index in different directions, which is measured as birefringence. If the benzene rings were oriented by simple tipping of their planes in the direction of applied stress, this would produce a positive birefringence effect, because the polarizability of the benzene ring is greatest in its plane. This is the effect which is observed in the glassy state, and indicates that elastic strains are taking place in the glassy state in a very localized way, since the benzene rings are oriented in the way they would be if they were independent entities not connected to a molecular chain.

In the rubbery state, the strain of the material clearly takes place in a completely different way on the molecular scale. Deformation takes place by the extension or uncoiling of molecular chains, and the chain backbones therefore tend to line up with the direction of stress or strain. When this takes place, the benzene ring side groups are brought into a lateral alignment; and since their planes are preferentially aligned in the sidewise direction, the birefringence observed is negative in sign. The stress is apparently transmitted through the molecular chains, rather than through a local continuum, and no specific stress is operating on the benzene rings themselves which would tend to orient them, as is true in the glassy state. An interesting additional observation is the fact that in flowing liquid styrene monomer, a positive birefringence is observed.[2] This indicates that the benzene rings are being oriented in the direction of flow because of the hydrodynamic drag characteristics of the monomer molecule when it is not incorporated in a polystyrene chain.

The large numerical magnitude of the stress-optical coefficient in the rubbery state is explained by the large decrease in elastic modulus (by a factor of about 1000) which takes place at the glass transition. As a result, the degree of strain (and consequent molecular orientation) produced by a given stress is much greater in the rubbery state than in the glassy state. Clearly the rubbery state is similar to that of a liquid, so that orientation of molecular chains can take place without the requirement of distortion of interatomic distances, bending of valence angles, etc., which take place in the glassy state. Many other polymers show a similar sign change in the stress-optical coefficient at the glass transition. The sign change can also be from negative in the glassy state to positive in the rubbery state, as in the case of polyvinyl chloride.[3]

Interesting insights into the details of the molecular distortion in the glassy state can also be obtained by the birefringence technique. Chemical structure can be systematically modified and the effect on the stress-optical coefficient measured. For example, the benzene ring can be moved away from the chain backbone in styrene-type polymers: starting with polymethyl styrene, the benzene ring can be moved away from the chain backbone by placing it at the end of an ester side group in polyphenyl methacrylate and polybenzyl methacrylate. These polymers are also glassy amorphous polymers at room temperature, and as the phenyl ring is moved away from the chain backbone, a marked increase in positive value of the stress-optical coefficient is obtained.[4] The chemical structures of the polymers and the behaviour of the stress-optical coefficient are shown in Fig. 8.4. These larger positive values indicate a greatly increased ease of orientation of the benzene ring when it is separated from the chain. A comparison of these values with the value for pure polystyrene indicates that there is considerable steric hindrance to the rotation of the benzene ring when it is attached directly to the chain backbone. This type of information on the freedom of side group rotation is very valuable in connection with interpretation of such phenomena as the molecular mechanism of secondary transitions.

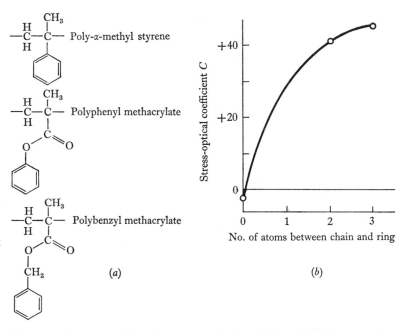

Fig. 8.4. Structure (*a*) and stress-optical coefficient C,(*b*) for a series of polymers having a benzene ring in the side group.

8.1.4 Secondary transitions in glassy polymers

One example of a secondary transition in a glassy polymer is the so-called beta transition in polymethyl methacrylate. This can be conveniently observed by dynamic (vibrational) mechanical measurements at low strains. In such measurements both a dynamic elastic modulus and the dynamic loss modulus or dissipation are usually measured. The presence of secondary transitions can be seen most clearly from the dynamic loss versus temperature curve (sometimes called the 'mechanical loss spectrum'). Polymethyl methacrylate shows a peak in mechanical loss not only at the glass transition at about

+120° C (the alpha transition) but also at the beta transition which is observed at about +40° C at low frequencies. This is illustrated in Fig. 8.5.

Dielectric constant and loss measurements as a function of temperature generally give very similar results to dynamic mechanical measurements and also show the presence of transitions. In dielectric measurements, the beta loss peak in polymethyl methacrylate shows up even more strongly than the

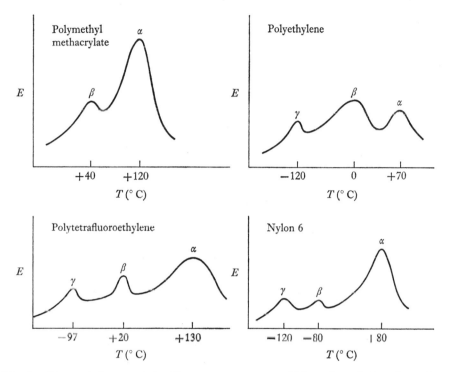

Fig. 8.5. Dynamic loss modulus E versus temperature T for polymethyl methacrylate, polyethylene, polytetrafluoroethylene and Nylon 6.

alpha peak or primary glass transition, which is the inverse of the behaviour observed in dynamic mechanical measurements. From this result, since the dipole in polymethyl methacrylate is located in the ester side group, the conclusion has been drawn that the molecular mechanism of the beta transition consists of a rotation of the ester side group, which would allow the dipoles to orient in the electric field.

In general, transitions in solid polymers have been interpreted in terms of motions of various parts of the molecular structure. For example, the principal

6

glass transition (or alpha transition) has been associated with the motion of chain segments of the molecular backbone. Secondary transitions have been associated with motions of smaller parts of the molecules, such as side groups. The details of the molecular mechanism of such loss peaks at transitions are not well understood at the present time. Unresolved questions still exist. For example, a beta transition and a dynamic mechanical loss peak are observed for polyvinyl chloride, where no side group rotation is possible. This casts some doubt on the interpretation of the beta transition in polymethyl methacrylate in terms of ester group rotation. An alternative type of interpretation would be in terms of the breakdown of a specific type of secondary bonding in the polymer rather than in terms of molecular motions. This would be similar to the transition in Nylon 6 which seems to be associated with hydrogen bonding and which will be discussed later. Dipole–dipole bonding would be possible in both polymethyl methacrylate and polyvinyl chloride, and it would seem quite possible that the thermal breakdown of such bonding could lead to an observable secondary transition.

8.1.5   Secondary transitions in crystalline polymers

As examples of the transitions in crystalline polymers, dynamic mechanical loss curves are shown for polyethylene, polytetrafluoroethylene, and Nylon 6 in Fig. 8.5. Although the polyethylene and polytetrafluoroethylene molecules are almost identical in structure, it is clear that the transitions are considerably different in nature. This shows that the dynamic mechanical loss spectra are not a direct result of the molecular structure, but reflect the details of the crystalline structure and morphology as well. The highest temperature secondary transition in polyethylene (alpha transition) has been attributed to the crystal phase, while the middle transition (beta transition) has been regarded as an amorphous transition. The lowest temperature transition, at about $-120°$ C, is called the gamma transition and is observed not only in polyethylene but in many other polymers which have a sequence of three or more methylene groups either in the chain backbone or in a side group. This transition can be observed, for example, in Nylon 6, as shown in Fig. 8.5. There is some evidence that this transition consists of two components, one amorphous and one crystalline. Most commonly this has been regarded as an amorphous phase transition.

The melting point of polytetrafluoroethylene is $327°$ C, and three widely spaced secondary transitions are observed at about $130°$ C, $20°$ C and $-97°$ C. The transition at room temperature is known to be a first-order crystalline phase transition. There are actually two component transitions in this room temperature region, one at 19° and the other at 30° C. The high-temperature and low-temperature transitions are both believed to be amorphous phase transitions, and the gamma transition at $-97°$ is probably the counterpart of the gamma transition at $-120°$ C in polyethylene. However, in polytetra-fluoroethylene the gamma transition is believed to be completely amorphous in nature.[5] These gamma transitions are believed to involve motions of very small segments of the molecule.

The transition observed at about room temperature in Nylon 6 (and in other nylon polymers) is found to be very sensitive to the presence of absorbed moisture. The transition is observed at temperatures from 50° to $-50°$ C depending on the moisture content of the polymer. This shift of the transition temperature to lower values is typical of the effect of a plasticizer on a glass transition, and from the fact that water is acting as such a powerful plasticizer, it is concluded that this transition is an amorphous phase glass transition involving the hydrogen bonds between adjacent molecules. Since neither polyethylene nor polytetrafluorethylene absorb water, their transitions show no such moisture sensitivity.

In connection with dynamic mechanical experiments, it is interesting to note that birefringence measurements can also be carried out in a dynamic way. A sinusoidal stress or strain is applied to the sample, as in the dynamic mechanical case, and the sinusoidal birefringence output is measured. One of the interesting results obtained from measurement of this type on poly-butene-1, which is a spherulitic polymer, is that the spherulites in such polymers show a time dependence in their deformation.[6] The spherulites are small spheres of oriented crystalline material; they deform almost instanta-neously from a spherical to an ellipsoidal shape, and this is then followed by internal relaxation effects within the spherulite which are time dependent and have a relaxation time of the order of a few seconds. This interpretation has been verified by the use of dynamic X-ray measurements.

8.1.6   Plasticity

It has been noted above that brittle and ductile polymers can be distinguished on the basis of their stress–strain curves. The same polymer can in fact show either brittle or ductile deformation depending on the deformation conditions (temperature, strain rate, etc.). This leads to a phenomenon known as the brittle–ductile transition, which is observed in many polymers as a function of temperature at constant strain rate, or as a function of strain rate at constant temperature. This transition as a function of temperature can be seen from data on the yield stress and fracture stress of polystyrene as a function of temperature, shown[7] in Fig. 8.6. These values are derived from stress–strain curves measured at a constant strain rate. The two stress values have a very different dependence on temperature. The yield stress shows a pronounced linear decrease with temperature which seems to extrapolate to zero stress slightly above the glass transition temperature of the polymer. The fracture stress is much less temperature dependent but shows a small linear decrease with temperature. The different slopes result in an intersection of these two

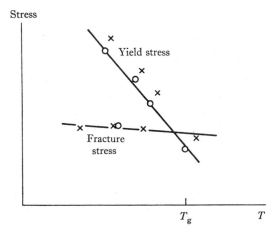

Fig. 8.6. Yield stress (for prestretched samples) and fracture stress (untreated samples) of polystyrene, as a function of temperature.

straight lines such that above the intersection point, when stress is increased on a sample, the yield stress is reached before the fracture stress, and the material shows plastic yield. Below the intersection of these two curves, as stress increases, the fracture stress is reached before the yield stress, and the material shows brittle fracture.

The intersection point in this case occurs at a temperature very close to the glass transition temperature. Plastic yield is not observed in the rubbery state; in this temperature range the material simply stretches homogeneously. In order to observe plastic yield, there must be some rigid structure which can collapse catastrophically.

(a) *Effect of molecular orientation*

It is interesting that the plastic yield process can be enhanced in polystyrene by pre-orientation of the polymer. This is accomplished by heating the material above its glass transition, stretching to orient the molecules, and then cooling again to freeze the molecular orientation produced. After such a treatment, polystyrene will show plastic yield over a much wider temperature range enabling measurement of the yield stress at lower temperatures (see Fig. 8.6). The reason for this effect is not known, but polystyrene can be converted from a brittle to a ductile material at room temperature in this way. It can be

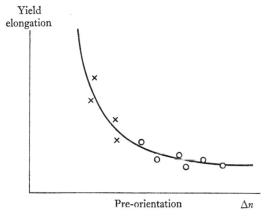

Fig. 8.7. Yield elongation or 'draw ratio' of polystyrene as a function of pre-orientation (expressed as birefringence Δn).

stated, however, that the effect is a result of raising the fracture stress, rather than lowering the yield stress.

The value of the yield stress of polystyrene does not seem to depend on the amount of pre-orientation introduced; it is a function only of temperature and strain rate. However, one characteristic of the drawing process is very significantly affected by the amount of pre-orientation; this is the 'draw ratio' or amount of elongation produced by the plastic yield. This shows a marked decrease as a function of degree of pre-orientation, as shown[7] in Fig. 8.7. This result suggests that the draw ratio is related in some way to the extensibility of the chains, and that during the pre-orientation process some of this potential extensibility has been used up.

(*b*) *Time effects*

Time effects, in the form of a dependence of the stress–strain curve on extension rate, are always observed in connection with the phenomenon of plastic yield. The dependence of the yield stress on both temperature and strain rate are shown[8] for polymethyl methacrylate in Fig. 8.8. The yield stress is always higher at higher strain rate but at any given strain rate shows the same linear dependence on temperature which was observed previously

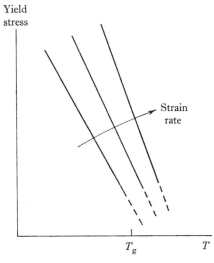

Fig. 8.8. Yield stress dependence on temperature and strain rate
for polymethyl methacrylate.

for polystyrene. The straight lines extrapolate to temperature values which are slightly different, but in all cases not far above the glass transition temperature of the polymer. The dependence of the yield stress on two other important molecular parameters, molecular weight and cross-linking, are shown[8] for polymethyl methacrylate in Fig. 8.9. For molecular weights covering a six-fold range, there seems to be essentially no effect of molecular weight. Similarly, the lightly cross-linked polymer shows behaviour which is almost identical with that for polymer containing no cross-links. These results indicate that there is no long-range molecular flow involved in the plastic yield process, because molecular flow (in a polymer melt, for example) is strongly dependent on molecular weight, being proportional to molecular weight to the 3.4 power, and true molecular flow is prevented by the presence of cross-links. This indicates that the plastic yield process takes place on a local scale, at the molecular level, and does not involve long range slippage of molecules past each other. This is a useful point to clarify since plastic yield is frequently referred to in the published literature as 'cold flow'.

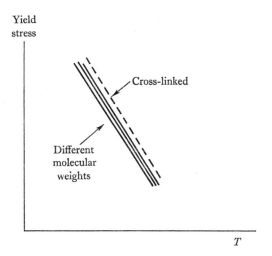

Fig. 8.9. Yield stress dependence on molecular weight and cross-linking, at a given strain rate and different temperatures, for polymethyl methacrylate.

(c) Creep experiments

Although plastic yield is usually studied by means of a stress–strain test in which the sample is extended at a constant rate, plastic yield can also be observed in a creep test in which the sample is subjected to a constant applied load.[9] This type of experiment provides some valuable added information on the nature of the yield process. In particular, the time effects involved are shown in a more interesting way than in stress–strain tests. In the creep experiment, the onset of plastic yield or neck formation does not take place immediately, but only after a so-called 'delay time' which may vary from seconds to many hours, depending on the temperature and stress level. If the temperature is held constant, the nature of the creep curves obtained at different stress levels is shown in Fig. 8.10. At low stress levels, the creep curve levels off at long times after relatively small strain and shows no plastic yield effect. At high stresses, after a certain period of normal creep, there is a relatively abrupt upswing of the creep curve corresponding to the formation and propagation of a neck in the sample. When the neck propagation is complete the creep curve flattens off again and no further marked changes are observed until final fracture. The time required for neck formation or plastic yield to initiate is a characteristic parameter of considerable interest. This type of creep behaviour has been studied for pre-oriented polystyrene over a range of stress level and temperature below the glass transition.[9] The results obtained are shown in Fig. 8.11; here a stress–temperature plane is plotted, and the presence or absence of plastic yield in the creep experiment is

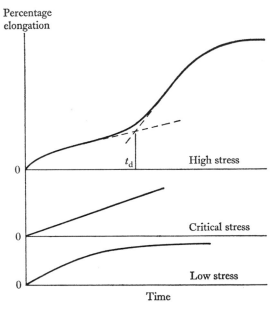

Fig. 8.10. Creep curves (elongation versus time at constant stress and temperature) for polystyrene. 'Delay time' for plastic yield is indicated by t_d.

noted for different stress–temperature combinations. It is found that there is a critical boundary in this plane, above which plastic yield takes place (at a rate depending on the temperature and load level), and below which plastic yield never takes place, no matter for how long a time the experiment is extended.

If the critical boundary is regarded as defining a critical stress level, it is clear that this critical stress depends strongly on temperature; it decreases with temperature in a roughly linear way, and extrapolates to zero stress at a temperature which is almost exactly the glass transition temperature of the polymer. In the higher-stress region where plastic yield occurs, the dependence

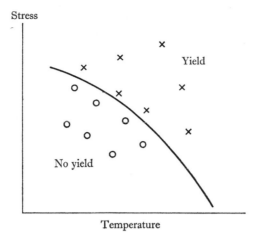

Fig. 8.11. Summary of creep curves for polystyrene. Conditions of stress and temperature leading to plastic yield indicated by ×, those without plastic yield by ○.

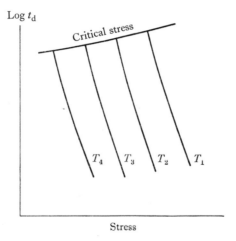

Fig. 8.12. Delay time (see Fig. 8.10) for plastic yield as a function of temperature and stress in polystyrene. No plastic yield occurs below the critical stress.

of the delay time on temperature and stress level (initial stress level, since the cross-section of the sample is changing during the experiment) is shown in Fig. 8.12. The logarithm of the delay time is a linear function of stress level at any given temperature, and shows also an approximately linear dependence on temperature. The curves for the different temperatures terminate at the critical boundary stress noted above. It is interesting that this critical boundary corresponds to something like a constant time value as well, which is in the neighbourhood of 10 hours. Thus it appears that if the plastic yield does not initiate within 10 hours, it never will.

This type of plastic yield under creep conditions is not restricted to polymers which show neck formation. Very similar behaviour is obtained

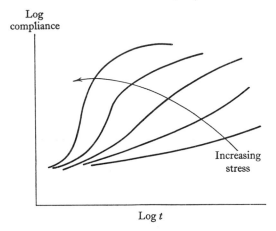

Fig. 8.13. Creep curves (compare Fig. 8.10) for polyvinyl chloride, at different stress.

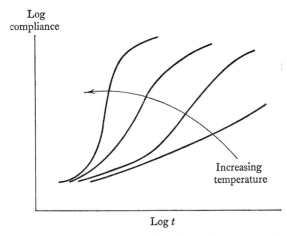

Fig. 8.14. Creep curves (compare Fig. 8.10) for polyvinyl chloride, at different temperatures.

with polyvinyl chloride films, prepared by solvent casting, which show yield without neck formation. Creep curves for these polyvinyl chloride films at different stress levels at a single temperature, and at different temperatures at a single stress level, are shown[10] in Figs. 8.13 and 8.14. The initial upswing of the curve is less easy to define in these curves at low temperature or low stress level, but it is interesting that the same pattern of curves is obtained for both stress variation and temperature variation. The curves show a generally sigmoidal shape between a lower and upper compliance value in both cases. A delay time for plastic yield can be defined in this case from the midpoint of the curves, half way between the lower and upper asymptotic values. This 'delay time' is found to decrease with increasing stress as was found for polystyrene.

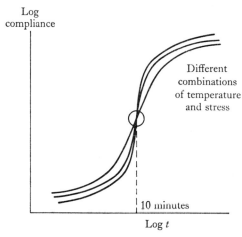

Fig. 8.15. Creep curves for polyvinyl chloride, after selecting combinations of temperatures and stress in such a way that the curves coincide.

(d) Stress–temperature equivalence

The similarity in the effect produced by variation of the stress and temperature parameters leads to a concept of stress–temperature equivalence which can be demonstrated in an interesting way. The stress level can be adjusted at different temperatures to give an essentially identical creep curve, as shown in Fig. 8.15, where the inflection point of the creep curve is seen in all cases at 10 min. If the corresponding stress and temperature values are plotted against each other, the result is as shown in Fig. 8.16. The stress decreases linearly with temperature and extrapolates to zero at approximately the glass transition temperature of the polymer. It would appear from this plot, and others shown previously, that the plastic yield phenomenon being observed here is in fact identical with the glass transition, and that the plastic yield process can consequently be regarded as a stress-induced glass transition.[10] The stress required to produce this glass transition (plastic yield) increases as temperature progressively decreases below the usual thermal glass transition value. Under stress the glass transition is produced by a combination of stress and temperature. This leads to a relationship for the glass transition temperature T_g

$$T_g = (T_g)_0 - C\sigma, \tag{8.2}$$

where σ is the stress, C is a constant, and $(T_g)_0$ represents the usual glass transition measured thermally on a sample which is not subjected to mechanical stress.

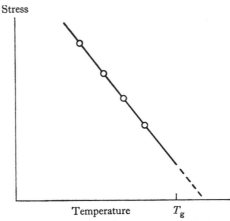

Fig. 8.16. Combinations of stress and temperature used in Fig. 8.15 to obtain coinciding creep curves.

(e) *Birefringence effects*

When plastic yield takes place, a change of birefringence is observed in the material at the same time. The sign of the birefringence produced is the same as that which is obtained by stretching the polymer in the rubbery state above the glass transition. This indicates that orientation of molecular chains is taking place in the same way as when the polymer is stretched in the rubbery state, even though the plastic yield is taking place in the glassy state below the usual glass transition temperature.

In polymers where the sign of the stress-optical coefficient changes at the glass transition temperature, the presence of a stress-induced glass transition can be observed in an interesting way by the use of birefringence measurements. When the stress on such a material is increased from zero progressively, at a temperature lying below the glass transition, birefringence (of the sign characteristic of the glassy state) will show an initial increase proportional to the applied stress in the early stages. However, as the yield stress is approached, the material will show a deviation from this glassy behaviour and a fairly abrupt change of sign of the birefringence in the direction of the sign characteristic of the rubbery state. This is illustrated for the case of unplasticized polyvinyl chloride in Fig. 8.17. Curves at several temperatures are shown here, and it is evident that the stress required to produce the abrupt reversal in birefringence decreases progressively as the temperature approaches the normal glass transition value.

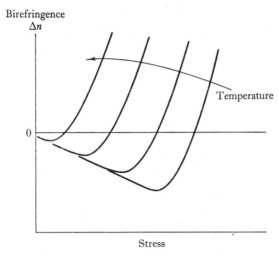

Fig. 8.17. Birefringence Δn as a function of stress for polyvinyl chloride, at various temperatures.

(*f*) *Interrupted stress–strain experiments*

A further very interesting demonstration of the time effects associated with the plastic yield process can be provided by an interrupted stress–strain curve; this type of experiment was first used by Vincent.[11] In this experiment, the sample is initially stretched at constant strain rate, following the normal procedure for a stress–strain test. However, when the sample has necked, and propagation of the neck has begun, the experiment is interrupted. This interruption can be carried out in two different ways. One method is simply to stop the motion of the cross-head of the testing machine and allow the sample to undergo stress relaxation at constant strain for a certain period of time, after which the constant strain rate is again resumed. In the second method of interruption, the motion of the crosshead is reversed and the sample is completely unloaded. After a suitable time of waiting under zero load, the extension of the sample is resumed at the original constant strain rate. The results obtained from these two different types of interruption are quite different, as is shown in Fig. 8.18. Following the interruption in which the sample is allowed to undergo stress relaxation, when straining of the sample is resumed, the stress–strain curve will go through another yield peak or stress maximum before constant steady-state drawing is again observed. Following an interruption in which the sample is completely unloaded, when straining of the sample is resumed no stress peak is observed; the sample shows a reduced modulus in the initial region of restraining and the stress asymptotically approaches the steady-state drawing value. This same type of behaviour is observed for both amorphous and crystalline polymers.

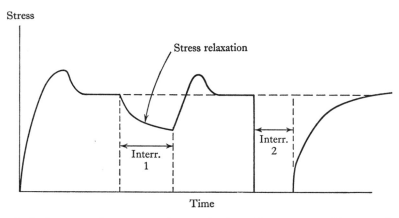

Fig. 8.18. Stress–strain test on a thermoplastic, with two types of interruptions.

When the effect of the interruption time period is examined, it is found that the load peak which is observed on renewed straining after an interruption of the first type builds up gradually with time to a final value. This indicates a hardening of the material with time, and the added rigidity must again be broken down by stress before steady-state drawing can be resumed. For the interruption by unloading, the modulus observed on restraining is found to first decrease with time to a minimum value and then to increase again, as a function of the elapsed time of the interruption. This indicates a temporary softening effect in the material, which disappears on further standing. The interrupted stress–strain experiment indicates that true hardening and softening of a solid polymer can be produced by loading and unloading, and these effects are clearly a significant part of the plastic yield process. The change of stress peak with interruption time in the first type of interruption is illustrated in Fig. 8.19, and the change of modulus with time in the second type of interruption is shown[12] in Fig. 8.20.

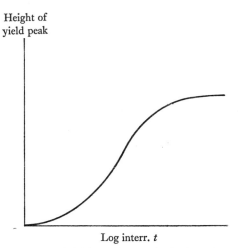

Fig. 8.19. Maximum stress (yield peak), reached after first type of interruption in Fig. 8.18, as a function of duration of interruption.

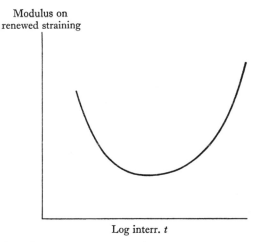

Fig. 8.20. Modulus observed, after second type of interruption in Fig. 8.18, as a function of duration of the interruption.

8.1.7   Theory of plastic yield

The conclusion was reached above that the plastic yield process in glassy amorphous polymers can be described as a stress-induced glass transition. The exact way in which the stress induces the glass transition still remains to be elucidated, however. The further clarification of this relationship will undoubtedly provide useful additional information on the nature of the glass transition itself. It is frequently assumed, for example, that the thermal glass transition is associated with the introduction of a certain amount of free volume into the material, as a result of thermal expansion. It is also known that the volume of a glassy polymer increases when it is subjected to a tensile stress. This corresponds to the fact that Poisson's ratio is less than 0.5 (typical values are in the range 0.30–0.35). It is not yet established whether the free volume introduced in this way up to the yield point is the same as the free volume introduced by thermal expansion, which produces the glass transition in a sample which is not subjected to stress. This can be established only on the basis of exact quantitative comparisons. It is difficult to make measurements of these volume changes produced by mechanical stress, since they are very small; also, it is difficult to make such measurements on a sample to which a mechanical stress is being applied. Some preliminary measurements on polystyrene under compression indicate that a volume increase does accompany the onset of plastic yield. This extra volume also seems to disappear when interrupted drawing of the first type (stress relaxation) takes place.

(a) *Yield criterion*

From the measurement of plastic yield under different types of mechanical stress, it can be deduced that the yield criterion corresponds to a maximum or critical shear stress plus a volumetric component.[13] The reason for the association of free volume with the glass transition effect is probably due to the convoluted shape of the molecules, such that molecular kinks, bends and side groups become trapped in each other in the glassy state. This entrapment can be regarded as an essential feature of the glassy rigidity. A certain amount of free volume (or space not occupied by the atoms of the molecules) is therefore needed to disentangle these kinks, so that the molecules can slip past one another in the way characteristic of the rubbery state or the plastic yield process. This can be visualized in a rough way by the sketch shown in Fig. 8.21. A very similar dilatational effect has been observed to accompany the deformation of particulate systems such as wet sand. The reason for this is very similar; such granular particles, when they are of irregular shape, can interlock with each other when very closely packed, and this effect can be overcome only by a slight volume expansion.

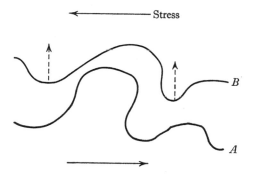

Fig. 8.21. Schematic view of two molecules A and B trying to slip past one another under stress; increased distance (dotted arrows) will help.

(b) Microscopic model for yield – glassy polymers

Some further speculations can be made as to the detailed nature of the yield process. When stresses are applied which are of a magnitude approaching the yield stress, because of the irregularity of the structure there are undoubtedly local spots where the tangled molecules are less interlocked than others and where molecular kinks become free at an earlier stage. Once such spots become free, it is probably easier for neighbouring regions of the molecules to become free also. These can be regarded as 'soft spots' in the structure which will grow with time. In these regions the secondary bonding is no longer localized, and the structure is very similar to that of a rubber or a liquid. When a large enough number of these spots have grown to a sufficient extent, the structure will become something like that of a sponge or foam in which only limited points of rigidity remain and where the bulk of the material is now capable of molecular slippage. This state during plastic yield is therefore not a true rubbery state and the glass transition involved must be considered as only a marginal glass transition, where the structure is similar to that obtained at the lower-temperature end of the thermal glass transition range. There are still local regions of rigidity which act as cross-links. This is indicated by the fact that the extensions obtained (draw ratio during plastic yield) are not as high as would be obtained from the same polymer in the rubbery state. Also, the relation between birefringence and strain, for the case of plastic yield, is different from that for rubbery extension,[10] indicating that the details of the molecular deformation and orientation are somewhat different in the two cases.

In an interrupted stress–strain experiment, where stress relaxation takes place, it can be assumed that these regions resolidify to some extent and that this is the reason for the reappearance of a stress peak when drawing is resumed. The reason for the softening of modulus when the drawing is interrupted by unloading of the sample is less clear. However, this unloading is undoubtedly accompanied by some retraction of the sample and while the sample is retracting it seems that the softening produced is still retained. Once this retraction phase is over, however, the material rehardens as noted previously. This rehardening process is probably taking place to some extent at all times, and the fact that the material softens sufficiently to allow plastic yield is a consequence of the competitive kinetics of the softening and hardening processes. The critical stress boundary for plastic yield, of the type shown in Fig. 8.11, is undoubtedly related to these kinetic features of the process, and represents the critical state where the softening process can become catastrophic. The kinetics of the softening process must be analogous to that of a branched chain reaction, which in chemical systems can lead to explosion and the very similar phenomenon of critical explosion limits.

(c) *Crystalline polymers*

The mechanism of plastic yield in crystalline polymers is still more speculative. It is clear that there must be differences of mechanism between amorphous and crystalline polymers. Measurement of the yield stress in polytetrafluoroethylene over a range of temperature including the gamma and room temperature transitions gives a curve of the sort shown in Fig. 8.22. This curve does not show a linear dependence on temperature, as was found for the glassy polymers, but rather closely parallels in shape the curve of elastic modulus versus temperature for this same temperature region.[14] The latter is shown in Fig. 8.23. The stress–strain curve is essentially an elastic straight line up to the yield point and these results therefore correspond to the fact that the yield begins at an approximately constant strain (about 5 % elongation) over this whole temperature range. It is interesting also that polytetrafluoroethylene does not show a load peak at the yield point but merely an abrupt change of slope, or knee, in the stress–strain curve. This type of yield point is frequently observed in textile fibres. The sample also shows no localized neck formation; the yielding process takes place homogeneously over the whole length of the tensile specimen.

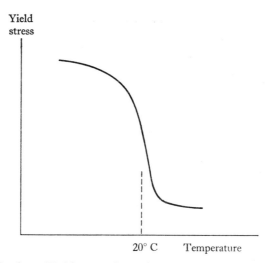

Fig. 8.22. Yield stress dependence on temperature for
polytetrafluoreoethylene.

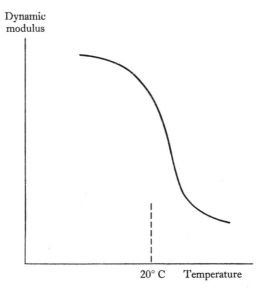

Fig. 8.23. Dependence of dynamic elasticity modulus on temperature
for polytetrafluoroethylene.

Studies of Peterlin and Sakaoku[15] using the electron microscope have shown that when yield and drawing take place in polyethylene the crystalline morphology undergoes some remarkable changes. Any spherulitic structure present in such a crystalline polymer breaks up when plastic yield takes place. In addition, folded chain lamellae break up into fragments which re-align into rows perpendicular to the stress axis. The mechanism of this process is not yet clearly understood.

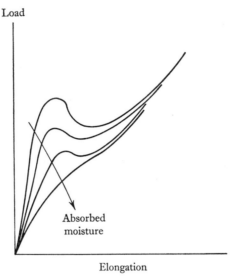

Fig. 8.24. Stress–strain curve (load versus elongation) for Nylon 6, as influenced by moisture.

(d) *Effect of moisture*

The effect of moisture on the stress–strain curve of Nylon 6 is very interesting, as shown in Fig. 8.24. As moisture content increases, the initial load peak gradually decreases and finally disappears. However, the later part of the stress–strain curve, at high elongations, remains almost unaffected by the moisture. This behaviour can be understood if it is assumed that the structure can be described as a two-phase system in which a crystalline lattice or network is embedded in an amorphous matrix. The plasticizing effect of the absorbed water can be assumed to be centred in the amorphous phase, and the initial yield peak can be assumed to represent a stress-induced glass transition of the amorphous matrix.[16] The later part of the stress–strain curve can then be regarded as the deformation of the crystalline lattice in the already softened matrix. This result emphasizes the point that in crystalline polymers, the mechanical behaviour must probably be interpreted as the mechanical behaviour of a composite system with supermolecular structure. The mechanical properties are therefore not simply those of a material but represent the properties of a structure as well. This is true also of such natural fibres as cotton and human hair.

8.1.8 Geometry of deformation

(a) Necking

Despite the fact that the deformation of solid polymers often occurs homogeneously throughout the material, very often special types of localization can be observed to take place. One example of this has already been discussed, that is neck formation. When neck formation takes place in a tensile stress–strain test, the load-elongation curve can no longer be given a simple interpretation after the yield point, since the sample is not stretching homogeneously and the neck is propagating over the specimen. This part of the curve therefore does not describe the stress–strain properties of the material itself. The true stress–strain properties of the material can be deduced from such an experimental curve, however, by simultaneously measuring photographically the diameter or cross-sectional area of the specimen at the initial necking point throughout the course of the stress–strain curve, and calculating the stress and extension of the sample at that point by the assumption that the volume or density of the specimen does not change as a result of extension.

When such a true stress–strain curve is derived, it is found that the load peak observed in the load–elongation curve is in fact an experimental artifact: the true stress–strain curve simply shows an abrupt flattening at that point and a gradual increase of stress on further elongation due to a 'strain hardening' effect in the material. Such a true stress–strain curve for polymethyl methacrylate is shown[17] in Fig. 8.25. This strain hardening may be associated with the increasing degree of molecular orientation as elongation increases.

(b) Deformation bands

Another type of localization of the yield process is the formation of deformation bands, at an angle approximately 45° to the axis of stress. These can be observed prominently in polystyrene in both tension and compression.[9, 18] The bands are very thin platelets a few microns in thickness which may extend over a considerable area. The yield strain within them corresponds to roughly 150% elongation. The bands show a birefringence of the sign which would be expected from the molecular orientation involved and the bands are recoverable by heating the specimen above the glass transition temperature. Such bands may form in great numbers and intersect with each other from opposite directions. They can be observed to form in advance of a propagating neck; and as the neck moves into their area, the bands coalesce with the neck and disappear, leaving homogeneously drawn material behind with no band structure.

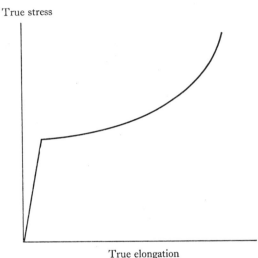

Fig. 8.25. True stress–strain curve for polymethyl methacrylate.

(c) Crazing

Another type of localization of yield deformation is observed in the phenomenon of crazing. A considerable advance in our understanding of this phenomenon has been made in recent years, particularly due to the work of Kambour.[18] It was at first believed that crazing represented some sort of internal cracking in the material. However, it has since been shown that these craze regions, which are thin and sheet-like, and have a silvery appearance by reflected light, are actually not empty but contain about 50% solid material within them. The other 50% consists of empty regions or voids. These crazes form with their planes perpendicular to the stress axis, and are therefore easily distinguishable from the 45° deformation bands described above. Crazes seem to represent a type of internal cavitation or separation of surfaces in which these surfaces are connected by strands of plastically yielded solid polymer.

(d) Yield and fracture

Even in the case of normal fracture, involving the propagation of a fracture crack, plastic yield may also play a significant role. In such cases the stress concentration at the tip of the crack appears to produce a local plastically yielded region which then forms a layer on the surface of the fracture crack as it propagates further. For this reason, the work of crack propagation of a polymer is often much higher than that which would be calculated theoretically from the surface energy of the crack itself, assuming that the work of fracture propagation consisted entirely in forming this new fracture surface. In the case of polymethyl methacrylate, the presence of such a yielded surface layer affects the refractive index of the material, and iridescent colours are often observed on fresh fracture surfaces of the polymer. This effect of yield on the energy of crack propagation has also been observed in ductile metals, and the deformation bands discussed above have an analogue in the Luders' bands which are observed in many cases when metal single crystals are subjected to tensile strain.

8.1.9   Modification of thermoplastics

(a) Plasticizers and stabilizers

Like other basic materials, thermoplastics must often be modified in various ways in order to obtain optimum properties. One such modification which has been used for many years is the incorporation of plasticizers into the polymer. Probably the earliest example of this was the use of camphor in celluloid. The use of this plasticizer was the discovery that made celluloid a commercially useful plastic. A very similar case is that of polyvinyl chloride, where the use of plasticizers made convenient fabrication and moulding of the polymer possible. The unplasticized polymer is a very hard, tough material which thermally degrades at temperatures where it becomes fluid. The use of chemical stabilizers to suppress this thermal degradation is also a very important part of the technology of polyvinyl chloride.

One of the most prominent effects of the incorporation of plasticizer is to lower the glass transition temperature of the polymer, thereby allowing moulding at lower temperature. This can be seen clearly from the shift to lower temperatures of the dynamic mechanical loss peak associated with the glass transition, as illustrated[20] for polyvinyl chloride in Fig. 8.26. In addition the toughness of the material is increased and the impact strength improved. In order to be suitable as a plasticizer, a compound must be compatible with or soluble in the polymer. However, many of the early plasticizers had disadvantages since due to their volatility they would gradually evaporate from the polymer and would also often migrate to the surface and spoil the surface appearance. These effects can be suppressed by the use of polymeric plasticizers (polymers which are soluble in or miscible with the base polymer). Another trend is to use 'internal plasticization', which consists

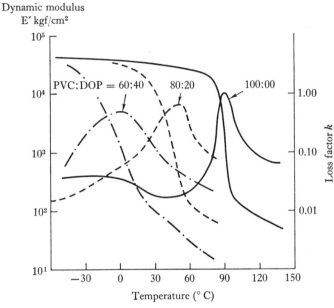

Fig. 8.26. Effect of plasticizer (dioctylphthalate) on the peak temperature of mechanical loss factor k in polyvinyl chloride. Corresponding E' curves also shown.

of making a copolymer with a comonomer which has a plasticizing effect, rather than adding a separate material as plasticizer. An important example of this is the vinyl acetate copolymer with vinyl chloride. This copolymer is now widely used for many applications such as phonograph records, plastic sheeting and vinyl tile, under the name 'rigid vinyl'. This material is also finding increased application as plastic piping, particularly under industrial conditions where corrosion of metal piping is a serious hazard.

An interesting illustration of a systematic variation in the internal plasticization effect is provided by the case of cellulose esters containing side groups with a progressive increase in length of the side groups (number of carbon atoms); this is shown[21] in Fig. 8.27 (*a*). As the side chains become longer, the mutual interaction of the cellulose molecules, in which hydrogen bonding is involved, becomes less and the cohesive forces contain a higher proportion of the weaker van der Waals' forces characteristic of paraffinic hydrocarbons, such as polyethylene. An interesting analogy can be drawn between this result and the effect of a similar series of organic acids with increasing chain length on the coefficient of friction, μ, between rubbing solid surfaces, as illustrated[22] in Fig. 8.27(*b*).

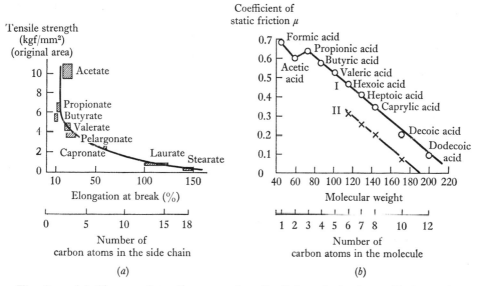

Fig. 8.27. (*a*) Change of tensile properties of cellulose derivatives with increasing length of the substituted side chains. (*b*) Decrease of the coefficient of friction μ with increasing chain length of the lubricant molecule. ○ curve for glass; × curve for steel.

(b) Fillers

Another type of modification which is widely used is the incorporation of solid fillers. These take many forms. Some are used as powders or very fine granules, such as chalk, talc, and diatomaceous earth. These materials, as well as such fibrous fillers as asbestos, are frequently added to polyvinyl chloride in the manufacture of vinyl floor tiling. These fillers provide greater rigidity, greater abrasion resistance, and in some cases also reduce cost. Some experimentation has also been done with the addition of glass fibers (short chopped fibres) to polyvinyl chloride compositions or polyethylene which can be used for injection moulding. This is an attempt to obtain some of the advantages of increased strength of the sort obtained in the fibreglass-reinforced thermo-setting resins such as polyesters and epoxies.

These compositions containing solid fillers are actually two-phase systems. Since the fillers added are generally rigid solids, the resulting composition is more rigid than the original polymer. Another interesting example of such a system is a mixture of asphalt and crushed rock of the type which is used for road surfacing. Asphalt itself can be regarded as a glassy polymer, and when it is 'melted' by heating it actually undergoes a glass transition. Some types of asphalt are themselves composite materials since they are suspensions of gel particles in an oily medium. In the liquid state, asphalt behaves as a linear viscoelastic material, but the composite does not show such a simple mechanical behaviour. When the composite is subjected to a compressive force, for example, and the resulting creep measured, an interesting pheno-menon of 'particle lock-up' can be observed. When this happens, no further creep takes place because the applied stress is being supported entirely by a structure of interlocked rock particles, and the asphalt itself therefore bears none of the applied load.[23]

(c) Foams

Another modification of thermoplastics which is finding increasing applica-tion is plastic foams. A considerable quantity of polystyrene and polyethylene foam is being manufactured which finds many applications as flotation or thermal insulation material, or as packing material, where high-temperature resistance is not required. Foams of this type can also be regarded as composite systems in which the two phases are solid and vapour. The mechanical properties of such foams depend considerably on whether the foam is an open-celled or closed-celled foam. In open-celled foams, there are channels through the material to the outside surface and they can therefore readily absorb liquids or transmit air in or out. Natural sponges are an example of an open-celled foam, and synthetic polymer sponges must therefore also be made with this open-celled structure. The detailed analysis of the mechanical response of such foamed and composite materials is just beginning to receive detailed attention. In all such cases, the response of the system depends not only on the properties of the polymer involved, but also on the geometry and nature of the overall structure.

(*d*) *Ionomers*

A new development in thermoplastics is the class of polymers known as 'ionomers', which are made by copolymerizing a small percentage of acrylic acid with ethylene to produce a modified polyethylene, and then reacting this material further with a metal ion to produce an organic salt between the metal and the acid groups (shown schematically in Fig. 8.28). The ion should be

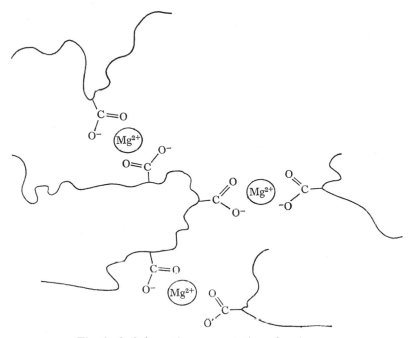

Fig. 8.28. Schematic representation of an ionomer.

polyvalent, so that an effective ionic cross-link is created. This type of cross-link provides the added strength and dimensional stability of a chemical cross-link at room temperature. However, these 'cross-links' have the special feature that they dissociate thermally when the polymer is heated and melted, so that the material is still a reversibly mouldable thermoplastic. The effect of these thermally labile cross-links is just the opposite of that of plasticization.

8.2   Resins (by R. Houwink)

8.2.1   Constitution

The oldest representatives of the group resins, characterized by globular, not interlinked, particles[24] are the natural resins, such as colophonium, dammar, shellac and mastix. Later materials like cumarone, a distillate from coal tar and some non-hardening resins such as cresol formaldehyde and urea formaldehyde condensates were added. Most of them are used in the paint industry since they are easily soluble. The changes in structure on cooling are pictured schematically in Fig. 8.29.

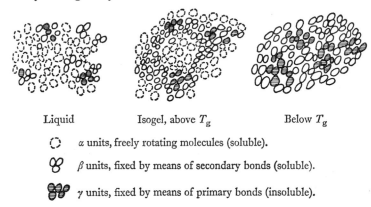

Liquid   Isogel, above T_g Below T_g

○ ⃝ α units, freely rotating molecules (soluble).

ℚ β units, fixed by means of secondary bonds (soluble).

🌢🌢 γ units, fixed by means of primary bonds (insoluble).

Fig. 8.29. Structural possibilities for a non-linear thermosplastic resin.

8.2.2   Elasticity

The figures mentioned in Table 8.1 for the modulus of elasticity are low, compared with glass, for which E is more than an order of magnitude higher, namely between 4000 and 8000 kgf mm^{-2}. The main cohesive forces in glass are primary bonds, whereas in resins they are secondary bonds.

TABLE 8.1. *Modulus of elasticity for some non-hardening resins*

Resin	Elasticity modulus at room temperature (kgf/cm²)
Mastix	93
Sandarac	150
Dammar	225
Colophonium	260

8.2.3. Viscosity

As to be expected, these materials show pure viscous flow above T_g. The temperature sensitivity of η, expressed by B in the formula,

$$\eta = Ae^{-B/T_m}$$

where A is a constant and T is absolute temperature, is very high compared with glass, as shown in Fig. 8.30.

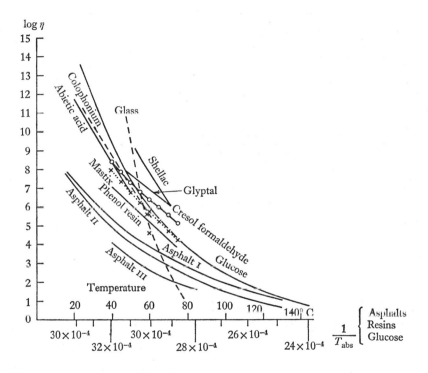

Fig. 8.30. Temperature sensitivity of η for some non-linear substances.

8.3 Asphalts (by H. K. de Decker)

8.3.1 Internal structure

Asphalts[25] are obtained as the residue of the distillation of crude petroleum oil; they are colloidal systems of which the most important constituents are maltenes and asphaltenes. The maltenes consist of the so-called asphaltic resins and oily constituents, and are soluble in low-boiling petrol; this solubility distinguishes them from asphaltenes.

The asphaltenes are hydrocarbons of high molecular weight, chiefly aromatic or hydro-aromatic in nature and kept dispersed in the oily medium by means of adsorbed protective bodies. They have a low hydrogen content, and are formed during the distillation process by polymerization and dehydro-genation. At low temperatures they will form gels with the oily constituents. Oily constituents rich in sulphur or aromatic in character should be solvents for the asphaltenes, the other oily constituents should be non-solvents. It is chiefly the behaviour of the maltenes acting as a solvent for the asphaltenes which determines the elastic and plastic properties of asphalt.

Pfeiffer and van Doormaal postulate that the aromatic or hydro-aromatic surface of the asphaltene micelles will make them peptize best in aromatic or hydro-aromatic maltenes. When the maltenes are removed or converted, for instance by blowing (leading to the so-called blown asphalts, in contrast to the steam-refined asphalts) there will no longer be complete peptization of the asphaltenes.

8.3.2 Elastic and plastic properties of asphalts

(a) *Asphalts without fillers*

Although some asphalts definitely show elastic reactions, it is difficult to determine a (static) elastic modulus, because the elastic deformation is invariably accompanied by viscous flow. By determining the elastic recovery after various times under a constant stress, and extrapolating back to the idealized case of pure elastic behaviour, shear modulus values in the range of 10^4 to 10^5 dyn/cm^2 are found in the literature. This is about 10^5 times smaller than values found by Houwink in rapid determination of elasticity of other thermoplastic resins at room temperature (see Table 8.1). A valid comparison of thermoplastics with respect to elastic modulus can only result from measurements at the same relaxation times (or the same frequency and amplitude in vibration experiments); such measurements have not yet come to our attention. The tensile strength is very low and is lower the softer the asphalt. A typical value is 0.5 kgf/cm².

Temperature sensitivity. The temperature sensitivity of flow resistance or 'viscosity' is usually a good indication of the internal structure and its relative strength (see Chapter 4). This is true for asphalts. A convenient measure of temperature sensitivity is the 'penetration index' or P.I., which is easily derived by means of a nomogram from the penetration ('softness') at room temperature, together with the softening temperature 'ring and ball' (that is the temperature at which a ball of given size sinks through a given ring filled with asphalt when the temperature is raised at a given rate; this temperature corresponds to a viscosity of 1.2×10^4 poise). A higher P.I. indicates less temperature sensitivity.

Pfeiffer distinguishes three types of asphalts according to their sol or gel structure between room temperature and their softening temperature.

Type I is the pure *sol type* in which the suspended particles (asphaltenes) are completely peptized; there is no structure to cause elastic reaction, nor is there any elasticity of the particles themselves contributing to the deformation. The behaviour is almost purely viscous, and the P.I. is typically -2 to 0.

Type II is a *sol type* in which there is an '*intra-particle*' *elasticity* during deformation. Thus the rheology is visco-elastic, but with a small elastic component; the P.I. is typically 0 to $+1$.

Type III is the *gel type* in which the asphaltene particles are not completely peptized and form a network structure which causes elastic reaction. The visco-elastic behaviour shows clear elastic as well as viscous components. The P.I. is typically $+1$ to $+5$.

Type I asphalts are generally 'straight-run' residues, obtained by vacuum-steam distillation of crude oil. Type III asphalts are typically 'blown', which means oxidized at high temperature by air. This treatment causes the asphaltenes and their protective layers to change to the point where they readily form network structures. In general, the blowing process also increases the proportion of asphaltenes; this is a second reason for increased network or gel formation. Examples are given in Table 8.2.

The P.I. as well as the elastic recovery increases if one moves from the straight run (type I) asphalt to the blown (type III); the asphaltene content also increases. No. 7 in the table refers to a type I asphalt in which rubber has been dissolved at 170° C. In this case, a network structure is formed in part by the rubber molecules.

Nearly straight lines are obtained when the logarithm of the penetration is plotted against temperature, compare Fig. 8.31. If the lines so obtained are

TABLE 8.2. *Physical properties of asphalts*

No.	Type	Production method	R. & B. (°C)	Pen. (25° C) 0.01 mm	P.I.	Elastic recovery	Asphaltenes (%)
I	I	Steam distillation	48	54	−1.5	—[25]	5.1
2	I	Steam distillation	54	57	0.0	−2.0[26]	—
3	II	Steam distillation	38	195	0.0	0.3[25]	12.2
4	II	Steam distillation	55	44	0.0	0.4[25]	15.5
5	III	Blown	65	37	1.0	5.6[26]	—
6	III	Blown	87	35	4.5	1.1[25]	28.9
7	III	No. 2 mixed with 5 % rubber	65	41	1.3	10.0[26]	—

R. & B: softening point by ring and ball method; Pen.: penetration by standard needle method; P.I.: penetration index, as defined in text; Elastic recovery: relative measurements by compression method[26] or shear method.[25]

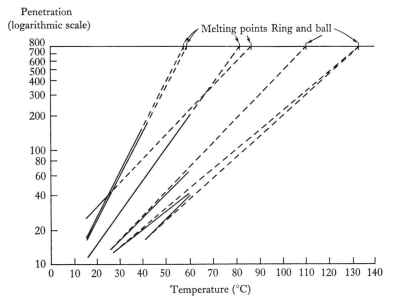

Fig. 8.31. Relation between penetration and temperature for various asphalts.

extrapolated to a penetration value of 800, the temperature at this point is the same as the ring and ball temperature.

For practical purposes, the 'ductility' of asphalts is often measured. This is the maximum length of thread obtained when a standardized test piece is pulled apart under specified conditions at 25° C. For material to be used for road covering, certain authors consider it desirable that this value should not be less than about 18 cm. Besides the length of thread at a given temperature, it is important to know how the length varies with temperature. The great temperature sensitivity of coal tar pitch ductility compared with that of

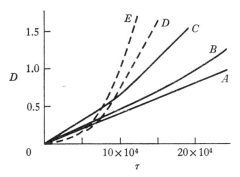

Fig. 8.32. *D–τ* curves for various asphaltic bitumens (from Saal, ref. 27.)

petroleum asphalts is very striking, a property which makes these pitches unsuitable in many practical applications.

The D–τ relationship. Steam-refined asphalts as a rule show pure flow at a sufficiently high temperature. This is analogous to the behaviour of the non-hardening resins. In contrast, blown asphalts show chiefly visco-elasticity, even at temperatures at which they are soft.

Fig. 8.32 shows some *D–τ* curves obtained with a rotating cylinder apparatus and a capillary–pressure viscometer at temperatures corresponding to a penetration of about 180 for each material. It is important to note that just as with the resins these *D–τ* curves can be expressed by the equation:

$$D = \frac{1}{\eta^*} \tau^n.$$

Values of *n* are tabulated below. *n* may be equal to 1, as for asphalt *A*, showing that pure flow is possible, though this is not observed for the blown asphalts. As for certain resins, *n* here also tends to decrease at higher temperatures; for asphalt *E*, $n = 1.5$ at 60° C and 1.3 at 150° C.

Kind of asphalt	Penetration at temperature of test	n
A: purely viscous	170 at 20° C	1
B: steam-refined, hardly visco-elastic	180 at 24° C	1.1
C: steam-refined, hardly visco-elastic	198 at 25° C	1.1
D: blown, very visco-elastic	190 at 25° C	1.5
E: blown, very visco-elastic	180 at 60° C	1.5

The sensitivity of η^* to temperature was found to be of the same order of magnitude as for other amorphous materials: η^* increases about 10 times per 10° C decrease in temperature.

It has already been pointed out that the properties of the maltenes is one factor governing the rheological properties of an asphalt. This supposition is confirmed by the following experiments. By means of gasoline, the asphaltenes were separated from the maltenes of three different bitumens A, B and C where:

A is a bitumen showing truly viscous flow;

B is a steam-refined bitumen showing a slight elasticity;

C is a blown bitumen showing decided elasticity.

The maltenes of these asphalts all proved to be purely viscous. If, however, any of the asphaltenes was dissolved in maltenes from bitumen A, it still yielded a purely viscous bitumen; maltenes from B always imparted a slight elasticity, and those from C a decided elasticity. It seems that in a bitumen where the maltenes act as lubricants for the asphaltenes, the nature of the lubricants is the pre-dominating factor in the rheological effect.

The following figures from Saal show that thixotropy plays a part in asphalts. He determined the viscosity of a blown asphalt (penetration 23 at 25° C) and repeated the experiment after the asphalt had been allowed to rest for varying periods of time. He found:

First determination	$\eta = 9.0 \times 10^8$ poise
Second determination immediately after the first	$\eta = 0.8 \times 10^8$ poise
Third determination (after 24 hours)	$\eta = 2.3 \times 10^8$ poise
Fourth determination (after 3×24 hours)	$\eta = 3.8 \times 10^8$ poise

The original viscosity was not recovered even after a total rest period of 72 h. The structure which was obviously destroyed by flow seems therefore not to have been re-established in that time.

(b) Influence of fillers

Asphalt is mixed with fillers to give the so-called asphalt cement, which is used with coarse aggregates of sand and stone to produce the 'asphalt concrete' compositions widely used on road surfaces.

Broadly speaking, one may say that up to a certain point the addition of fillers raises the melting point, increases the hardness and tensile strength and

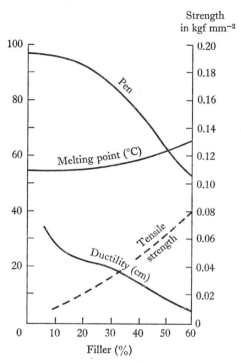

Fig. 8.33. Influence of limestone on the mechanical properties of asphalt bitumen (from Evans, ref. 28).

decreases the ductility. This behaviour is illustrated clearly in Fig. 8.33 which demonstrates the effect of limestone as a filler. Fillers having a still more pronounced influence on the mechanical properties of asphalt are known. For example, the so-called micro-asbestos has an effect not unlike that of rubber (see above). In an investigation similar to that illustrated by Fig. 8.33, a tensile strength of 20 kgf/cm² was obtained with 60 % of micro-asbestos as a filler.

The favourable effect of fillers can probably be ascribed to the fact that the adhesion between asphalt and filler is stronger than the cohesion in the bitumen itself. The tensile strength increases with increasing content of filler to a maximum value with optimum filler content. This fact is explained schematically in Fig. 8.34.

In Fig. 8.34(*a*) relatively too much bitumen is present to be bound at the surface of the filler; it acts as a lubricant and keeps the tensile strength low. On the other hand, when the bitumen content is too low (Fig. 8.34*c*), not all the filler particles will be glued together, so that the tensile strength in this

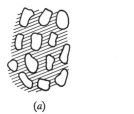  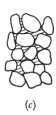

(*a*)   (*b*)   (*c*)

Fig. 8.34. Bitumen filler mixtures with varying content of filler. (*a*) Too little filler content: small tensile strength. (*b*) Correct filler content: maximum tensile strength. (*c*) Too much filler: small tensile strength.

case will also be small. Furthermore this type of material will be rigid because a stress will be communicated via the relatively few points at which the filler particles are in contact, and the result will be high local stresses. As soon as all the holes are filled with asphalt, this phenomenon will disappear, and the ideal situation of Fig. 8.34(*b*) will be attained. It could be calculated that for these optimum conditions the bitumen layer around the particles would be between 0.2 and 0.6 μm in thickness. The material could already be easily moulded when this layer was 0.1 to 0.3 μm in thickness.

A practical asphalt concrete for road surfacing contains stone, sand, and filler; the smaller particles fill the voids between the larger ones. The overall range of particle diameters may be from 1 to 10^{-3} cm. The actual asphaltic binder constitutes perhaps 5 % of the total material (this is analogous to the cement hydrate binder in cement concrete). It is noteworthy that asphalt concretes show visco-elastic behaviour (the road surface has a certain resilience!) even if the binder is a purely viscous (type 1) asphalt.

References to chapter 8

1   R. D. Andrews & J. F. Rudd, *J. Appl. Phys.* **28**, 1091 (1957).
2   D. Vorländer & J. Fischer, *Ber. Deutsch, Chem. bes.* **65**, 1756 (1932).
3   R. D. Andrews & Y. Kazama, *J. Appl. Phys.* **39**, 4891 (1968).

4   J. F. Rudd & R. D. Andrews, *J. Appl. Phys.* **31**, 818 (1960).
5   R. W. Gray & N. G. McCrum, *J. Polymer Sci.* A-2, **7**, 1329 (1969).
6   R. S. Stein, chapter 6 in *Rheology* **5** (Ed. F. R. Eirich; New York, 1969).
7   S. L. Cooper, W. Whitney, N. S. Schneider & R. D. Andrews, M. I. T. (unpublished work).
8   S. J. Kurtz, G. Langford, R. D. Andrews & N. S. Schneider, M.I.T. (unpublished work).
9   D. H. Ender & R. D. Andrews, *J. Appl. Phys.* **36**, 3057 (1965).
10 R. D. Andrews & Y. Kazama, *J. Appl. Phys.* **38**, 4118 (1967).
11 P. I. Vincent, *Polymer* **1**, 7 (1960).
12 W. Whitney, Sc. D. Thesis, M.I.T. (Nov. 1964).
13 W. Whitney & R. D. Andrews, *J. Polymer Sci.* **C16**, 2981 (1967).
14 G. P. Koo & R. D. Andrews, *Polymer Eng. and Sci.* **9**, 268 (1969).
15 A. Peterlin & K. Sakooku, *J. Appl. Phys.* **38**, 4152 (1967).
16 J. Rubin & R. D. Andrews, *Polymer Eng. and Sci.* **8**, 302 (1968).
17 S. W. Allison & R. D. Andrews, *J. Appl. Phys.* **38**, 4164 (1967).
18 W. Whitney, *J. Appl. Phys.* **34**, 3633 (1963). A. S. Argon, R. D. Andrews, J. A. Godrick & W. Whitney, *J. Appl. Phys.* **39**, 1899 (1968).
19 R. P. Kambour & G. A. Bernier, *Macromolecules* **1**, 190 (1968).
20 F. Zinhardt, *Kunststoffe* **53**, 18 (1963).
21 M. Hagedorn & P. Moeller, *Cellulosechemic* **12**, 29 (1931).
22 W. B. Hardy & T. Doubleday, *Proc. Roy. Soc. A*, **100**, 550 (1922).
23 E. M. Krokosky, E. Tons & R. D. Andrews, *A.S.T.M. Proc.* **63**, 1263 (1963).
24 R. Houwink, *Elasticity, plasticity and structure of matter*, pp. 138–45 (Cambridge, 1937).
25 J. Ph. Pfeiffer, *The properties of asphaltic bitumen* (Elsevier, 1950).
26 H. K. de Decker & H. A. W. Nijveld, *Proceedings of the Third World Petroleum Congress*, section VIII, 496 (The Hague, 1951).
27 R. N. J. Saal & G. Koens, *J. Inst. Petrol. Techn.* **19**, 176 (1934).
28 A. Evans, *J. Inst. Petrol. Techn.* **18**, 957 (1932).

9 Elastomeric materials (rubbers)

B. B. Boonstra†

9.1 Rubber-like elasticity

The definitions of rubber-like elasticity are usually based on the description of the behaviour of vulcanized natural rubber at room temperature. A strip of good vulcanized rubber without fillers (pure gum vulcanizate) can be extended to seven or eight times its original length and on release immediately retracts to practically that same original length. Many rubber-like materials show a similar behaviour although their extendability may not be as large and the retraction is not as rapid or as complete. Extension to at least twice the original length (100% elongation) has been suggested as a definition but the description is arbitrary. The natural rubber pure gum vulcanizate is kept in mind as the model for rubber-like high elasticity. It is only recently that a synthetic rubber (linear polybutadiene) has become available, which exhibits higher elasticity than natural rubber; for all types of rubber this high elasticity refers to the vulcanized product.

Elasticity or resilience is the ratio of the energy regained on release divided by the original energy input, by stretching, compression or shear. In the case of natural rubber this can be as much as 85 to 90%.

As explained in Chapters 1, 4 and 8, rubber-like (or high extension) elasticity is based on the kinetic movement of the segments of the long chain molecules and is similar to the apparent elasticity of a volume of gas in a cylinder that one is trying to compress under a piston. This is in contrast to the elasticity exhibited by billiard balls on impact or a steel coil spring on extension. These are based on the deformation of bond angles between the atoms and the change in interatomic distances and are quite different in character. Whatever the cause of the elasticity, its numerical rating is still given by the ratio, regained energy/input energy.

Vulcanization or cross-linking is the process that brings out the truly elastic properties by reducing viscous flow. Unvulcanized rubber demonstrates a pronounced viscous or plastic flow, particularly after it has been mechanically

† With contributions on chemistry of vulcanization by G. R. Cotten (Cabot Corporation).

198

plasticized by mastication. This mastication is necessary to make the incorporation of the various ingredients into the rubber possible.

The visco-elastic behaviour of rubber is dependent on its chemical constitution and the degree and type of vulcanization or cross-linking.

These two subjects will be discussed first, after which the visco-elastic behaviour of unvulcanized and vulcanized (cross-linked) rubber will be considered using stress–strain diagrams.

9.2 The chemical structure of rubber and its changes during vulcanization

9.2.1 The chemical structure of natural rubber

Natural rubber is a polymer built from isoprene units joined in a heat-to-tail manner. More recently the molecular weight of raw rubber has been obtained from light-scattering measurements[1] and found to be 1.85×10^5, that is, a

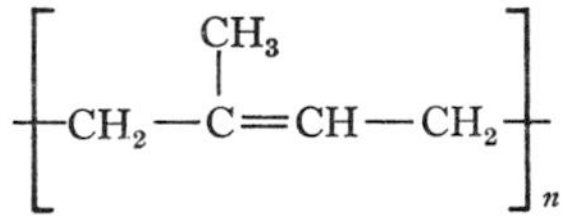

$$\left[-CH_2 - \underset{\underset{CH_3}{|}}{C} = CH - CH_2 - \right]_n$$

Fig. 9.1. Isoprene units in natural rubber.

single rubber molecule consists of approximately 30000 isoprene units. In normal processing the rubber is considerably broken down during mastication and the extent of this breakdown depends on the specific conditions employed, but on the average the molecular weight is reduced nearly ten-fold. This degradation is desirable in order to achieve visco-elastic properties suitable for subsequent mixing and extruding operations. The shearing forces during the mastication process break the carbon–carbon bonds homolytically,[2] and the free radicals thus formed are then mostly stabilized by reaction with atmospheric oxygen.

The polyisoprene chains can take one of two stereo-configurations according to the relative positions of pendant —CH_3 groups (see Fig. 9.2). The natural rubber is the *cis*-isomer with a molecular repeat distance of 8.1 Å, while β-gutta-percha is the *trans*-isomer with a molecular repeat distance of 4.7 Å as determined by X-ray crystallography. The two isomers have markedly different physical properties, in particular the *trans*-isomer permits closer molecular chain packing and, hence, a higher degree of crystallinity at room temperature. Thus natural rubber has a melting point of about 28° C and specific gravity of 0.92 at 20° C, while gutta-percha has a melting point of 56–65° C and specific gravity of 0.95 at 20° C. In the unstretched state, rubber is to a great extent amorphous at room temperature, the molecules being distributed at random. During the process of stretching, molecular alignment takes place and crystallization is initiated, leading to the characteristic S-shape form of the stress–strain curve for uncured natural rubber, not found with non-crystallizing synthetic rubbers (e.g. styrene–butadiene copolymer).

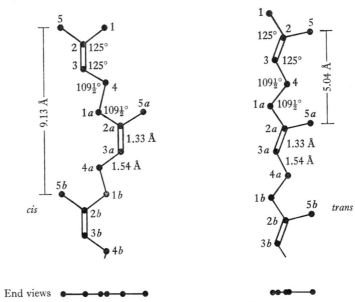

Fig. 9.2. Calculated planar polyisoprene units.

In recent years, it has been possible to prepare synthetic *cis*-polyisoprene by the use of stereo-specific catalysts and the product so obtained is physically and chemically identical with natural rubber. This gave the final proof that our structural representation of natural rubber is correct.

9.2.2 The reaction mechanism of vulcanization

The conversion of linear polymers (e.g. polyisoprene) into a three-dimensional network, characteristic of rubbery materials, is achieved by cross-link formation using either sulphur or peroxides as the curing agents. There are, however, some exceptions to this, for example, in the case of a recently developed styrene–butadiene block-copolymer, where chemical cross-links have been replaced by glassy regions of polystyrene segments.

The beneficial effects of curing natural rubber with sulphur have been discovered independently by Goodyear and Hancock. In 1906 it was discovered that anilines and other amines accelerated vulcanization of rubber with sulphur, and around 1920 it was found that a combination of zinc oxide and stearic acid formed an effective activator for accelerated sulphur vulcanization of rubber. Since those times the number of commercially used accelerators have increased enormously, and they can be classified into the following main groups of compounds: (*a*) mercaptobenzthiozoles, (*b*) guanidines, (*c*) dithiocarbamates, (*d*) various other chemical compounds.

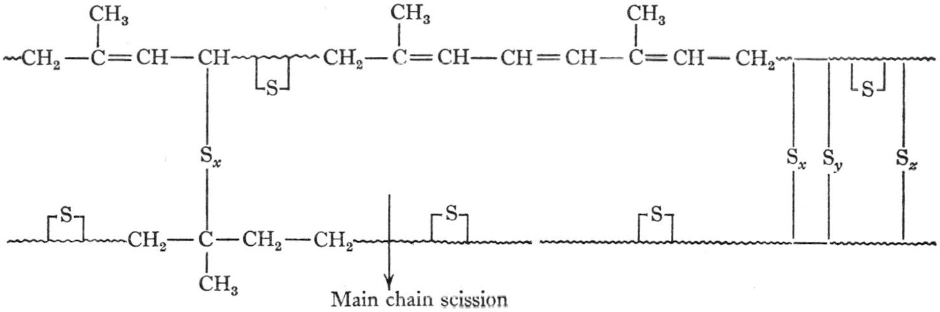

Fig. 9.3. Structural features of an unaccelerated sulphur–natural rubber vulcanizate network.

Much of the mechanism of unaccelerated sulphur vulcanization has been elucidated by a group of English workers[3] using model olefinic compounds. In natural rubber, it is believed that sulphur can take part in a number of different substitutions and addition reactions which may result in mono- or polysulphidic (S_x) cross-links where the number of sulphur atoms per cross-link may be 2, 3, or more. In addition some intramolecular reactions may also occur, leading to monosulphidic linked structures. A general case is illustrated in Fig. 9.3. By using chemical probes[4] the relative proportion of polysulphides, disulphides and monosulphides can be established.

In unaccelerated sulphur systems, approximately 40–55 sulphur atoms are combined with the rubber for each cross-link. This suggests a very complex network structure which includes a larger number of cyclic sulphides as shown schematically in Fig. 9.3. The efficiency of the cross-linking reaction is considerably improved by use of accelerators and activators. A typical industrial recipe would have a cross-link efficiency of approximately 5–6 sulphur atoms per cross-link[4] and this would increase to roughly 3 sulphur atoms per cross-link on longer cure times.

The exact nature of the cross-links influences the tensile properties of the rubber. Since the bond strength of the different types of cross-link can be represented as follows:

$$-C-S_{n+2}-C- \; < \; -C-S_2-C \; < \; -C-S-C \; < \; -C-C-$$

it has been suggested that under stress the polysulphidic cross-links break and then form anew at some new position, allowing a more homogeneous distribution of stress throughout the system. This view has been supported in particular by studies of creep[5] and permanent set. More recent evidence favours the view that internal tensions occur already during vulcanization and become relieved by re-arrangement in the case of S_x links during cooling, whereas the S links and C—C bonds do not have this possibility. The stronger carbon–carbon and monosulphide cross-links are actually less desirable for practical rubber compounds requiring high tensile strength, since they cause breakage of the main polymer chains under stress at relatively low elongations and lead gradually to catastrophic rupture. However, these bonds will result in lower permanent set and better ageing than the polysulphide links. We may summarize, therefore, that modern curing systems are designed to avoid production of cyclic sulphides and facilitate formation of polysulphidic cross-links with an average efficiency of 3–8 sulphur atoms per cross-link.

The rate of development of cross-links may be conveniently followed by a number of methods that utilize the relationship between the increase in modulus, or viscosity, and the number of cross-links introduced into the system. Recently some specialized instruments have been developed such as the Vulkameter, Curometer and Viscurometer. All these new methods permit measurement of viscosity and shear modulus, of the stock while it is being cured, by recording the force required to subject the rubber to small angle oscillations (usually 3°) in shear. From such a recording trace, a number of important parameters can be directly read off. In particular, it is possible to determine initial viscosity, scorch time, and optimum cure time of the material.

Most investigators believe that rate of cross-link formation follows first order reaction kinetics and therefore a plot of $\ln\left[(L_\infty - L)/(L_\infty - L_0)\right]$ versus (time of cure) is widely used in the literature. The letter L stands here for some physical property; say modulus, Curometer torque, or an amount of combined sulphur, that can be related directly to the cross-link density by a proportionality constant. Subscripts ∞ and o represent infinite and zero time respectively.

A typical Curometer curve and its representation replotted on a logarithmic scale is shown in Fig. 9.4.

From such plots it is possible to derive first order rate constants and an induction period for the cross-linking reaction so that one can examine them in terms of other variables such as concentration of accelerator, sulphur, zinc stearate (activator) etc.

As a result of studying the effect of free radical scavengers (benzoquinone, hydroquinone, diphenyl-picryl-hydrazyl) on various curing systems[6] it was

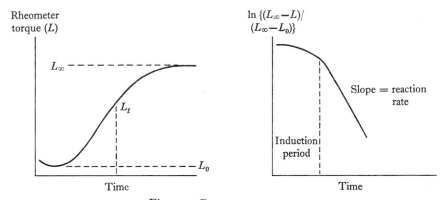

Fig. 9.4. Curometer curves.

suggested that unaccelerated sulphur cure and sulphur cure accelerated by guanidine proceed via polar mechanisms. Other accelerators produce mixed polar and free radical reactions. Peroxide cures are entirely of a free radical nature.

The actual cross-link density in vulcanizates without any fillers (pure gum vulcanizates) can be derived either by measurement of the equilibrium modulus, or the equilibrium swelling on the vulcanizates. The cross-link density can be obtained from the equilibrium modulus by utilizing the statistical theory of rubber elasticity[7] while the results from equilibrium swelling measurements may be interpreted in terms of Flory's theory.[8] Unfortunately, both of these methods suffer from the disadvantage of not differentiating between chemical cross-links and 'trapped entanglements',[9] that is, molecular entanglements trapped between two chemical cross-links. The magnitude of the correction, in a specific case of natural rubber cured with t-butyl peroxide, has been shown to be approximately one entanglement per 250 isoprene units.[10] The long and cumbersome procedure required to obtain the equilibrium modulus may be avoided by measurement of the compression modulus of swollen vulcanizates[11] since there the equilibrium is established very rapidly.

9.3 The elastic behaviour of rubber

Depending on the application, a higher or lower level of elasticity is desirable for a rubber part. In tyres, elastic bands, gaskets, V-belts and packings, high elasticity is wanted. In shock absorber engine mounts and torsion-springs (Torsilastic), and in acoustics, low elasticity and high damping are required.

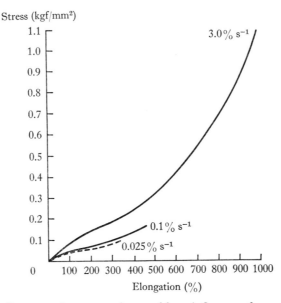

Fig. 9.5. Stress–strain curves of raw rubber: influence of speed of test.
(Stress calculated for actual cross-section.)

9.3.1 The stress–strain curve

Many experiments have been made on the elastic behaviour of unvulcanized rubbers though only a few can be mentioned here. The stress–strain curve for raw natural rubber is given in Fig. 9.5 at various rates of straining.

The effect of raising the temperature is to reduce the tensile strength, but increase the breaking elongation; both are caused by the decrease of the cohesive forces. The effect of a slower rate of stretching is quite similar and will be explained later with the help of the Williams, Landell and Ferry (WLF) superposition principle.

These effects are much smaller in vulcanized rubber, but they still exist (Fig. 9.6).[3] The effect of increasing amounts of cross-linking by sulphur is shown in Fig. 9.7. It is clear that increased sulphur-bonding causes stiffening, that is, higher forces are necessary for a given elongation. Raw rubber shows great extensibility which is lost when the rubber is masticated,

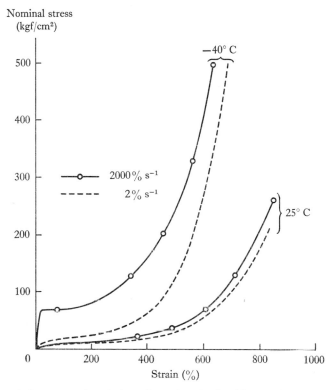

Fig. 9.6. Stress–strain relations for vulcanized rubber at different rates and temperatures. (Stress calculated for actual cross-section.)

but reappears when the rubber is vulcanized. In commerical vulcanizates which use accelerators, the amount of bound sulphur is much lower, of the order of only 1 %, and the tensile strength is considerably higher because the cross-linking reaction is more efficient and less degradation of the rubber takes place during the short reaction time than during the long curing times with sulphur only.

For the sample with 2 % of bound sulphur a force of only 3 kgf/mm² is necessary to stretch it 850 %;† for the same with 5 % of bound sulphur the necessary force is 13 kgf/mm². We have also reproduced in Fig. 9.7 one of the curves for raw rubber together with a curve for masticated rubber. The raw rubber appears to be very extensible (the extension is for a great part plastic); masticated rubber has lost its extensibility to a large degree, but as soon as vulcanization sets in it reappears. The more sulphur bound, the smaller the

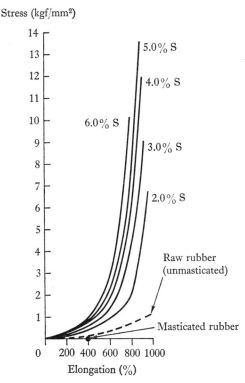

Fig. 9.7. Stress–strain curves for a rubber–sulphur mixture with increasing degree of vulcanization. (Stress calculated on actual cross-section.)

extensibility, however. This takes place at first gradually, until more than 5 % of sulphur has been bound, then there is a sharp falling off of extensibility to the point (shown in Fig. 9.8) where it is only a few per cent and the rubber is hard rubber or ebonite $(C_5H_8S)_n$, containing 40 to 50 % of bound sulphur.

† The stiffness of rubber is often expressed in this way instead of by the modulus of elasticity, since the latter cannot be determined for rubber in a reliable way. (Hooke's law does not hold.)

In the range which contains between 10 and 25 % of bound sulphur, the rubber has a leathery consistency.

If these observations are summarized as in Fig. 9.8 there may be said to be two maxima in the curve representing the extension at rupture plotted against the percentage of bound sulphur: one with raw rubber, and the other when there is about 2 % bound sulphur. The rupture strength, on the other hand, shows three maxima: (*a*) one for raw rubber, (*b*) the next for soft vulcanized rubber with about 5 % of bound sulphur, and (*c*) a third for ebonite. After the process of vulcanization has passed the 'optimum cure' the rubber is often

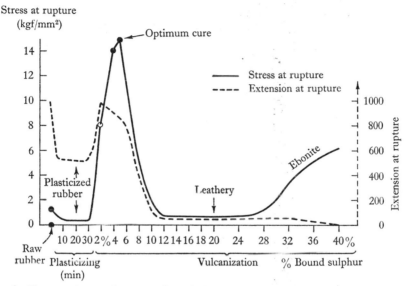

Fig. 9.8. Changes in tensile strength and elongation at rupture in rubber during the whole manufacturing process (rubber–sulphur mixture without fillers or accelerators). (Stress at rupture calculated on actual cross-section during test.)

spoken of as 'over-vulcanized'. An example of an accelerated formulation for pure gum natural rubber is, in parts by weight: smoked sheets, 100; zinc oxide, 3; stearic acid, 1.5; benzthiazol disulphide, 0.6; sulphur, 2.5.

The optimum cure time at 140° C is about 30 minutes. The shape of the stress–strain diagram as actually measured with the testing machine (i.e. not corrected for decreasing cross-section) is that of an S and continues through the origin into the compression curve which approaches 100 % compression asymptotically, as shown in Fig. 9.9. Fig. 9.10 shows that at the beginning of the extension, very little force is developed but after about 400 % elongation, the force starts to rise much more rapidly, particularly when the 'optimum cure'

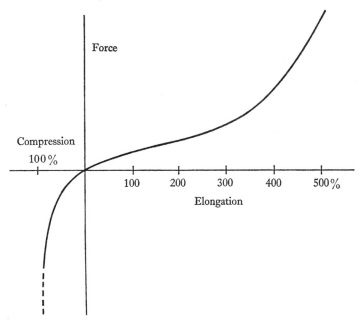

Fig. 9.9. Deformation curve for vulcanized rubber, under compression and elongation.

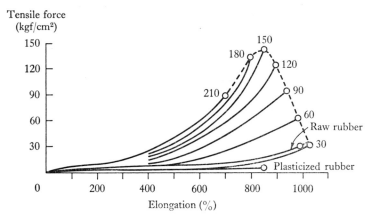

Fig. 9.10. Effect of increasing vulcanization on the stress-strain curve of natural rubber. (Mixture: 100 parts of rubber, 8 parts of sulphur.) Numbers along the curve are minutes of vulcanization time.

is reached (in this case 150 minutes). This increase in force is caused by the chain molecules becoming oriented in the direction of stretch when approaching the fully extended state.

9.3.2   Effect of cross-linking and types of cross-links

An increase in strength with time of vulcanization, during which time sulphur bridges are formed between the macromolecules, is evident in Fig. 9.10. The effect of these cross-links is to reduce the unlimited flow of macromolecules past each other so as to make it possible for them to become oriented in the direction of stretching. These oriented molecules interact more strongly with each other and have greater strength in the direction of orientation as is the case with fibres and high polymers oriented by cold drawing. The interaction can be strong enough to cause crystallization as in the case with natural rubber. Therefore the tensile strength of vulcanized rubber depends greatly on the secondary forces between the macromolecules. These should be weak enough not to reduce flexibility and mobility but also strong enough to allow build-up to sufficient tensile strength. During the stretching process, the individual macromolecules slip along each other resulting in a more even stress distribution among the molecules; this is the main reason for greater strength in oriented polymers. It has been explained in §2.3 that the theoretical strength is many times higher than the experimental one, basically because the loads carried by the various bonds vary over a wide range. Therefore, they break one after another. By equalizing these stresses to a narrow distribution a closer approach to the theoretical strength is obtained. Pre-stretched rubber, cooled to $-180°$ C, falls apart in a bundle of fibres when smashed.

The slippage process, which occurs in rubber during stretching, absorbs energy which is not regained on retraction. This loss in energy is measured as hysteresis. Increasing the number of cross-links beyond the optimum prevents slippage necessary for molecular alignment and results in the reduction of strength known as over-cure.

In §9.2 it was demonstrated that one may distinguish three main types of cross-links:

(1) Polysulphidic bridges $-\overset{|}{\underset{|}{C}}-S_x-\overset{|}{\underset{|}{C}}-$. The majority are produced by vulcanization with sulphur and an accelerator of the diphenyl guanidine or mercaptobenzothiazole type. These bonds are relatively unstable, they break and reform again at a different part of the polymer molecule.

(2) Mono- and disulphidic bridges $\overset{|}{\underset{|}{C}}$—S—$\overset{|}{\underset{|}{C}}$ and $\overset{|}{\underset{|}{C}}$—S—S—$\overset{|}{\underset{|}{C}}$. This type is considerably more stable and less subject to rupture and reformation. The monosulphide is formed by vulcanization with tetramethyl thiuram disulphide (TMT).

(3) Carbon to carbon bonds $\overset{|}{\underset{|}{C}}$—$\overset{|}{\underset{|}{C}}$ which are the most stable and obtained by peroxides or by radiation.

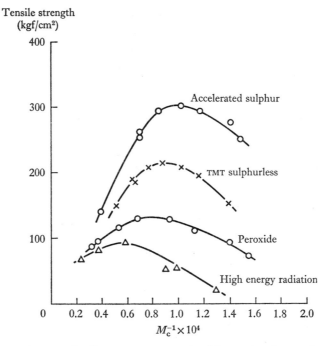

Fig. 9.11. Tensile strength of pure gum natural rubber vulcanizates plotted against $1/M_c$ for various vulcanizing systems.

Going from type (1) via type (2) to type (3) in vulcanizates, one finds higher resilience (lower hysteresis), lower permanent set, better ageing and lower creep.

It is interesting to note that the tensile strength decreases in the same sequence, the less stable cross-links allowing more stress equalization. This is illustrated by Fig. 9.11 where tensile strength of natural rubber is plotted versus the cross-linking density obtained from stress–strain data.

9.3.3 Effect of molecular weight

As shown in Fig. 9.12, there is a strong dependence of the tensile strength on molecular weight before vulcanization. Mastication reduces the molecular weight and it is evident from Fig. 9.12 that excessive mastication will result in low tensile strength.

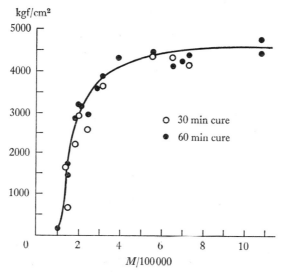

Fig. 9.12. Dependence of tensile strength on initial molecular weight (M) of butyl rubber.

9.3.4 Elastic memory

This remarkable effect was described by Kohlrausch in 1876. A rubber thread (Fig. 9.13) is twisted $2 \times 360°$ to the right and held in that position for 18 hours (*a*). It is then released and twisted $45°$ beyond the position of equilibrium to the left and held so for 30 s (*b*). An elastic after-effect is observed first from (*b*) past the equilibrium position to (*c*) for example, then the direction of the after-effect is changed and the thread twists to the left until a final position at (*d*) is reached. The behaviour of the thread does indeed give the impression that it possesses a 'memory'.

9.3.5 The influence of temperature

As early as 1925 le Blanc and Kröger pointed out the fact that a stress–strain curve for raw rubber at $-50°$ C can be recorded which is similar to that for well-vulcanized rubber. For the sake of comparison several sets of stress–

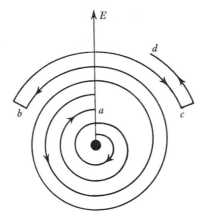

Fig. 9.13. 'Memory-capacity' of a rubber thread.

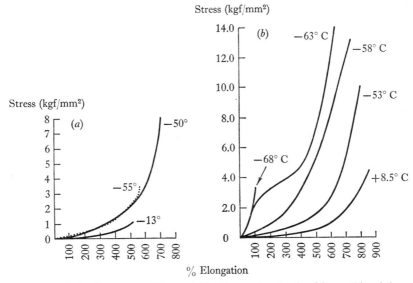

Fig. 9.14. (*a*) Cooling raw rubber. (*b*) Cooling vulcanized rubber. 5 % sulphur, vulcanized for 80 min. (Stress calculated for actual area.)

strain curves observed at various temperatures for both raw and vulcanized rubber are reproduced in Fig. 9.14.

Considering first Fig. 9.14(*a*) it will be seen that at −13° C the values of both the tensile strength and the elongation at rupture of raw rubber are very small, but that they increase markedly upon further cooling until a temperature of about $-50°$ C is reached. At lower temperatures than $-50°$ C a decrease in the values of the above-mentioned mechanical properties again takes place. Furthermore a slight stiffening has occurred due to the low temperature. For vulcanized rubber (Fig. 9.14(*b*)) the effect of cooling is very similar as regards stiffening and change in tensile strength; the elongation at

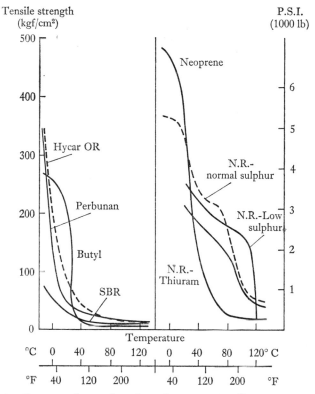

Fig. 9.15. Tensile strength as a function of temperature for pure gum vulcanizates.
Hycar OR contains 40% Acrylonitrile + 60% butadiene (NBR).
Perbunan contains 25% Acrylonitrile + 75% butadiene (NBR).

rupture, however, decreases steadily. These last effects are represented comprehensively in Fig. 9.17, which will be discussed on the succeeding pages.

Many investigators considered cooling to be a 'physical vulcanization'. In our opinion this concept is of great importance if we consider that it assumes that cooling brings about an increase in the secondary forces acting between the macromolecules which may be compared to an increase in the number of cross-links.

The change of tensile strength with temperature is illustrated[12] for a number of rubbers in Fig. 9.15. In comparing the two graphs, it should be remembered that the data are for tensile strength on original cross-section, not on actual cross-section.

It is interesting to note that the crystallizing rubbers, natural rubber, and butyl and neoprene rubber lose their strength at higher temperature than the non-crystallizing SBR and NBR. The general trend of tensile strength and elongation is shown in Fig. 9.16 for a natural rubber vulcanizate containing 50 parts of reinforcing carbon black per 100 parts of rubber by weight.[13]

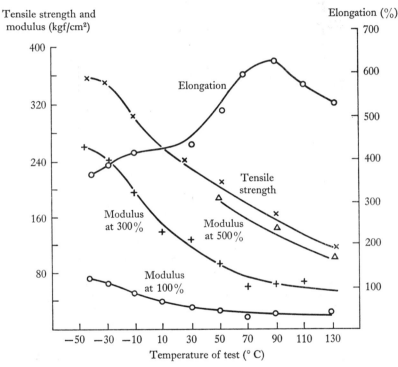

Fig. 9.16. Effect of temperature on physical properties.
(Natural rubber tread mixture.)

9.3.6 Effect of rate of deformation (time)

Of great interest is the superposition of tensile and breaking elongation data at various temperatures by the method of Williams, Landell and Ferry[14] developed for viscous deformation (rate processes). This principle states that the same tensile and breaking strain can be obtained at different temperatures if the rate of strain at the higher temperature is sufficiently larger. The relationship between temperature difference and difference in rate is given by the shift factor α_T. A small correction in the tension is necessary because of the increase in tension at higher temperature required by the kinetic theory of elasticity. If plotted on a double logarithmic scale all values can be collected on one master curve, as shown[15] in Fig. 9.17.

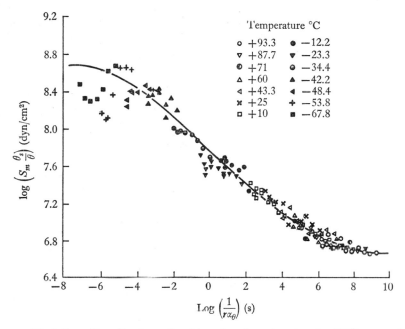

Fig. 9.17. Variation of tensile strength with reduced strain rate r for SBR gum rubber, r is rate of extension in s^{-1}, S_m is the tensile strength, θ_s is a reference temperature $50°$ C above the glass transition temperature, and θ the temperature of measurement.

The shift factor can be written in the general form:

$$\log \alpha_t = \frac{A(T-T_0)}{B+T-T_0}; \quad A = 8.86, \ B = 101.6.$$

A and B are constants with numerical values dependent on liquid parameters and the free volume (available for molecular movement) at the glass transition temperature. It is remarkable that Williams, Landell and Ferry's equation is so generally applicable, which indicates that the constants are characteristic of the glass–liquid transition rather than of the particular polymer. The fact that this theory applies to the tensile and breaking elongation indicates that a rate process (viscous flow) is involved.

9.3.7   Effect of fillers; reinforcement

For most applications rubbers contain fillers, particulate solids, dispersed in the polymer. These fillers are used in rubber either for cheapening or for improvement of properties. The first category is that of the so-called inactive fillers, the second that of active or reinforcing fillers. The last play such a dominating part in rubber technology that they merit some extra discussion. In Fig. 9.18 the influence of both types of fillers on the stress–strain curve is demonstrated. In the case of the inactive filler barytes, there is no significant effect on the shape of the stress–strain curve, the tensile strength becomes

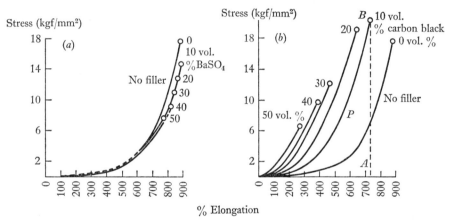

Fig. 9.18 (*a*) Influence of an inactive filler (BaSO$_4$) on vulcanized rubber. (*b*) Influence of an active filler (carbon black) on vulcanized rubber. (Stress calculated for actual cross-section.)

smaller with increasing loading which can be expected, since there is less rubber per cross-section; the elongation decreases slightly. Carbon black, the most universal active filler, raises the whole stress–strain curve to a higher stress level and increases tensile strength up to a maximum loading, after which it decreases again. The elongation also becomes rapidly shorter as the black is added in appreciable amounts beyond the optimum. A criterion for activity for a filler has been defined by Wiegand; it is the total energy necessary for rupture which can be found from the stress–strain curve on the original cross-section, as the area between the curve and the elongation co-ordinate

$$e_\mathrm{r} = \int_0^{\lambda_\mathrm{b}} \sigma \mathrm{d}\lambda.$$

It is evident that the shortening of the stress–strain curve by increasing loading with barytes leads to lower e_r values whereas the raising of the stress level by carbon black will increase the energy at rupture. Typical values for energy at rupture are: for a well cured pure gum vulcanizate, 380 kgf.cm/cm³; for the corresponding vulcanizate with 20% by volume of carbon black, 580 kgf.cm/cm³. The energy at rupture has been related to the heat of wetting or heat of adsorption. The integral heat of wetting per gram of black decreases with higher loading of black, and by extrapolating to zero concentration a value of 11 g cal per gram of black was found.

The increase in stiffness as expressed by Young's modulus, due to the presence of wetting fillers of spherical particles, according to Guth[16] is

$$E = E_0(1 + 2.5C + 14.1C^2), \qquad (9.1)$$

where C is volume concentration of filler. This relation is valid up to 30 vol. per cent of thermal black but for higher reinforcing blacks the equation should be replaced by:

$$E = E_0(1 + 0.67fC + 1.62f^2C^2), \qquad (9.2)$$

where f is the shape factor related to length over diameter ratio. These equations do not take into account the effect of particle size and surface activity which are important for the contribution of the filler particle as a 'multiple cross-link'. Their general validity is therefore questionable, particularly when they are applied to stress at, for instance, 300% instead of to Young's modulus. A more complete treatment is given in Chapter 4 on Rheology.

It is of interest here to note that the carbon black surface is very inhomogeneous and contains sites of various adsorptive activities. Differential heats of adsorption were determined by Smith and co-workers[17] and by Atkins and Taylor.[18] A typical curve is shown in Fig. 9.19.

A reinforcing black has a surface area of about 100 m²/g so that at optimum loading (50 parts per 100 rubber by weight) there is about 34 m² per cm³ of compound. This factor is an overriding one, and covers a very wide range, from 1 m²/g to over 200 m²/g. Coarse fillers with much lower surface area 1–5 m²/g show little reinforcement.

The specific activity of the surface

As shown in Fig. 9.19 the carbon black surface contains a small percentage of sites with an adsorptive energy of about 14 kcal/mol for propane. This value is much higher than the heat of liquefaction of propane. For carbon black in rubber, secondary bonds between carbon and rubber at these active sites are stronger than rubber to rubber bonds. This is particularly so when the surface area of the filler is large (small particle size) and is the reason for the higher stress level of the stress–strain curve; the particles form bridges between the rubber molecules. Plate 9.1 (facing p. 320) shows an electron micrograph of carbon black particles.

The particles of non-reinforcing fillers have interaction forces which are easily broken since they are weaker than the rubber to rubber forces.

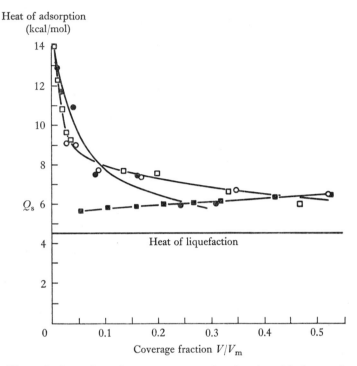

Fig. 9.19. Heat of adsorption of propane per gram of carbon black as a function of coverage fraction (V/V_m). V = volume of propane adsorbed per g of black. V_m = volume of propane absorbed to form a monolayer.

□ ISAF black	V_m = 14.90 cm³/g
○ Low structure ISAF black	V_m = 11.45 cm³/g
■ Graphitized channel black	V_m = 9.40 cm³/g
● EPC channel black	V_m = 12.30 cm³/g

One way to determine the strength of the bonds between rubber and filler is by measuring the swelling of the vulcanizate in a good solvent. Fillers with little adhesion will separate from the swelling rubber and pockets of solvent will form around each particle. It appears that increased loadings of such a filler also increase the swelling of the rubber matrix. On the other hand, strong adhesion bonds between filler and rubber will not separate on swelling but restrict the swelling. Increased carbon black loading in vulcanizates causes a decrease in the swelling equilibrium;[19] see Fig. 9.20.

Most filler particles are not ideal spheres but consist of branched and irregular shaped aggregates (see Plate 9.1, left). The higher the branching of these aggregates the higher is the structure of a filler. High structure fillers will cause high stress levels in the rubber vulcanizates provided that the specific surface activity is high enough to have stronger adhesive forces than the rubber to rubber force and provided the surface area is large enough.

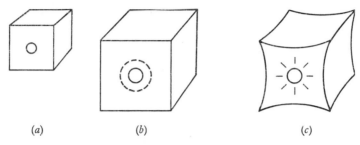

(a)   (b)   (c)

Fig. 9.20. (a) Original vulcanizate element with filler particle. (b) Swollen vulcanizate with non-adhering filler and pocket of solvent. (c) Swollen vulcanizate with adhering filler restricting swelling.

9.3.8 Stress softening and reinforcement; interaction

One of the typical phenomena observed with filled vulcanizates is that of stress softening and often referred to as the 'Mullins' effect.[20] This is particularly noticeable with carbon blacks. It is an indication of breakdown of the structure during deformation, in other words, a thixotropic effect, and it has been most useful in explaining the interaction between filler and elastomer.

When the stress–strain curve of a vulcanizate is determined after the sample has been elongated one or more times previously, the values for the stress are lower than for the original unstressed sample. The effect is illustrated in Fig. 9.21. The lowering of the stress is only appreciable at elongations below the pre-stress elongation. In unfilled rubbers this reduction of stress is relatively small, probably due to disentanglement, breakage of some bonds and orientation or crystallization.

Several explanations have been put forward for the large effect in filled rubbers, the most probable one being that of slippage of the adsorbed hydrocarbon chains along the filler surface.[21] The hydrocarbon segments are adsorbed onto the carbon black surface by forces that allow two-dimensional mobility. This is the concept of mobile adsorption.[22] Under the influence of stress, those chains in the network that carry the highest load will slip along the filler surface and alleviate this high stress resulting in a more homogeneous stress distribution.

This slippage process dissipates a certain amount of energy which represents a contribution to the tension, and raises the resistance to tear and to abrasion.

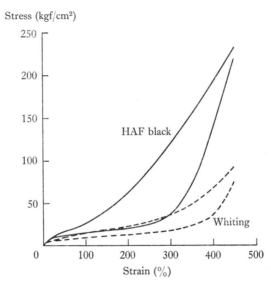

Fig. 9.21. Stress–strain curves during first and second extensions on rubbers containing different fillers (50 parts per hundred parts of rubber) (after L. Bateman, ref. 3.)

A schematic picture of the slippage process is given in Fig. 9.23. After retraction the chains between particles do not immediately resume their original position (see Fig. 9.23 *b*). During slippage they are stretched to about the same length and when elongated for the second time the energy which was dissipated by slippage does not have to be provided again. The stress at the second extension cycle is therefore smaller. Repeated cycling results in only a further slight reduction of the stress. The observation that the stress at the second cycle is lower only up to the maximum elongation of the first cycle is in agreement with this picture. Beyond that elongation the behaviour of the vulcanizates at the first and the following cycles is practically the same. The ultimate tensile strength of both is approximately identical, therefore it is evident that no major portion of chains is broken. In time, part of the loss in stress in the first cycle of extension is recovered, especially at higher tempera-

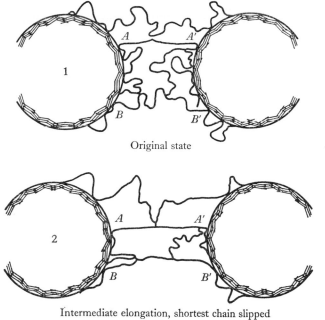

Original state

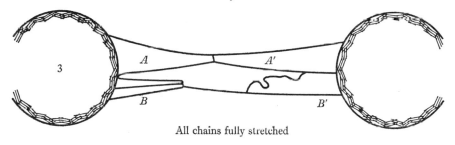

Intermediate elongation, shortest chain slipped
distances beyond *A–A'*

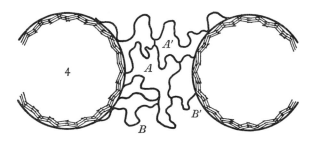

All chains fully stretched

After retraction all chains have equal lengths between particles

Fig. 9.22. Molecular slippage model of reinforcement mechanism.

ture; the network is striving to regain its original configuration. However, a small part of the loss is permanent, and this can be due to actual breakage of chains or cross-links. If the elongation is carried out close to the breaking point the permanent softening becomes a larger fraction of the total stress softening.

The concept also provides a logical explanation of the reinforcing effect of active fillers in elastomers, especially in non-crystallizing rubbers such as copolymers of butadiene with styrene or acrylonitrile. The vulcanization reaction introduces cross-links at random resulting in a broad distribution of

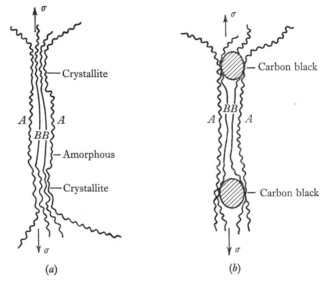

Fig. 9.23. (*a*) Locally overstressed molecules *B* in crystallized rubber. (*b*) Locally overstressed molecules *B* in carbon black reinforced rubber.

network chain lengths between cross-links. If no slippage occurs during elongation, the chains that carry the highest load will break one after another with continued extension of the sample. Finally only a few chains carry the ultimate breaking load. The attachments to the carbon black surface are of a mobile nature and allow the highest loaded chains to dissipate the stress by slippage along the surface (Fig. 9.22). A more homogeneous distribution of stress results, so that a larger percentage of chains carry the load to the ultimate breaking. In the case of SBR this leads to an improvement in tensile strength at room temperature by a factor of 10–15. In crystallizing rubbers such as natural rubber the improvement is much smaller because the crystallites already fulfil a function similar to that of the carbon black particles, and thus these rubbers are self-reinforcing. This is shown in Fig. 9.23. Crystallization will be discussed in §9.7.4.

9.3.9 Visco-elastic phenomena and hysteresis

During the elongation of vulcanized rubber, a process which is reversible to a high degree, the segments of the chain molecules slide along each other. However, this flow is limited by the linkages which exist between these molecules at certain points, so the system is both viscous and elastic. As a consequence of the interaction of adjacent molecules during their viscous flow on stretching, frictional heat will be developed. This is the irreversible part of

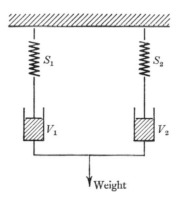

Fig. 9.24. Dashpot model for the rheological behaviour of vulcanized rubber.

the energy of stretching, that is the hysteresis. Keeping the rubber stretched for a longer period of time results in additional viscous flow, manifested by a relaxation of tension. By the same mechanism, a vulcanizate elongated under constant load will creep, that is, slowly increase its elongation.

This behaviour can be represented (Fig. 9.24) by a mechanical model of a system of springs and dashpots for which Kuhn[23] derived:

$$E_t = E_{01} e^{-t/\lambda_1} + E_{02} e^{-t/\lambda_2} + E_{03} e^{-t/\lambda_3} + \dots.$$

In this equation E_0 and E_t are the stress at time o and time t, and λ is the time of relaxation following from Maxwell's theory of relaxation. It is the time after which the tension would have decreased by a factor $1/e$.

The relaxation spectrum explains the phenomenon of elastic memory previously mentioned in §9.3.4. The segments with the short relaxation times will give an immediate response while those with long relaxation times give a later response. For natural rubber Haegel[24] determined:

$$E_{01} \quad 10^6-10^7 \text{ dyn/cm}^2 \qquad \lambda_1 = \infty$$
$$E_{02} \quad 10^2-10^6 \qquad\qquad \lambda_2 = 10-10^4 \text{ s}$$
$$E_{03} \quad 10^6-10^8 \qquad\qquad \lambda_3 = 10^{-3}-10^{-4} \text{ s}$$
$$E_{04} \quad 10^{11}-10^{12} \qquad\quad\; \lambda_4 = 10^{-13} \text{ s.}$$

The first term shows that the high elasticity modulus E_{01} remains constant after the elongation is attained and does not decrease with time. The stress

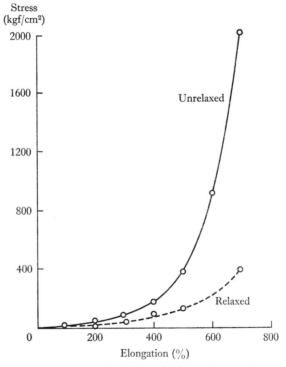

Fig. 9.25. Effect of relaxation (4h, 70° C) on the tension on actual cross-section of natural rubber (pure gum vulcanizate).

relaxation in natural rubber is complicated by phenomena stemming from its tendency to crystallize. Fig. 9.25 shows the magnitude of the stress relaxation in natural rubber. The temperature was raised to 70° C to accelerate the process. The decay of stress with time follows the empirical equation:

$$\sigma = a - b \log t, \tag{9.3}$$

in which a and b are constants. This equation can be derived from Kuhn's equation if a very broad relaxation time spectrum is assumed. It can be seen from Fig. 9.26 that the relaxation is greater at higher elongations. This is due to the crystallization which occurs at high elongation and which increases the stiffness and stress. On relaxation, particularly at higher temperature, these crystals melt and recrystallize thus reducing the stress to a much lower value. Similarly the phenomenon of creep or flow (the increase in length at constant load) is also influenced by crystallization. The general equation for flow is:

$$\alpha = a + b \log t, \tag{9.4}$$

$$v = \frac{\alpha_2 - \alpha_1}{\log t_2 - \log t_1},$$

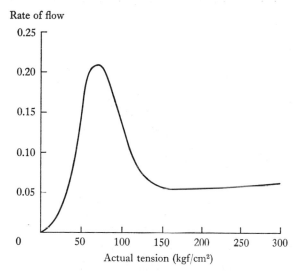

Fig. 9.26. Tension–flow diagram of vulcanized natural rubber at 20° C.

where v rate of flow or creep, α elongation ratio l/l_0, t time and a and b are constants.

Fig. 9.26 shows that the rate of flow of vulcanized rubber rises up to a certain tension and then falls. The drop in the rate of flow is due to crystallization which decreases the mobility of the rubber molecules in respect to each other.

Flow and relaxation in a rubber vulcanizate is not only a viscous re-orientation or recrystallization phenomenon but is also strongly affected by the fact that rubber after elongation does not immediately reach its final value of crystallization as, for instance, is shown by measurements of double

8

refraction.[26] This after-crystallization strengthens the rubber structure and reduces the force needed to keep the rubber at a certain elongation. If the tension is kept constant, the rubber will elongate further.

Chemorelaxation is a term introduced by A. V. Tobolsky[27] for the decay of stress that occurs in vulcanized rubber, kept at constant elongation at high temperatures of $70°$–$130°$ C, and due to oxidative scission of rubber chains. It follows the equation

$$\sigma_t = \sigma_0 e^{-kt}, \tag{8.5}$$

where k is the rate constant. From the temperature dependence of this rate constant, the activation energy of the oxidative chain scission can be determined. For natural rubber a value of 30 kcal was found. In the absence of oxygen, the relaxation rate is reduced by a factor up to 1000.

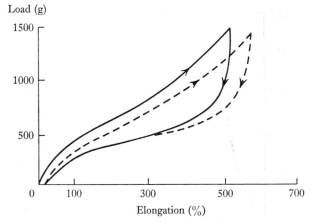

Fig. 9.27. Hysteresis curves for vulcanized rubber (according to Bouasse and Carrière) (ref. 25).

The important phenomenon known as hysteresis is due to energy expended as internal friction during deformation of a rubber specimen, this energy is not recovered on retraction. As a consequence the return curve of the stress–strain curve lies below the stress level of the original. This is demonstrated by Fig. 9.27. The energy of hysteresis is represented by the area between the stressing and the releasing curve. Repeated cycles tend to narrow this difference but it remains a measurable value. Hysteresis is a most important property of rubber since it is responsible for the heat build-up (during frequent deformation) which is detrimental to abrasion resistance but is desirable when the rubber is to be used for vibration damping, or for its high coefficient of friction.

As explained in §9.3.7, fillers will increase hysteresis losses. According to Schallamach[28] in his theory of abrasion resistance, it is hysteresis that reduces the movement of a rubber tyre over the road. In this respect, high hysteresis improves abrasion resistance but it also causes the rubber to heat and abrasion resistance decreases rapidly as the temperature rises. If this heat can be rapidly dissipated, the final effect of hysteresis on abrasion resistance is favourable. In a practical sense, high hysteresis rubber shows good skid resistance and bad abrasion.

9.4 The plastic behaviour of rubber

9.4.1 General principles

Raw rubbers exhibit many properties common to all thermoplastic polymers of equivalent molecular weight. Their long molecular chains give rise to elastic behaviour, which is particularly noticeable at short deformation times when the molecules do not have sufficient opportunity to slide past each other. At long deformation times these molecules exhibit viscous flow since they are not cross-linked in a three-dimensional network. In Chapter 4 it was shown that viscous flow in a polymer is related to the jump frequency of the molecular segments which varies with temperature. Below the glass transition temperature the jump frequency is negligibly small but it increases rapidly with increasing temperature. This interaction between time and temperature for viscous liquids above the glass transition temperature forms the basis for the superposition principle and for the Williams, Landell and Ferry equation[29] which is often used to relate measurements made at different temperatures to those made at some standard temperature and hence obtain relaxation spectra over a very wide interval of time. An example of such a procedure[30] is given in Fig. 9.28.

9.4.2 Measurement of viscosity

Many of the manufacturing processes (e.g. extrusion, calendering, and moulding) depend largely on the ability of raw rubber to deform and flow under applied force. In common with all other visco-elastic materials, rubber may be considered a non-Newtonian fluid, since its viscosity, η, decreases with increasing rate of flow (shear rate, D). The general derivation of rheological equations relating shear stress, shear rate and apparent viscosity as well as the methods available for determination of these parameters, have already been discussed in §1.1. A typical plot of shear stress versus shear strain for an SBR rubber[31] is given in Fig. 9.29 and it may be noted that the

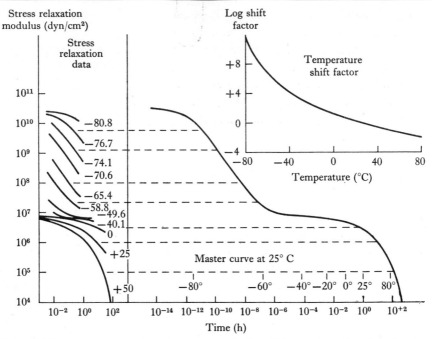

Fig. 9.28. Time–temperature superposition principle illustrated with polyisobutylene data. The reference temperature of the master curve is 25° C. The inset graph gives the amount of curve shifting required at the different temperatures.

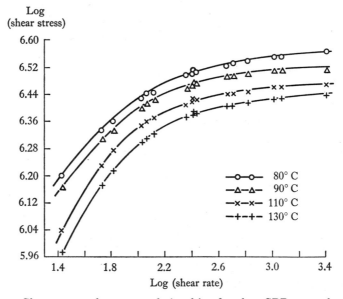

Fig. 9.29. Shear stress–shear rate relationships for clear SBR 1712 elastomer.

apparent viscosity of the material, that is, the ratio (shear stress)/(shear rate), decreases with increase in shear rate.

In crystallizing rubbers, such as natural rubber, the apparent viscosity may increase very rapidly due to shear-induced crystallization.[32] The point at which this occurs depends on the molecular weight of the material (Fig. 9.30), its chemical composition and temperature.

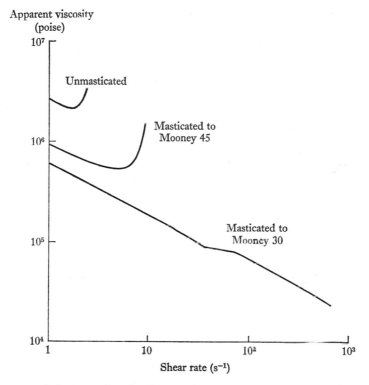

Fig. 9.30. Influence of molecular weight on shear-induced crystallization of natural rubber (ref. 32).

The viscosity at very low shear rates, n_i is very dependent on the molecular weight, M, of the polymer and for amorphous, linear, high molecular weight polymers these are related by the equation[33]

$$n_i = K.M^{3.5}, \tag{9.6}$$

where K is a constant. In industrial processing, rubber is degraded by milling until the material reaches the desired viscosity, and can be processed further with relatively low expenditure of work. The degree of degradation and the fluidity of compounded stocks is customarily measured using a shearing disc (Mooney) viscometer, although other older instruments such as Williams' plastometer are also being used.

The shearing disc viscometer (Fig. 9.31) consists of a cylindrical cavity maintained at the desired temperature and a rotor which is made to rotate

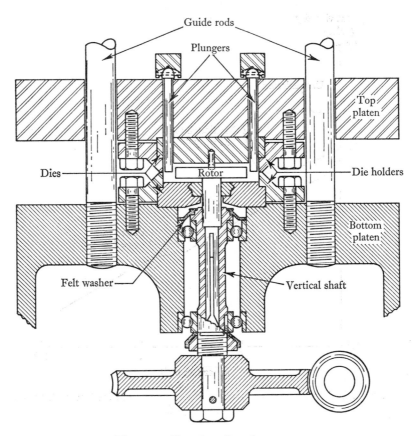

Fig. 9.31. Shearing disc viscometer.

within it at constant speed. The sample of raw rubber is placed on both sides of the rotor, and the chamber is closed under pressure. The torque required to maintain constant rotor speed is measured on a calibrated dial and recorded as Mooney Viscosity Value. The Mooney Viscosity is useful for predicting the behaviour of the materials during extrusion, or calendering, although it is measured at a much lower shear rate (*ca.* 1 s^{-1}) than those encountered in commercial processes:

Process	Range of shear rates (s^{-1})
Compression moulding	1–10
Calendering	10–100
Extrusion	100–1000
Injection moulding	1000–10 000

9.4.3 Extrusion experiments with raw rubbers

One of the methods of studying the visco-elastic behaviour of raw rubbers under large stresses is by means of extrusion tests. From such experiments it is possible to determine the viscous parameters of the material by application of the rheological equations discussed already in Chapter 1. However, in addition to these, raw rubber exhibits other important characteristics associated generally with any high molecular weight, linear polymers. The two most important of these phenomena are extrusion shrinkage and melt fracture.

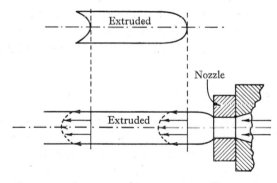

Fig. 9.32. Recovery phenomena during tubing (Barus phenomenon).

The post-extrudate swelling was observed first in 1893 (Barus), when it was noted that extruded marine glue slowly relaxes and changes its shape on standing, eventually producing a bulge in the direction opposite to that of the original flow (Fig. 9.32). A related effect which has since received much attention is 'die swell', or 'extrusion shrinkage'.

The die swell, S, may be defined by:

$$S = (A_{\mathrm{j}}/A_{\mathrm{c}}) - 1 \tag{9.7}$$

where A_{j} and A_{c} stand for the cross-section areas of the extrudate and the capillary (die) respectively. Similarly the percentage extrusion shrinkage, P, is defined by:

$$P = 100S/(S+100). \tag{9.8}$$

In practice the die swell plays an important role, since it controls the dimensions of all extruded rubber goods. Several independent mechanisms causing the die swell have already been proposed in the literature[34] in which the die swell is considered to be a relaxation process due to pseudo-elastic character of the melt. Newman and Trementozzi[35] have considered the tendency towards flatter velocity distribution within the capillary (plug flow). Some recent data[36] indicate that die swell of filled rubbers is mainly caused by molecular orientation. High structure fillers, for example carbon black, decrease the overall orientation of polymer–molecules occurring during the capillary flow through particle–particle interaction which occurs at filler loadings in excess of approximately 20 % by volume.

Melt fracture, another important phenomenon, is recognized by visual examination of the surface of an extrudate; it normally produces either fine crazing of the surface, or the cross-section of the extrudate becomes uneven giving the so-called 'bamboo effect'. Melt fracture produces discontinuity in the relationship between shear stress and shear rate. In practice, the sensitivity of any rubber compound to melt fracture is determined using the Garvey die[37] and observing the contour as well as the edge of the extrudate (Fig. 9.33).

9.4.4   Effect of fillers on viscosity

In rubber compounding two general categories of fillers are recognized. The non-reinforcing fillers (e.g. whiting, chalk, barytes) are characterized by relatively large particle size (average diameter greater than 0.5 μm) and a low surface area (5 m²/g or less). The reinforcing fillers (e.g. furnace and channel blacks, fine silica) range in particle size from 10 to 200 nm and in surface area from 10 to 300 m²/g.

The effect of fillers on viscosity of suspensions (§4.3) led, in moderately concentrated suspension, to equations of the type:

$$\eta/\eta_0 = (1 + D_1 V + D_2 V^2),$$

where V is volume concentration. In the Guth–Gold equation $D_1 = 2.5$ and $D_2 = 14.1$. It describes accurately the viscosity of rubber containing non-reinforcing, spherical fillers (even MT grade of carbon black) at moderate concentrations. Additional constants have to be employed, however, for non-

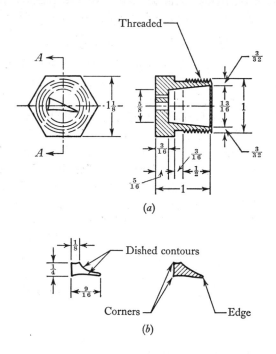

(a)

Threaded

Dished contours

Corners ——   —— Edge

(b)

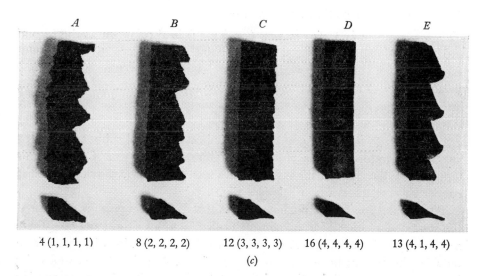

| A | B | C | D | E |

4 (1, 1, 1, 1)   8 (2, 2, 2, 2)   12 (3, 3, 3, 3)   16 (4, 4, 4, 4)   13 (4, 1, 4, 4)

(c)

Fig. 9.33. Garvey die. (*a*) Section on *AA*. (*b*) Details of die opening. (*c*) Type samples showing evaluation of tubing properties.

spherical and fine particles. The influence of the particle shape has been evaluated by introducing the shape factor f into the equation, namely:

$$\eta/\eta_0 = 1 + 0.67f.V + 1.62f^2.V^2.$$

See similar equations (9.1) and (9.3) applied to the modulus. In all practical systems, the value of f is normally adjusted to give the best possible agreement with experimental observations. An independent method of evaluating the shape factor of fillers, based on statistical analysis of electron microscope data, has been suggested[38] and applied to a series of carbon blacks. The authors were able to differentiate between anisometry and bulkiness of filler particles. At higher concentrations, above a filler volume fraction of 0.3, all the equations disagree with experimental findings and the viscosity of such suspensions increases very rapidly as the particle–particle interaction increases.

The viscosity of rubbers containing reinforcing fillers depends to a great extent on the polymer filler interaction. A Guth–Gold type equation successfully predicts the viscosity of such stocks up to not too high concentrations, providing that the assumption is made[39] that all the 'bound rubber' (i.e. the absorbed rubber shell around each filler particle) is also counted as filler volume.

9.5 Thermodynamics of rubber

9.5.1 The glass transition temperature

This temperature (T_g) of polymers (see §4.5.2) is also known as the second order transition point because at this point second derivatives of several thermodynamic capacitive variables such as specific volume and entropy (heat capacity or specific heat) as a function of temperature tend to become infinitely large. Plots of these variables versus temperature show discontinuities at T_g.

When the specific volume or the specific heat of natural rubber are plotted versus temperature, the discontinuity is found at about $-72°$ C. Below this temperature the slope of the line becomes less steep and at the same time the rubber becomes hard and brittle (see Fig. 9.34).

The explanation of this phenomenon is that as the temperature is lowered the rotational motion of the chain molecules becomes slower and at T_g these motions become insignificant compared to the deformations imposed from outside, the rubber becomes brittle; the 'free volume' between molecules becomes a very small fraction of the total volume. The change in specific heat indicates that several degrees of freedom of rotational molecular motion have 'fallen asleep'. Vibrational motion is still continuing. Evidently kinetic (rubber-like) elasticity is only possible above T_g.

9.5.2 Thermodynamics of rubber extension and contraction

Much progress has been made in the last decades towards understanding the thermodynamics of rubber and the fundamental concepts of kinetic elasticity. It has been known for a long time that raw rubber warms up when stretched so quickly that a minimum of heat is lost to the environment; about 10 cals/g

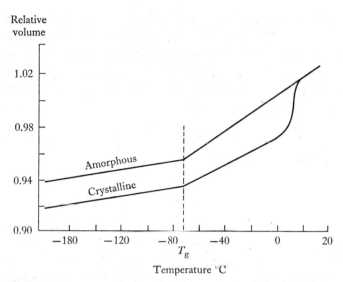

Fig. 9.34. Second order transition point of natural rubber. Relative volume of amorphous and partially crystalline rubber hydrocarbon as a function of temperature.

can be developed. Although we now know that most of this is heat of crystallization, it led Lord Kelvin to conclude that rubber must contract when heated, in contrast to metals, glass and solids, which was confirmed quantitatively by experiment.

9.5.3 Theory[40]

The thermodynamic free energy equation for any material can be written

$$G = E - TS - Fl, \qquad (9.9)$$

assuming the volume change in stretching to be negligible and where G is free energy, E is internal energy, S is entropy, T is temperature, F is extension force *per unit area*, and l is length.

For a reversible extension at constant temperature $G = 0$ so that:

$$dE - T(\partial S)_T - F(\partial l)_T = 0, \tag{9.10}$$

from which:
$$F = \left(\frac{\partial E}{\partial l}\right)_T - T\left(\frac{\partial S}{\partial l}\right)_T.$$

Using a property of total integrals, this can also be written:

$$F = \left(\frac{\partial E}{\partial l}\right)_T + T\left(\frac{\partial F}{\partial T}\right)_l. \tag{9.11}$$

Measuring F at constant length and different temperatures equation (9.11) allows calculation of $(\partial E/\partial l)_T$, the change in internal potential energy due to stretching. For rigid materials like metals $(\partial E/\partial l)_T$ is considerably larger than zero and $T(\partial F/\partial T)_l$ is practically negligible so that $F = (\partial E/\partial l)_T$. In words: the force is caused by change in potential energy, that is, larger distance between atoms or deformation of valence angles. The deformation causing the force is usually completely reversible.

In the case of rubber-like elasticity the term $(\partial E/\partial l)_T$ is almost negligible in comparison to the term $T(\partial F/\partial T)_l$; hence

$$F = T\left(\frac{\partial F}{\partial T}\right)_l = -T\left(\frac{\partial S}{\partial l}\right)_T. \tag{9.12}$$

This means that the force excited by the extension is due to a decrease in entropy. The entropy, in turn, can be expressed in terms of the total probability of the system, W, by Bolzmann's law: $S = k \ln W$. In other words, the elastic force is caused by the molecular segments not occupying the most probable position.

Since the tension is held by the molecular chains themselves, there must also be a small change in interatomic distance or bond angle, which contributes a small portion to the force as was the case with a rigid material such as steel. An analytical picture of the contributions to the elastic force of entropy and potential energy for vulcanized natural rubber is given by Fig. 9.35). Up to about 300 % elongation both terms are small but the entropy term $T(\partial F/\partial T)$ is increasing rapidly to reach a maximum at about 550 % elongation. The potential energy term $(\partial E/\partial l)$ in the case of natural rubber is negligible only at less than 300 % elongation, it goes through zero at about 350 % and then becomes strongly negative reaching its minimum at about 550 %.

The effect is due to crystallization and release of the latent heat of crystallization, which is a loss to the system and therefore negative. At greater than 550 % elongation, the increase in crystallization with extension is less rapid

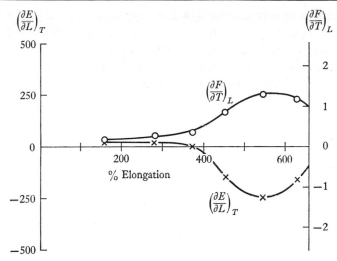

Fig. 9.35. Thermodynamical analysis of the elastic tension of natural rubber.

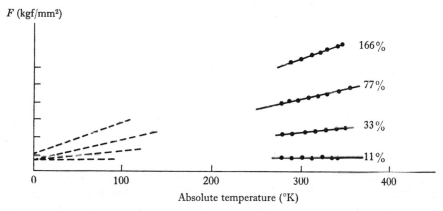

Fig. 9.36. Elastic force at constant length as a function of absolute temperature.

and the $(\partial E/\partial l)$ curve swings upward again as bond angles and atomic distances change as the maximum elongation is approached.

The plot of force versus absolute temperature should be a straight line which, extrapolated, will go to the origin if $F = T(\partial F/\partial T)_l$, or in other words $(\partial E/\partial l)_T = 0$. It can easily be shown that the intercept at $T = 0$ actually indicates the value of $(\partial E/\partial l)_T$. However, because of the short temperature range in which tension can be measured and the large temperature span that has to be covered by the extrapolation, the values of $(\partial E/\partial l)$ are not very precise (see Fig. 9.36).[41]

The reduction in elastic tension in Fig. 9.37 occurring below the temperature of point B (about 36° C) where the slope of the line changes abruptly, is due to crystallization. Assuming the degree of crystallinity to be proportional to the relative reduction in elastic tension, it is possible to estimate the percentage crystallinity. Percentages estimated this way agree well with X-ray determinations.

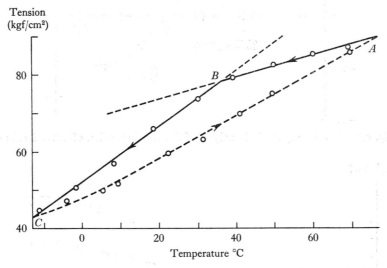

Fig. 9.37. Elastic tension as a function of temperature with natural rubber (pure gum vulcanizate) at 357% elongation.

Extrapolated to $T = 0$, the line to the left of point B in Fig. 9.37 will reach a point well below the horizontal axis; this indicates a negative $(\partial E/\partial l)_T$ corresponding to the heat of crystallization of that portion of the vulcanizate. This is, of course, again a measure of the percentage crystallized rubber. These findings showing the rubber elasticity to be a kinetic phenomenon have been supplemented by statistical considerations.

9.5.4   Statistical theory of elasticity

Assume a rubber molecule to consist of freely rotating chain segments so that for very long chains an unlimited number of configurations is possible. The principle is shown in Fig. 9.38.

It can be shown that for chains with n such mobile segments (and assuming a Gaussian distribution) the most probable distance between the ends is given by

$$r = l \sqrt{\left\{ n \left(\frac{1 + \cos \alpha}{1 - \cos \alpha} \right) \right\}},$$

where l is the length of the chain segment and α is the semi-angle of the cone of rotation. When rubber is stretched the chain molecules are in less probable states and thermal motion tries to restore the most probable state. This is the cause of the elastic tension (see Fig. 9.38(c)).

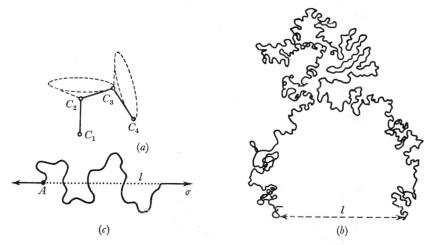

Fig. 9.38. Configurations in chain molecules. (a) Some possible configurations due to free rotation in a carbon chain. (b) Distance r between the chain ends in a 1000-link paraffin chain according to the statistical theory. (c) Tension σ in a chain molecule.

From the probability function itself, which will not be given here, the elastic force $F = -T(\partial S/\partial l)_T$ (see equation (9.12)) can be calculated. Thus $F = NkT(\alpha - 1/\alpha^2)$ where N is the number of chains molecules per cm³, k is Boltzmann's constant and α is the extension ratio.

This equation can be rewritten as:

$$F = \rho \frac{RT}{M_c} \left(\alpha - \frac{1}{\alpha^2} \right)$$

where M_c is the average molecular weight of the free chain or the chain between cross-links, and $R = k \times 6 \times 10^{23} = 1.98$ cal/° C.

Flory has corrected this formula, pointing out that loose chains ends do not contribute to the elastic tension, so that the entity $1/M_c$ should be multiplied by $(M - 2M_c)/M$ where M is the molecular weight of the chain molecules before cross-linking and M_c the molecular weight between cross-links. When M is large compared to $2M_c$ the correction is small.

This theory does not take into account the fact that at higher elongations the chain ends do not follow a Gaussian distribution curve. In Fig. 9.39 the

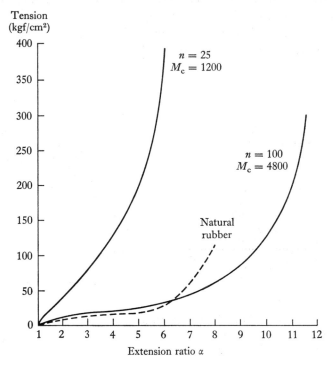

Fig. 9.39. Theoretical and experimental stress–strain curves for simple elongation.

theoretical stress–strain curves with corrections by Treloar[42] for non-Gaussian distribution is compared with the experimental curve for pure gum natural rubber. The agreement is very reasonable and provides convincing proof that the basic assumptions of the theory of kinetic elasticity are correct.

9.6   Chemical constitution and mechanical properties

9.6.1   Chain regularity

Isoprene with its conjugated set of double bonds can undergo addition reactions in four different ways.

Polymerization can take place at the 1, 3 or 3, 4 double bonds so that there will be unsaturated side groups pendant on the main chain as shown.

CH$_3$
H$_2$ |
C=C—C=C $\longrightarrow$ ~C—C~   or   ~C—C~
H H$_2$ H | H H
                C=CH$_2$ |
                H   C
                                    CH$_3$ CH$_2$

1,2 polymerization   3,4 polymerization

If the addition is 1, 4 the double bond stays in the backbone and the polymer will show *cis–trans* isomerism. Apart from this, head–head and head–tail junctions can be envisaged.

cis-1,4-polyisoprene   *trans*-1,4-polyisoprene
natural rubber   gutta percha

In the case of natural rubber (*cis*-1, 4 configuration) and gutta percha (*trans*-1, 4 configuration) the monomer units are arranged in a completely regular manner indicated as head to tail addition:

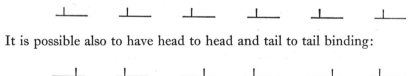

It is possible also to have head to head and tail to tail binding:

or a random arrangement giving a very irregular chain structure. In the head to tail arrangement complete regularity is obtained and this leads to crystallization. In the case of natural rubber (the *cis* isomer), the crystallization in the raw polymer occurs between about 12° and 28° C and in the vulcanizate only after stretching above 300 % elongation.

In gutta-percha (the *trans* isomer) the tendency to crystallize is so strong that the normal state of the polymer (at room temperature) is a hard horny crystalline material which melts at about 60° C and then obtains a degree of rubber-like elasticity which, however, is less than that of natural rubber. This is due to the stronger interchain interaction in gutta.

The *trans* form of an organic molecule usually has the higher melting point and this is borne out in the case of gutta-percha versus natural rubber. Chains resulting from 1, 2 or 3, 4 addition of isoprene do not have as good rubber-like properties as the *cis*-1, 4 addition product, due to the lesser mobility of the chain backbone and the bulkiness of the vinyl side groups. The chemical reactivity of the latter also makes such polymers subject to oxidation.

9.6.2 Stereo regularity, *d–l* isomerism, tactics and atactic polymers

In vinyl polymers a further type of isomerism is possible. In a polymer such as polystyrene the carbon atoms in the chain carrying the phenyl side group are asymmetric since the two attached ends of the chain are not equal. Therefore the spatial configuration of the four dissimilar groups around these C atoms can be in two different ways, leading to *d–l* isomerism as in the structure of sugars. The arrangement of asymmetric C atoms in the chain can be represented as

where * denotes the asymmetric carbon.

According to a nomenclature introduced by Natta[43] who did much of the pioneering work in this field, a polymer chain consisting entirely of asymmetric C atoms in an identical configuration:

$$d—d—d—d—d \quad \text{or} \quad l—l—l—l—l—l,$$

is called isotactic. If *d* and *l* alternate *d—l—d—l—d—l—d—l* it is syndiotactic and if there is just random placement the polymer is said to be atactic.

Stereo-specific polymerization, pioneered by Ziegler[44] and Natta,[45] has been the most important development in polymer chemistry of the last ten years and has given a much clearer insight into the effect of chemical configuration on mechanical properties.

The melting point of the isotactic polymers, which are crystalline, is much higher than the softening temperatures of the atactic species which are mostly amorphous. Isotactic polystyrene has a melting point of 240° C whereas the atactic polymer softens around 80° C without a sharp melting point. Solubilities also are in general much lower for the isotactic polymers. The regularity of the steric structure leads to a denser packing of the molecules into an ordered three-dimensional structure which crystallizes readily and, in general, to a tougher material.

9.7 Composition of different elastomers

9.7.1 Organic elastomers

Synthetic rubbers, manufactured by emulsion polymerization, have developed rapidly since World War 2, Since 1956, solution polymerization with metal-organic catalysts has proceeded to the point where stereo-regular polymers such as polybutadiene and polyisoprene (corresponding to natural rubber), can be made industrially, which even surpass natural rubber in resilience.

The nomenclature has undergone many changes. The best known rubbers arc

Polychloroprene (Neoprene) CR
Good oil and weather resistance
Specific gravity = 1.23

Ethylene sulphides ('Thiokol) (Perduren)
Excellent oil and weather resistance
Specific gravity = 1.2–1.31

Styrene–butadiene copolymer (SBR)
General-purpose rubber
Specific gravity = 0.97
This polymer is made by both
emulsion and solution polymeriza-
tion.

$$\frac{m}{n} = 0.1-0.9$$

Acrylonitrile–butadiene copolymers
NBR
Good oil resistance
Specific gravity = 0.98–1.04

$$\frac{m}{n} = 0.2-0.4$$

Polyisobutylene–isoprene copolymers
Butyl rubber IIR
Low gas permeability
Specific gravity = 0.92

$$\frac{m}{n} = 0.02-0.03$$

Cis-1, 4-*polybutadiene* BR
Good low-temperature resilience
Specific gravity = 0.92

Cis-1, 4-*polyisoprene* IR
Practically identical with natural rubber
All-purpose rubber
Specific gravity = 0.92

Chlorosulphonated polyethylene
Hypalon CSM
Excellent ozone resistance
Specific gravity = 1.10–1.28

Ethylene–propylene terpolymer
EPDM
The third monomer has an un-
saturated group in the side chain
which makes cross-linking possible
Good cracking resistance
Specific gravity = 0.86

$$p \ll m < n$$

Fluorinated hydrocarbon rubbers
Viton A FPM
High temperature and chemical resistance
Specific gravity = 1.82

Urethane rubbers AU and EU
Good strength properties
Specific gravity = 1.08–1.22

(R has a molecular weight of about 1000–5000)

Polymethyl siloxanes Si
Silicone rubbers for high temperature
application
Specific gravity = 0.98

$$\left[\begin{array}{ccc} CH_3 & CH_3 & CH_3 \\ Si-O-Si-O-Si-O \\ CH_3 & CH_3 & CH_3 \end{array} \right]_n$$

and fluoro derivatives: FSi

$$\left[\begin{array}{c} CH_3 \\ | \\ Si-O \\ | \\ CH_2 \\ CH_2 \\ CF_3 \end{array} \right]_m \left[\begin{array}{c} CH_3 \\ | \\ Si-O \\ | \\ CH_3 \end{array} \right]_n$$

$$\frac{m}{n} = 0.075$$

Polyepichlorohydrin and copolymers (Hydrin rubbers) ECO
Excellent oil resistance and low-temperature flexibility
Specific gravity = 1.27–1.36

$$\left[\begin{array}{c} H\ \ H \\ C-C-O \\ | \\ CH_2 \\ Cl \end{array} \right]_n \ \text{and} \ \left[\begin{array}{c} H\ \ H \\ C-C-O \\ |\ \ \ H \\ CH_2 \\ Cl \end{array} \right]_n \left[\begin{array}{c} H\ \ H \\ C-C-O \\ H\ \ H \end{array} \right]_m$$

9.7.2 Special high-temperature elastomers

In addition to the silicone rubbers, partly fluorinated silicone rubbers and fluorinated hydrocarbons (Viton type), some exotic polymers for use at extra high temperatures have been developed recently. These are:

Poly-m-carborene siloxanes of various types based on decaborene; one type can be represented as below:

$$\sim \left[\begin{array}{cccc} CH_3 & & CH_3 & CH_3 \\ | & & | & | \\ Si-C\equiv(B_{10}H_{10})\equiv C-Si-O-Si-O \\ | & & | & | \\ CH_3 & & CH_3 & CH_3 \end{array} \right]_n$$

The tensile strength is about 500 lbf/in² (36 kgf/cm²) at an elongation of about 150 %. These elastomers have been tested at about 310° C and retain 25–40 % of the tensile strength at that temperature.

Poly-2,4-perfluoroalkylene-perfluoroalkyltriazines. These elastomers have tensile strengths of 6–36 kgf/cm² at elongations of about 100 %. Rapid decomposition occurs at 420° C.

Future developments may be in the direction of ladder polymers which have a double parallel chain and many of which are very heat resistant, but no rubber-like types have yet been developed.

9.7.3 Inorganic elastomers

All the materials showing rubber-like properties which have been considered, so far, could be classified as organic since they all contain carbon either as part of the chain or in the side groups. There are, however, a number of inorganic polymers[46] some of which have elastic properties. Generally speaking, the mechanical strength of these elastomers is much less than that of the organic rubbers discussed in the previous chapter. Elemental sulphur which normally consists of rings of eight sulphur atoms shows a large increase in melt viscosity at 159° C, which points to ring opening and polymerization to high molecular weight chains. Maximum viscosity is found at about 190° C and if the hot melt is cooled rapidly a metastable dark brown material results which has elastic properties as well as viscous flow.

Another group of elastomeric organic materials are the so-called phosphonitriles (I), better named phosphonitrilic dichlorides, and their derivatives such as the fluoro elastomers (II).

The polymer (I), prepared by heating the cyclic trimer or tetramer, has properties similar to uncured natural rubber and has been suggested in the patent literature for limited special applications such as flame proofing textiles, brake lining bondings and oil additives. The polymer (II) is not entirely inorganic, but is resistant to almost all chemicals, decomposes above $300°$ C and has a T_g of $-77°$ C.

$$
\begin{array}{cccc}
CF_3 & CF_3 & C_3F_7 & C_3F_7 \\
| & | & | & | \\
CH_2 & CH_2 & CH_2 & CH_2 \\
| & | & | & | \\
O & O & O & O \\
| & | & | & | \\
\text{\textasciitilde}P{=}N{-}P{=}N{-}P{=}N{-}P{=}N\text{\textasciitilde} \\
| & | & | & | \\
O & O & O & O \\
| & | & | & | \\
CH_2 & CH_2 & CH_2 & CH_2 \\
| & | & | & | \\
C_3F_7 & CF_3 & CF_3 & C_3F_7
\end{array}
$$

$$
\left[-P{=}N{-} \begin{array}{c} Cl \\ \diagup \\ \diagdown \\ Cl \end{array} \right]_n
$$

(I)   (II)

9.7.4 Crystallizing and non-crystallizing rubbers

There is a considerable difference between crystallizing and non-crystallizing rubbers. The first, natural rubber, neoprene, butyl rubber are not strengthened nearly as much by fillers as are the non-crystallizing rubbers, SBR, NBR, EPDM, emulsion-polyisoprene and emulsion-polybutadiene. This effect is shown for tensile strength in Table 9.1.

TABLE 9.1. *Reinforcing effect of carbon black on various rubbers*

Rubber	Tensile strength (kgf/cm²)		Tensile reinforcing effect of carbon black
	Unloaded vulcanizate	Loaded vulcanizate	
First group (non-crystallizing)			
Polybutadiene (Na)	15–25	120–180	5 to 12 times
SBR	20–30	150–200	5 to 10 times
NBR	20–40	150–200	4 to 10 times
Polyisoprene	15–20	100–150	5 to 10 times
Polymethylbutadiene	20	80–120	4 to 6 times
Second group (crystallizing)			
Natural rubber	200–300	300–325	1 to 1.6 times
Neoprene	150–200	200–250	1 to 1.7 times
Polyisobutylene (Butyl rubber)	250–350	300–350	1 to 1.4 times

The reason for this effect can be easily understood. The emulsion-polymerized copolymers cannot crystallize because the components occur in random order in the molecular chain so there is no regular pattern in adjacent molecules. In these copolymers, such as styrene–butadiene, acrylonitrile–butadiene, ethylene–propylene, the effect of about 20 volume per cent of reinforcing fillers is an increase in tensile strength of more than 10 times. As explained in the section on reinforcement, the filler (carbon black) particles take over the function of crystallites which occur in the regular chain polymers such as natural rubber, polychloroprene, Butyl rubber, and the stereo-regular synthetic polyisoprene, polybutadiene and others.

In these rubbers the effect of the same 20 volume per cent of reinforcing filler is to increase the tensile strength by a factor of only 1.5 to 2.5, evidently because the crystallites that form on stretching are already fulfilling a function similar to that of a reinforcing filler.

The processing behaviour of synthetic rubbers is different from that of natural rubber in so far as they are chemically different. Natural rubber breaks down very rapidly under mastication on a roll mill or in an internal mixer, and by exposure to heat. This behaviour is due to reaction with atmospheric oxygen which can cause both cross-linking and scission of the macromolecular chains. In natural rubber the scission always predominates; in SBR scission takes place first under the mechanical stresses of milling, but simple heating will harden it so that here cross-linking is more prevalent. This is also true in the case of NBR.

Stereo-regular *cis*-1,4-polybutadiene hardly breaks down at all during milling below a temperature of 149° C. For this reason it is difficult to process and is often used in blends with natural rubber and SBR or oil extended SBR. However, the difference in viscosities between the components of the blend may cause unequal distribution of reinforcing filler in the blend; and the common curatives may cause different states of cross-linking. This again is reflected in the mechanical properties of the final vulcanizate.

9.7.5 Thermoplastic elastomers

Thermoplastic rubbers behave as elastomers up to temperatures of 60–70° C, but when heated over 110° C, they become thermoplastic polymers.

These rubbers have alternate hard glassy polymer chains and soft rubbery ones in regular sequence in the macromolecule. Polystyrene is an example of the hard glassy polymer A and polybutadiene represents a rubber-like B. The complete molecule can be represented as:

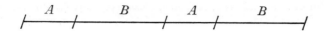

The hard segments A of adjacent chains will coil together to form hard regions similar to reinforcing filler particles or crystallites and acting as cross-links between the rubbery chains B. The net result is a rubber with a stiffness that depends on the relative amounts of A and B. When heated above the softening temperature of the A region the cross-links no longer hold and the polymer behaves as an unvulcanized rubber. The technical advantages and limitations of these materials are evident; no reinforcing fillers or vulcanizing ingredients are required, no vulcanizing step is necessary and the scrap and overflow can be remoulded. A vulcanizing rubber cannot be used more than once; although it is possible to re-use such rubber scrap, this requires an elaborate process of softening and chemical breakdown and the regenerated rubber has very inferior properties compared to the original material. In contrast, thermoelastics can be used over again to manufacture articles; scrap, overflow and remnants are simply heated again, fused together and remoulded into useful articles. Of course the finished articles cannot be used near or above the softening temperature of the hard regions; this constitutes a severe limitation in their usefulness.

References to chapter 9

1   G. V. Schulz, K. Altgelt & H. Canton, *Macromol. Chem.* **21**, 13 (1956). K. Altgelt & G. V. Schulz, *Macromol. Chem.* **36**, 209 (1960).
2   G. Ayrey, C. G. Moore & W. F. Watson, *J. Polymer Sci.* **19**, 1 (1956).
3   L. Bateman, *The chemistry and physics of rubber-like substances* (New York, 1963), chapter 15.
4   C. G. Moore & B. R. Trego, *J. Appl. Polymer Sci.* **8**, 1957 (1964).
5   W. Cooper, *J. Polymer Sci.* **28**, 195 (1958).
6   R. Shelton & E. T. McDonel, *Rubber Chem. Tech.* **33**, 342 (1960).
7   P. J. Flory, *Chem. Revs.* **35**, 51 (1944).
8   P. J. Flory, *J. Chem. Phys.* **18**, 108 (1950).
9   L. Mullins, *J. Appl. Polymer Sci.* **2**, 1 (1959).
10 C. G. Moore & W. F. Watson, *J. Polymer Sci.* **19**, 237, (1956).
11 L. D. Loan, *Techniques of polymer science*, Society of Chemical Industry Monograph No. 17 (London, 1962).
12 B. B. Boonstra, *India Rubb. World* **121**, 300 (1949).
13 J. M. Buist in *Applied Science of Rubber* (Ed. W. J. S. Naunton; London 1961), p. 709.
14 M. L. Williams, R. F. Landell & J. D. Ferry, *J. Amer. Chem. Soc.* **77**, 374 (1955).
15 Thor L. Smith, *J. Polymer Sci.* **32**, 99–113 (1958).
16 E. Guth, *Proc. 2nd Rub. Tech. Conf.*, p. 353 (London, 1948). *J. Appl. Phys.* **16**, 20 (1945).

17 W. R. Smith & R. A. Beebe, *Ind. Eng. Chem.* **41**, 1431 (1949). W. D. Schaeffer, M. H. Polley & W. R. Smith, *J. Phys. & Coll. Chem.* **54**, 277 (1950).

18 J. H. Atkins & G. L. Taylor *J. Phys. Chem.* **70**, 1678 (1966).

19 G. Kraus, *Rubber World* **35**, 67, 254 (1956). B. B. Boonstra, *Rubber Chem. Tech.* **38**, 943 (1965).

20 L. Mullins, *J. Rubber Res.* **16**, 275 (1947). *J. Phys. Coll. Chem.* **54**, 239 (1950).

21 R. Houwink, *J. Polymer Sci.* **4**, 763 (1949). *Rubber Chem. Tech.* **29**, 888 (1956). B. B. Boonstra, *Reinforcement of elastomers* (Ed. G. Kraus; New York, 1965) p. 559. E. M. Dannenberg & J. J. Brennan, *Rubber Chem. Tech.* **39**, 597 (1966).

22 E. M. Dannenberg, Lecture to the I.R.I. London (Oct. 1966). S. Ross & J. P. Oliver, *On physical absorption* (London, 1964).

23 W. Kuhn, *Naturw.* **26**, 661 (1938). *Z. Phys. Chem.* **B42**, (1939). *Helv. Chim. Acta* **30**, 487 (1947).

24 G. Haegel, *Helv. Chim. Acta* **30**, 487 (1947).

25 R. Houwink, *Elastomers and plastomers*, p. 258 (New York, 1950).

26 P. A. Thiessen & W. Wittstadt, *Z. Physil. Chem.* **B41**, 33 (1938).

27 A. V. Tobolsky *et al.*, *J. Appl. Physics* **15**, 380 (1944).

28 A. Schallamach, 'Abrasion and tyre wear', in ref 3.

29 J. D. Ferry, *Viscoelastic properties of polymers*, chapter 11, (Wiley & Sons, 1961).

30 F. Castiff & A. V. Tobolsky, *J. Polymer Sci.* **19**, 111 (1956).

31 T. C. Einhorn & T. B. Turezky, *J. Appl. Polymer Sci.* **8**, 1257 (1963).

32 P. P. A. Smit & A. K. van der Vegt, International Rubber Conference, Brighton, 1967.

33 F. Bueche, *Physical properties of polymers* (Interscience, 1962).

34 S. Middleman & J. Gavis, *Phys. Fluids* **4**, 355, 963 (1961). R. S. Spencer & R. E. Dillon, *J. Colloid Sci.* **3**, 163 (1948); E. B. Bagley *et al.*, *J. Appl. Polymer Sci.* **7**, 1661 (1963).

35 S. Newman & Q. A. Trementozzi, *J. Appl. Polymer Sci.* **9**, 3071 (1965).

36 G. R. Cotton, *Rubber Age* **100**, 51 (1968).

37 B. S. Garrey, M. H. Whitlock & J. A. Freese, *Ind. Eng. Chem.* **34**, 1309 (1942).

38 A. I. Medalia, *J. Colloid Interface Sci.* **24**, 393 (1967). A. I. Medalia & F. A. Heckman, *Carbon* **7**, 567 (1969). A. I. Medalia & F. A. Heckman, *J. I.R.I.* **3** (2), 66 (1969).

39 J. J. Brennan, T. E. Jermyn & B. B. Boonstra, *J. Appl. Polymer Sci.* **8**, 2687 (1967).

40 See for a comprehensive treatment: L. R. G. Treloar, *The physics of rubber elasticity* (Oxford, 1949).

41 K. H. Meyer & C. Ferri, *Helv. Chim. Acta* **18**, 570 (1935).

42 L. R. G. Treloar, *Trans. Far. Soc.* **42**, 77, 83 (1946).

43 G. Natta, *Angew. Chem.* **68**, 393 (1956). *Chem. Ind.* 520 (1957); *J. Polymer Sci.* **16**, 143 (1955).

44 K. Ziegler *et al.*, *Angew. Chem.* **67**, 541 (1955). See C. E. H. Bawn & A. Ladwith, *Quarterly Rev.* **16**, 361 (1962).

45 G. Natta & P. Corradini, *J. Polymer Sci.* **39**, 29 (1959).

46 *Inorganic polymers*, International Symposium at Nottingham 1961 (Chemical Society, Special Publication No. 15, London, 1961).

10 Fibres

R. H. Peters

10.1 Introduction

Fibres are characterized by their shape and dimensions and are defined as 'a unit of matter characterized by flexibility, fineness and a high ratio of length to thickness'. They are in general composed of linear polymer molecules which are, with one or two exceptions (e.g. glass), oriented preferentially in the direction of the fibre axis; such an alignment of chains allows a considerable measure of contact, conditions in which interchain forces of the non-polar or polar type may operate over considerable lengths of the molecules, thereby developing adequate strength (see Figs. 2.6, 2.7 and 2.8). To achieve adequate contact between chains and avoid slippage under load, the fibres must be long (i.e. of high molecular weight); in fact no fibres of technical merit have been made from polymers of molecular weight less than 4000–5000. Below this value, a weak brittle product is formed; with increase in molecular weight, the mechanical properties improve though with diminishing effect as the size increases. For synthetic fibres, a limit is set by the ease of fabrication which can be impeded by the large viscosities encountered in spinning using compounds of high molecular weight.

To develop the interchain forces and hence achieve maximum strength, the molecules or segments of the chain must be able to fit alongside each other, that is, possess elements of geometric periodicity and shape symmetry. However, due to the thermal movement of the segments, this alignment can be disturbed to an extent dependent on the flexibility of the chains. Chains containing groups such as —O—, —NR—, etc. are very flexible whereas rigid units such as

$$\text{—} \bigcirc \text{—} \quad , \quad \text{—HC} \underset{CH_2-CH_2}{\overset{CH_2-CH_2}{\diagdown}} CH\text{—} , \text{ etc.,}$$

restrict segmental movement. Restricted segmental movement will result in an improved resistance to disruption; this property is most easily shown by the melting point. For example, polyethylene terephthalate with a rigid unit has a much higher melting point (264° C) than the corresponding aliphatic compound polyethylene adipate (56° C) which has a flexible chain. From the practical point of view, increase in stiffness is an advantage provided it is not

taken to excess; thus the very stiff ring system $-\left[-\left\langle\bigcirc\right\rangle-\right]_n$ is too rigid

for fibre formation. In general, stiff parts of the molecules must be connected through more flexible units which can allow the chain to be oriented.

10.2   Morphology of fibres

In certain regions, the packing of the chains in many polymers is so regular that the arrangement assumes a crystalline character. This is clearly shown by X-ray pictures which also indicate that other regions present are amorphous in character. These observations have led to the picture of the structure as one in which crystalline regions are embedded in an amorphous matrix; moreover the size of the former is smaller than the length of the molecule, which may extend through several crystalline regions thereby holding the structure together.

The detailed shape of the crystalline regions is open to some controversy. Thus disintegration of fibres (e.g. polyamides, cellulose, etc.) with acids, ultrasonic vibration, etc., yields crystalline residues which are fibrillar in character. These observations have led to the picture given in Fig. 10.1a in which the crystalline regions are fibrils joined together by the chain molecules which run through more than one fibril.

However, studies of molten polymers suggest that when the chains crystallize slowly from solution, they do so in a folded pattern to yield a plate-like or lozenge structure. On cooling a molten polymer, crystallization appears to originate from a centre or nucleus from which radiate a collection of fibrils; in the latter the chains crystallize in the plate-like structures which are themselves oriented at right angles to the direction of the fibril. This suggests that the fibrils are made up of a set of these lamella structures stacked one upon the other. The structure is held together in a manner similar to that indicated for the fibrillar model with any one chain being incorporated in the folds of more than one crystalline region. This case is indicated in Fig. 10.1b.

In assessing the detailed structure, cognizance must be taken of the origin of the fibre. In natural fibres such as wool and cotton, the arrangement of the molecules is a consequence of the growth process whereas in a synthetic fibre the melt is quickly cooled on extrusion to form the fibre. It is therefore quite possible for some fibres to have a detailed structure of the form indicated in Fig. 10.1 *a* while others will have folded chains.

(*a*)   (*b*)

Fig. 10.1. (*a*) Fringed fibril structure. (*b*) Modified fringed micelle structure (from Hearle, 1967).

10.3 Importance of crystallinity

Crystallinity in a fibre is an advantage, since in general crystalline materials show better mechanical properties and better retention of useful properties with increasing temperature compared with amorphous materials. However, although crystallinity contributes to the mechanical properties, it is not indispensible; in fact synthetic fibres can be divided into two classes, the relatively highly crystalline, such as nylon, polyester, etc. and the more or less amorphous ones such as acrylic fibres. It must be added that with the latter some alignment of the chains will occur by virtue of the strongly polar character of the side chains (CN) even though the chains can not sterically approach sufficiently close to form a crystalline arrangement.

10.4 Chemical constitution

Fibres are derived from chemical, vegetable and animal sources.

10.4.1   Chemical

(a) *Hydrocarbon fibres*

Many hydrocarbons have been produced for fibres but the only important commercial ones are polyethylene and isotactic polypropylene.

Depending on the method of manufacture, polyethylene fibres are of two kinds, one of low and the other of high density, depending on the number and size of alkyl side groups which the chain carries. The high density product is mainly linear and has a high melting point.

As a fibre, polypropylene is favoured compared with polyethylene because of its higher softening point ($155°$ C compared with $110°$ C for high density and $90°$ C for low density polyethylene).

(b) *Hydrocarbon polymers containing substituents*

Fibres of commerical value are usually copolymers because of deficiencies in the homopolymers. Thus Saran or Tygan is a copolymer of vinyl and vinylidine chloride; polyvinyl chloride, although used to a small extent as a fibre, suffers from a low softening point, whereas polyvinylidine chloride is a horny material which is not easily transformed into fibres. The copolymer has the advantage that the softening point is higher than that of polyvinyl chloride but at the same time low enough for fibres to be melt spun without decomposition.

The most important fibres of this type are derived from acrylonitrile and various monomers such as methyl acrylate, itaconic acid, etc.; they have been classified according to the quantity of acrylonitrile units as acrylic (at least 85%) or modacrylic (35–85%). As a group these fibres are atactic and amorphous or at best in a quasi-crystalline state. However, the interaction energy of the CN dipoles is of the order of 8.5 kcal/mol, a value which exceeds that of hydrogen bond formation in polyamides and in polymers containing hydroxyl groups. The detailed constitutions of the acrylic fibres have not been completely disclosed; only a few are known: e.g. Dynel has 40/60 acrylonitrile vinyl chloride, Verel and Teklan have 60/40 acrylonitrile/vinylidine chloride.

One other fibre may be mentioned: namely the hydroxyl-containing polyvinyl alcohol, a product of interest because of its ability to absorb water. The polymer carries some side chains, but although atactic, is able to crystallize because of the small size of the hydroxyl groups. Fibres made from polyvinyl alcohol vary in their properties, notably water sensitivity, according to their chemical history as well as the conditions of manufacture. Usually, to ensure that the fibres are insoluble in water they are cross-linked.

(c) Polyesters

Fibre-forming polyesters are of two main types, namely those of formula

$$-[\!-O.R.CO-\!]_n$$ which are derived actually or notionally from the hydroxy-carboxylic acid, $HO.R.COOH$ and those of formula

$$-O.R.O.COR'.CO-_n$$

obtained from the diol $HO.R.OH$ and the dicarboxylic acid

$$HOOCR'.COOH.$$

However, although many suitable polyesters are possible, only two have any commercial significance as fibres. The more important, polyethylene terephthalate, has $R = CH_2CH_2$ in the diol and $R' = \langle\bigcirc\rangle$ in the dicarboxylic acid; the other is Kodel in which

$$R = CH_2\!-\!\!CH \overset{\textstyle CH_2-CH_2}{\underset{\textstyle CH_2-CH_2}{\diagup\diagdown}} CH\!-\!CH_2\!-\!- \quad \text{and} \quad R' = \langle\bigcirc\rangle$$

(d) Polyamides and related materials

These are derived actually or notionally from (a) amino acids of the form $NH_2(CH_2)_n COOH$ or (b) from diacids of formula $HOOC(CH_2)_m COOH$ and an amine of the type $NH_2(CH_2)_n NH_2$. They give repeat units of

$$-[NHR.CO]- \quad \text{or} \quad -[NHRNHCOR'CO]-.$$

Two important fibres are made commercially, namely Nylon 6 with repeat unit $-[NH(CH_2)_5CO]-$ and Nylon 66 with repeat unit

$$-[NH(CH_2)_6NH.CO(CH_2)_4CO]-.$$

Other examples of this type of substance are the synthetic polypeptides of formula

$$\overset{\textstyle R}{-NH.CH.CO-}$$

of which

poly (γ-methyl-L-glutamate) ($R = HOOC.CH.CH_2-$)
$$CH_3$$

may be mentioned since fibres have been prepared from it.

The melting points of these polymers are quit high (Nylon 6, 215° C and Nylon 66, 264° C) because of the high energies of interaction operative in the formation of hydrogen bonds between amide groups.

(e) *Carbon fibres*

Whiskers, that is thread-like mono-crystals, can be grown from certain metals, inorganic compounds and carbon. These substances are valued because their very high strength in the longitudinal direction makes them valuable materials for reinforcement. Carbon fibres are of particular interest since these are the strongest filaments available, being stronger even than glass. They are prepared by heating polyacrylonitrile fibres which initially blacken but when heated to temperatures up to 3000° C graphitize to give a fibre composed entirely of carbon.

(f) *Elastomeric fibres*

Recoverable extensibilities of conventional fibres are limited to only a few per cent of their original lengths. Textile requirements for elastic-like yarns used to be met by the use of composite threads made of rubber cores with spirally wound coverings of normal fibres that served to control the stretch and preserve the core from abrasion.

Nowadays, synthetic fibres with reversible extensions of several times their original lengths are available. Early examples include random copolymers of ethylene terephthalate with large proportions of aliphatic groups such as azelate or sebacate and the partly substituted nylons. In the former, the introduction of the second acid of formula $HOOC(CH_2)_nCOOH$ reduces, in a large measure, the ability to crystallize, but more important perhaps replaces the rigid phenyl group by a flexible aliphatic methylene chain, thereby giving a polymer whose molecules, in part at least, have the possibility of assuming the coiled configurations found in rubbers. In the latter, substitution of the amide groups by a bulky group such as methyl separates the chains and removes the possibility of hydrogen bonding. Thus where substitution has occurred, the larger interchain forces are reduced as well as the crystallinity, allowing the possibility of chain coiling. The effects are accentuated as the degree of substitution increases, when the melting point drops, the modulus decreases and the ease of stretching increases (Fig. 10.2). Ultimately the material becomes liquid.

Unfortunately, such materials are technically unsatisfactory since they have low melting points, poor solvent resistance and marked stress decay. A more satisfactory product is formed by constructing a polymer from sections or blocks of rubber-like and fibre-like polymers. Thus the introduction of large aliphatic polyoxyalkalene units into polyethylene terephthalate (up to 60 % by weight) produces fibres with long range elasticity combined with low stress decay and higher melting points.

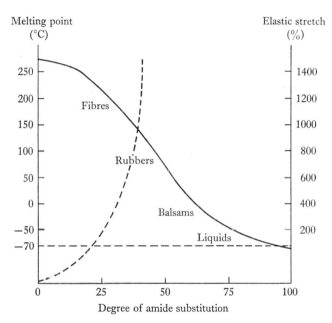

Fig. 10.2. Effect of progressive reduction in hydrogen bonding on the properties of Nylon 66.

   The optimum properties were obtained with a molecular weight of *ca.* 4000 in the polyether block. These products have minimum disruption of the crystalline regions and hence a minimum depression of the melting point of the parent polyester. It may be noted in this context that the change in melting point is proportional to the molar and not the weight fraction of the second species, which is present so that a large amount by weight of the rubber-like block may be introduced without a significant drop in melting point.

H E M

Unfortunately, for various technical reasons not connected with the mechanical properties, these particular fibres are not commercially satisfactory; nevertheless their molecular construction is the basis of commercial elastomeric yarns which essentially consist of 'hard' and 'soft' sections arranged alternately.

(g) Bulked and texture yarns

With the object of imparting bulk and a measure of springiness or recoverable elasticity, fibres are crimped. Crimp is found in natural fibres such as wool and may be produced in synthetic filaments by the action of heat and mechanical forces. For example, in the Agilon process, the fibres are drawn under tension around a heated knife which compresses them at the point of contact and stretches them at the opposite side of the cross-section. On rapid cooling, the shape imparted to the fibre is set-in giving an undulating or sawtooth configuration. There are many processes of this type used to produce bulky stretch yarns.

An alternative technique to obtain crimp is to form a composite filament; to do this, polymers of different characteristics are extruded side by side so that they coalesce longitudinally to give a fibre rather like a bimetallic strip. The different properties of the two components cause the development of crimp. Thus, Orlon (Suyelle) is an acrylic bicomponent filament or heterophile whose components differ in hydrophilic character so that the composite acquires a crimp when dry but straightens out when wet. With Cantrece bifilaments, Nylon 66 is conjoined with Nylon 6, the different behaviour of the two components yielding a crimped fibre.

10.4.2   Vegetable fibres

Two main sources of textile fibres are seed hairs (e.g. cotton) and the stems of plants – the bast fibres (e.g. flax). To these must be added a third source, namely wood, which yields fibres which are short in length and suitable for making paper.

Chemically the important fibrous component of these materials is cellulose, a polymer formed in nature by the joining of anhydroglucose units through oxygen bridges across the 1, 4 positions (formula (1)).

In this polymer, it is important that the anhydroglucose units are in the β configuration since such an arrangement yields a linear chain. When the anhydroglucose units adopt the α configuration, the direction of the bonds is

Cellulose $n \geqslant 1500$

I

such that the molecule adopts a helical shape. This shape is found in starch. In formula (1), the glucose units are arranged in the *trans* configuration, an arrangement expected in the crystalline regions. However, rotation around the oxygen atoms joining the units together is possible and is likely to occur in the amorphous regions.

The cellulose in these materials is associated with various impurities from which the fibre is isolated. Thus with cotton, the wax and other impurities may be removed by solvents followed by treatment in water or by a more vigorous treatment with caustic soda solution. Flax is treated more gently, the extent of removal of the non-cellulosic impurities (e.g. lignin) being determined by the end use. Bast fibres such as flax are retted, that is, treated under moist conditions, so that bacterial action breaks down the stem into fibre 'bundles'. Subsequent treatments with mild alkali and bleaching agents are then used to obtain the desired removal of the impurities. Wood, on the other hand, is vigorously cooked with sodium sulphide or sulphite solution in order to remove the very large quantities of impurities which hold the fibres together.

The length of the fibre depends on the source. Cotton is of length $\frac{1}{2}$–$2\frac{1}{2}$ in according to variety. Flax gives fibres which are 'ultimates' of length 1.0–1.25 in joined together by residual intercellular cements (lignin, hemi-cellulose, etc.) giving a fibre 'bundle' of length from 6 to 40 in. With wood, the fibre length ranges up to 3–4 mm.

The most important textile fibre is cotton. This has the appearance of a long, irregular, twisted, flattened tube tapering at its tip (Plate 10.1). The outside layer of the fibre is composed of a thin cuticle – the primary cell wall; running along the centre is the lumen, a narrow tube, which supplies the biological substances necessary for growth but which collapses when the seed hair dries out. The cross-section of the fibre is shown in Fig. 10.3. The bulk of the fibre, the secondary wall, lies between the cuticle and the lumen and is deposited on the inside of the cuticle in a series of growth rings. Each layer which is laid down is made up of a large number of fibrils which spiral around the fibre axis at an angle of 30° but whose sense reverses in coincidence with the external convolutions.

10.4.3 Animal fibres

Of the many animal fibres available, wool and silk are the most important. These are proteins composed of units of amino acids in the L-form; eighteen different individual acids have been isolated by hydrolysis in acid solution. These units are joined together through amide (or peptide) linkages yielding a linear polymer

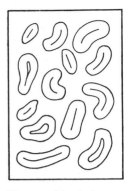

Fig. 10.3. Typical cross-sections of mature cotton fibres.

$$-\mathrm{CH.CO.NH.CH.CO.NH.CH.CO.NH-.}$$
$$\underset{R_1}{\qquad}\underset{R_2}{\qquad}\underset{R_3}{\qquad}$$

The side groups R_1, R_2, etc. vary in size (e.g. H, CH_3, benzyl, etc.) and in chemical character from non-polar to acidic or basic (e.g. CH_2COOH or $(CH_2)_4NH_2$). Two of the acids differ in structure from the rest and merit particular note. Cystine is a diacid (II) in which two amino acids are joined together by a disulphide bridge. This acid may be incorporated in two adjacent chains thereby forming a cross-link which can contribute to the mechanical properties and render the wool insoluble in solvents which do not break the disulphide bond.

$$\begin{array}{cc}
\underset{\mathrm{NH_2}}{\overset{\textstyle\mathrm{COOH}}{\mathrm{CH}}}-\mathrm{CH_2}-\mathrm{S}-\mathrm{S}-\mathrm{CH_2}-\underset{\mathrm{NH_2}}{\overset{\textstyle\mathrm{COOH}}{\mathrm{CH}}} & \begin{array}{c}\mathrm{H}\\ \mathrm{N}\\ \mathrm{H_2C}\qquad\mathrm{CHCOOH}\\ \mathrm{H_2C}-\mathrm{CH_2}\end{array}\\
(\textsc{ii}) & (\textsc{iii})
\end{array}$$

The other acid is proline (III). When this is incorporated into the chain, the angles at which the chemical bonds are disposed because of the rigid nature of the ring, cause the chain to change direction.

Wool is of particular interest, since to a considerable extent, the polypeptide chain forms a helix with the substituents protruding and the adjacent coils of the helix held together by the hydrogen bonds between the amide groups. This configuration is normally referred to as the α-form.

When viewed under the microscope, the wool fibre is seen to be covered with overlapping scales, the cuticle. Through the centre runs a canal, the medulla, which in many fibres is very small. The bulk of the fibre, the cortex, is made up of cells rather spindle-like in shape. These cells are of two kinds,

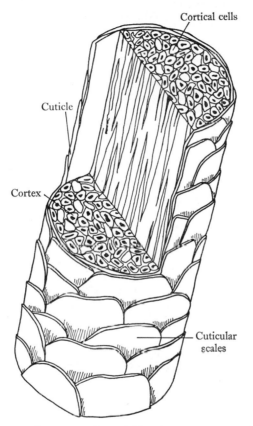

Fig. 10.4. Structure of wool fibre (para-cortex shaded).

ortho and para; in the important merino wool, these cells are arranged bilaterally rather like a heterophile filament (Fig. 10.4).

This structure is probably responsible for the crimp which exists in wool fibres since the less swellable form, the para-cortex, is always found on the inside of the curvature so that the two structures twist around one another.

Silk is morphologically more simple than wool, being essentially a filament formed by extrusion from the gland of the silk worm. By suitable degradation of the silk protein (fibroin) with the help of enzymes, a crystalline portion has been isolated which contains mainly amino acids carrying small substituents, namely the amino acids glycine (R=H), alanine (R=CH$_3$) and serine (R=CH$_2$OH). This portion of the fibroin molecule being different from the remainder suggests that the silk molecule may be regarded as analogous to a block polymer made up of two segments, one of which is crystalline and the other not, since the latter contains the bulky side groups (e.g. tyrosine,

R=CH$_2$⟨◯⟩OH). The arrangement in the crystalline regions in silk is

different from that of wool, the chains being in an extended β-form giving a structure which may be described as pleated sheets.

10.4.4   Mineral

Glass fibres need only be mentioned briefly (§11.1). Glass is an inorganic product of fusion which has cooled to a rigid condition without crystallizing. From the fibre point of view, it is notable because it is a three-dimensional random network, and hence, is isotropic.

TABLE 10.1. *Typical compositions used in glass-fibre manufacture*

Glass type	Composition (wt %)					
	SiO$_2$	Na$_2$O	B$_2$O$_3$	Al$_2$O$_3$	MgO	CaO
Sheet glass	72.5	13.5	—	1.5	4.0	8.5
E glass	54.6	0.6	8.0	14.8	4.5	17.4
		(Na$_2$O + K$_2$O)		(Al$_2$O$_3$ + Fe$_2$O$_3$)		
19 XE glass	65.0	14.5	—	2.0	7.5	11.0

The composition of some glasses is shown in Table 10.1. Of these, sheet glass is ordinary window glass and under the name of alkali glass is used for making continuous filaments. The composition is chosen to give satisfactory melting, good weathering properties and freedom from crystallization during fabrication. E glass is also used for continuous filaments; it has good electrical insulation properties and a high degree of resistance to attack by moisture. The alkali content is kept low and a proportion of B$_2$O$_3$ is incorporated to make the glass easier to melt. 19 XE glass is used for the manufacture of glass wool. The requirements for this glass are less stringent than the E type and the soda, lime and magnesium contents are comparatively high for ease of melting.

10.5 Spinning

10.5.1 Man-made fibres

The conversion of a polymer into a filament suitable for textile uses may be carried out in three ways: namely, wet, dry and melt spinning.

(a) *Wet spinning*

Regenerated cellulose filaments are made by treatment of highly purified wood pulp with caustic soda solution followed by carbon bisulphide. Chemical reaction occurs and some of the hydroxyl groups are converted to the xanthate.

$$\text{Cellulose} . \text{O} . \text{Na} + \text{CS}_2 \rightarrow \text{S}{=}\text{C}\begin{smallmatrix} \diagup \text{S} . \text{Na} \\ \diagdown \text{O} . \text{Cellulose} \end{smallmatrix}$$

The cellulose xanthate so formed dissolves to give a viscous liquid – viscose. During this process, a controlled but considerable amount of degradation takes place, the degree of polymerization (i.e. number of anhydroglucose units in the polymer chain) being reduced to a value between 250 and 600. In principle for filament formation, this liquid is pumped at a constant rate through spinnerets whose orifice diameters are in the range 0.05–1 mm. The spinneret is submerged in an acid bath containing salts (e.g. sodium and zinc sulphates); the filaments on emerging into this bath coagulate and the cellulose is regenerated from decomposition of the xanthate by the acid.

The structure of the final filament is dependent on the conditions of regeneration. In the bath, the salts cause coagulation whereas the acid causes decomposition. These processes do not occur at the same rate. By changing the temperature, time of immersion, amount of electrolyte, etc., the relative rates at which these two processes occur can be changed and so control the time which the polymer has to form its structure. In addition, as the fibre is pulled through the bath, the filaments are stretched, thus orienting the molecules and increasing the strength of the fibre.

A result of this interplay of variables is that more than one structural arrangement is possible in the final filament. In the production of the fibre known as 'regular' viscose, the xanthate solution, on first meeting the bath, forms a skin through which reagents diffuse; as regeneration of the inside or core occurs, there is a substantial loss of volume and the filament shrinks, collapsing the skin so that the cross-section of the resulting fibre is serrated or crenulated. Clearly, regeneration of the skin and core take place at different rates and the tensile forces on the two portions differ. Different physical structures result; in the skin the molecules are more oriented, contain smaller crystalline regions and have superior mechanical properties as compared with the 'core'.

This last point has led to the development of fibres which contain more 'skin'. Essentially, to achieve this, the rate of coagulation is slowed down by modifying the bath, in particular by the addition of retardants (e.g. aliphatic amines), so that the period in which the filament is held in a plastic state is extended; this gives a greater period of time for the structure to develop and enables a greater degree of stretch to be imparted. Concomitant with the increase in the ratio of 'skin' to 'core', the cross-section of the filament changes from crenulate to circular and the strength of the filaments is much improved. Finally, in addition to the advantages obtained by process modification, improved products ('polynosics') have more recently been manufactured where the cellulose xanthate used has a higher degree of polymerization than usual (500–700).

(b) Dry spinning

In this process, the polymer dissolved in a suitable solvent is extruded into fairly long heated tubes in which the solvent evaporates; this is accompanied by a reduction in volume of the extruded filament by 15–35 % (Fig. 10.5).

The conditions under which the filament is extruded determine the shape of the cross-section. Thus if the rate of evaporation from the surface is less than the diffusion rate of the solvent from the centre of the filament, the solvent concentration will be the same across the filament and a circular cross-section will result. On the other hand, if the converse is true, a surface skin is formed and the filament is more like a rubber pipe containing liquid. In this situation, solvent diffuses out through the skin and the volume of polymer solution decreases causing the cross-section to distort and become flat or 'dog-boned'.

Two of the most important fibres which are dry-spun are cellulose triacetate and cellulose secondary acetate. The former is made by treatment of cotton linters or highly purified wood pulp with acetic anhydride until the three hydroxyl groups in each anhydroglucose unit are acetylated. The derivative so formed is spun from solution in methylene chloride. Secondary cellulose acetate fibres are spun from acetone: this polymer is formed by partial hydrolysis of the cellulose triacetate to a degree of substitution (i.e. fraction of the hydroxyl groups of the anhydroglucose units which are acetylated) of *ca.* 2.3.

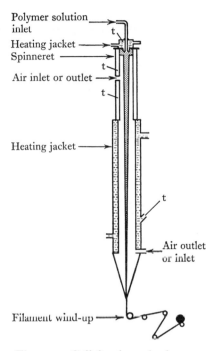

Fig. 10.5. Cell for dry spinning.

(c) *Melt spinning*

The molten polymer is fed to the spinnerets at a constant rate by a metering pump. The liquid streams emerge vertically downwards, rapid cooling occurs and the filaments solidify. The production details of the different fibres vary from polymer to polymer but in general the diameters of the filaments depend on the rate at which the polymer passes through the orifice, the nature of the polymer and the wind-up velocity, rather than on the diameter of the spinneret itself. The filaments, as extruded, have high extensibility but low strength and are stretched or drawn in a second process.

10.5.2   Glass fibres

On melting, glass changes from an elastic solid to a free flowing liquid. For sheet glass the viscosities must lie in the range 100–1000 poise for fibre formation. In order to form fibres, the molten glass is extruded through an orifice, when it cools rapidly and is drawn down and wound on a drum. Speeds of formation may be as high as several thousand feet per minute and the final diameter of the attenuated filament a few thousandths of an inch. The filament is attenttuated on leaving the orifice and reaches a final diameter a few centimetres below. Because the heat transfer between the centre and the surface of the fibre is quite good, there is no appreciable thermal gradient and hence no permanent mechanical stress in the fibres.

10.6   Mechanism of spinning

In order to obtain satisfactory spinning, it is important that the magnitude of the viscosity be within a certain range, since if this is too small droplets form, and if too high the large tensile force required for wind-up will result in cohesive failure of the filament. The situation however, is, complicated by the fact that polymer solutions or melts often exhibit structural viscosity, the apparent viscosity decreasing with shear rate so that the output exceeds that calculated by Poisseulle's equation to an extent determined by the shear (Fig. 10.6). On emerging from the spinneret some of the shear stresses are released and the visco-elastic polymer stream frequently swells. The extent to which this die swell occurs is modified by the tensile force applied from the take-up device and varies from polymer to polymer. Thus the die swell decreases with winding speed for melt-spun polystyrene but remains nearly constant for polyamides and polyesters.

As the jet emerges from the spinneret it is smooth, but can become rough and irregular and even break up into individual fragments if the shear rate exceeds a critical level. Although the mechanism of this melt fracture is not known, the phenomenon may be regarded as an elastic instability.

When the fluid emerges from the orifice and passes through the transition region in which the die swell effect is observable, the jet is subjected to a tensile stress causing extension. In the upper zone of the spinning path (50–150 cm), the diameter of the filament decreases and the major part of the fibre structure is formed, molecular orientation and solidification taking place. The changes in velocity of the polymer along the spinning path are shown in Fig. 10.7 where it may be seen that the velocity of the filament changes in a

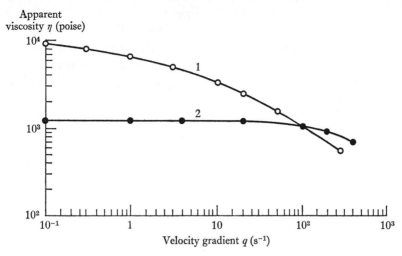

Fig. 10.6. Velocity-gradient dependent viscosity of fibre-forming melts. (1) Polypropylene; (2) polyethylene terephthalate.

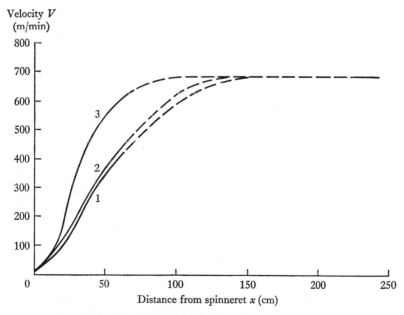

Fig. 10.7. Velocity distributions along the spinning path for melt-spun fibres. (1) Polycapronamide; (2) polyethylene terephthalate; and (3) polystyrene.

sigmoid fashion reaching a limiting value at some distance from the spinneret. As the jet is subjected to axial tension, orientation increases and to an extent which varies with the distance from the spinneret (Fig. 10.8).

Concomitant with the orientation of the polymer molecules is the solidification process. In most processes of fibre formation, a phase transition occurs. Many polymers crystallize on cooling from the melt, and the degree to

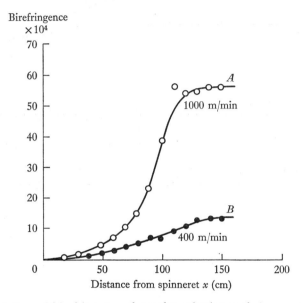

Fig. 10.8. Variation of birefringence along the spinning path in running melt-spun polyester fibre (ρQ) = const. = 2.6 g/min. (A) V_L = 1000 m/min; (B) V_L = 400 m/min. Courtesy of Dr Shiro Nishiumi, Toyo Rayon Co. Ltd.

which the 'as spun' fibre is crystalline varies very markedly with the nature of the polymer. Thus a slowly crystallizing material such as polyethylene terephthalate is amorphous, whereas melt-spun polyamides and polypropylene are crystalline. Those wet-spun polymers which are slowly precipitated (cellulose, polyvinyl alcohol) almost reach an equilibrium condition. With melt-spun fibres, the T_g is important since molecular mobility is only possible in the temperature range T_g to T_m. In the conditions of spinning, the temperature of the filament may pass through T_g before crystallization has got under way. If this is so, the molecular mobility ceases and a glassy fairly stable fibre results. A common example of this is polyethylene terephthalate with a T_g of 67° C. In other instances, if T_g is below room temperature and crystallization is incomplete in the time available for filament formation, further crystallization can occur and the properties of the fibre may change. An example of this is polyvinylidine chloride with a T_g of $-17°$ C. Analogous complications arise in dry and wet spinning.

This short discussion is sufficient to indicate that there are many variables to be controlled in fibre spinning and that the products themselves are usually metastable since equilibrium cannot be reached in the time available.

10.7 Drawing and stretching

In all fibre spinning, the filament is attentuated and stretched in order to achieve a high degree of orientation in the final product. With viscose rayon, stretching may be applied at the same time as regeneration; with other fibres, the stretch is applied in a second process. In general, the polymer structure may be modified when its temperature is at least equal to the glass transition temperature of the polymer. Some transition temperatures are quoted in Table 10.2.

TABLE 10.2. *Glass transition temperatures of textile polymers*

Polymer	Temperature (° C)
Cellulose acetate	$>$ 180
PVC	70–80
Polyamides	-30 to $+45$
Polyesters	60 to 80
Polyacrylonitrile	80 to 100

In stretching dry-spun fibres, which are often amorphous in character (c.g. acrylics), the elongation of the stretched fibre decreases but is accompanied by an increase in strength (Fig. 10.9). With melt-spun fibres, the filaments are relatively weak in the 'as spun' condition. A typical load extension curve is given in Fig. 10.10 where it may be seen that as the stress is increased, a point is reached at which the fibre undergoes considerable extension with little change in stress: as elongation is increased, another stage is reached where the stress rises sharply and the fibre is fully 'drawn'. During this large irreversible extension the crystalline structure is modified, or if the filament is amorphous crystallization may be induced; however, probably what is more important, the molecules become oriented along the fibre axis thereby imparting greater strength to the fibre. In practice, the drawing is carried out continuously, the filament passing from one roller to another revolving at a higher speed.

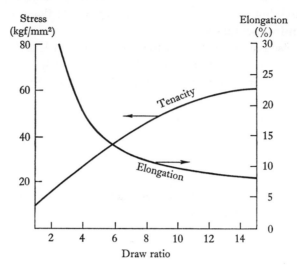

Fig. 10.9. Effect of draw ratio on tenacity and elongation of a typical fibre.

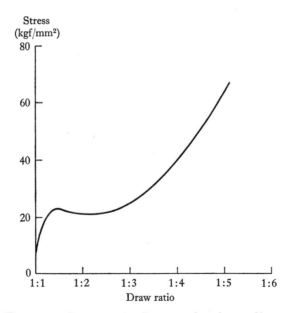

Fig. 10.10. Stress–strain diagram of undrawn fibre.

The extent to which a fibre may be drawn is dependent on the structure of the spun filament. Thus if more orientation is introduced into the threadline, the draw ratio is decreased (Fig. 10.11). Increased orientation improves the tensile strength but decreases elongation at the breaking point of the undrawn fibres.

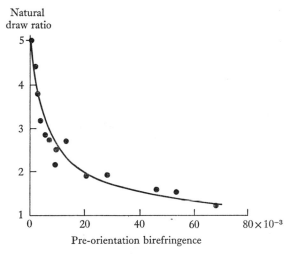

Fig. 10.11. The variation of the natural draw ratio with degree of pre-orientation.

10.8 Mechanical properties

The desirable properties of textile materials derive from both the intrinsic physical properties and the manner in which the fibres and yarns are put together. For example in a non-woven fabric, a web of fibres is stuck together with an adhesive so that the material is rather coherent, and if the adhesive is strong, shows the high initial modulus expected of the fibre. This situation may be contrasted with that in knitted or woven fabrics; when a load is applied, the structure deforms very easily but after the initial deformation, it tightens up and becomes strong. This behaviour means that the fabric has a very low initial modulus which under strain becomes very high. However, although the properties rely markedly on the construction of the fabric, the mechanical properties of the fibres themselves determine not only the strength and recovery from deformation but play an important role in processing: thus in opening a mass of tangled fibres prior to spinning, fibre breakage is of importance; in winding, permanent straining must be considered and in weaving, relaxation effects must be taken into account. Special care must be observed in spinning and weaving glass fibres because of their brittleness and inflexibility. In this section, the mechanical properties of fibres are described rather than those of composites, for example ropes, fabrics, etc.

There are several aspects of the mechanical properties of fibres which should be discussed. Of these, the stress–strain relationships, the change of strain under constant stress or stress at constant strain are the most important.

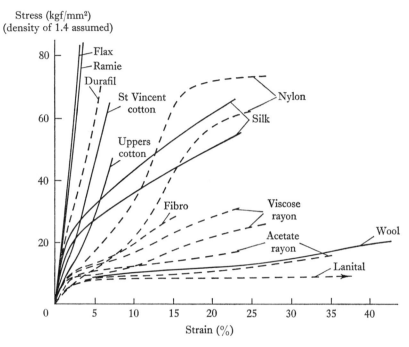

Fig. 10.12. Stress–strain curves of various fibres tested at 65 % relative humidity, 20°. (Note: Durafil is the Lilienfeld rayon, from 1945).

10.8.1   Stress–strain curves

Typical results are given in Fig. 10.12. The detailed shapes of the curves vary widely due to the differences in morphology, degree of polymerization, inter-molecular forces, etc. The mechanical behaviour may be represented dia-grammatically as in Fig. 10.13. These curves show an initially straight portion A followed by a portion B in which extension is easier and finally a region where extension is more difficult until the fibre breaks. A simple explanation of these portions may be made. Portion A represents a slight stretching of the chain molecules themselves and a straightening up of the molecular network particularly in the non-crystalline regions with some straining of the 'bonds' holding the molecules together. This is an elastic deformation and the fibre returns to its original length on removal of the load. With larger loads, the stress on the cross-linkages will become big enough to break some of them (portion *B*).

Rupture may occur for one or both of two reasons. As the structure is strained, the cross-linkages may break and the molecules slip over each other or the molecules themselves may break first. Recovery is now incomplete, due to rupture and reformation of the cross-linkages in new positions. In the later stages of this process (*C*), the possibility of further extension of the network becomes more difficult and the slope of the curve increases. Finally,

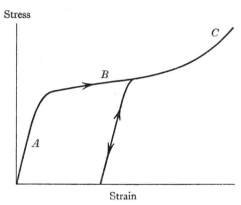

Fig. 10.13. Typical fibre stress–strain curve.

the force on some of the molecules is sufficient to cause them to slip or break, leading to cumulative breakdown of the whole fibre. This simple picture fits most theories of fine structure that have been proposed. It may be noted that once the structure had yielded, the load–extension relation on reduction of the load does not follow the original path. This will be discussed later. The properties of some specific fibres will be discussed now.

Natural fibres have great molecular length so that any slippage of the molecules over one another is likely to be negligible and fibre breakage is expected to arise from breakage of bonds.

(*a*) Cotton is an exceptional fibre in that its molecules spiral round the axis of the fibre at an angle of about 30° so that some extension results from stretching out the spiral rather like a spring, and hence the modulus is lower than might be expected.

The stress–strain curve for cotton is slightly concave to the extension axis and there is no obvious yield point. In general, finer cottons have higher strengths or tenacities and initial moduli than the coarser cottons. The breaking extension varies from 5–10 % but is not related to fineness. However, better correlation is obtained between tenacity and molecular orientation as measured by the birefringence (Fig. 10.14), the latter being closely related to the spiral angle.

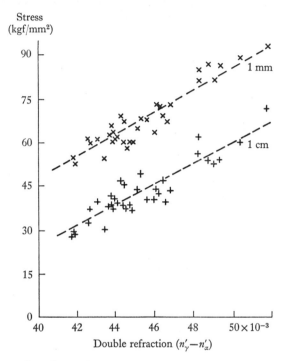

Fig. 10.14. Correlation between tenacity and birefringence of cotton.
Test lengths of 1 mm and 1 cm.

(*b*) The bast fibres (e.g. flax) in which the molecules are very nearly parallel to the fibre axis show a higher strength, a lower breaking extension and a lower work of rupture. These fibres are the strongest but least extensible of organic fibres.

(*c*) Wool fibres with coiled molecules are characterized by low strength but great extensibility. The nature of the keratin molecules means that there is a reversible straining, but above the yield point the molecules will unfold giving a very large extension. At not too large extension, removal of the force allows the molecule to return to its normal conformation. When extension takes place, the molecular conformation is changed and in steam it is possible to extend the α helical arrangement to give the extended β conformation. This transformation is shown up by the corresponding changes in the X-ray diagram.

(*d*) The tensile properties of synthetic fibres depend to a considerable extent on the molecular weight of the polymer and the conditions of spinning and drawing. The general tendency in synthetic fibres (Fig. 10.12) is for

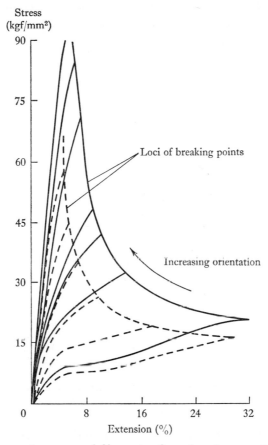

Fig. 10.15. Stress–strain curves of filaments of varying degrees of orientation. The dotted curves are acetate, and the full curves are cellulose fibres regenerated from acetate. The lowest curve in each set is for unoriented material.

moderately high strength to be combined with moderately high breaking extension resulting in a tough fibre, this being open to modification through the amount of drawing.

(*e*) The stress–strain curves of viscose and acetate fibres show an initial rapid rise with a marked yield point, followed by a nearly flat portion rising more rapidly again as breakage is approached. The curves vary widely for different types of rayons, the biggest difference being between stretched and unstretched fibres: the stretched fibres have good molecular orientation and show high strength and little extensibility similar to bast fibres whereas unstretched ones are weaker but more extensible. This behaviour is shown by cellulose and cellulose acetate of varying degrees of orientation (Fig. 10.15).

One comparison may be made between viscose fibres and cotton. With long molecules, interchain forces will be large and the chance of slippage reduced.

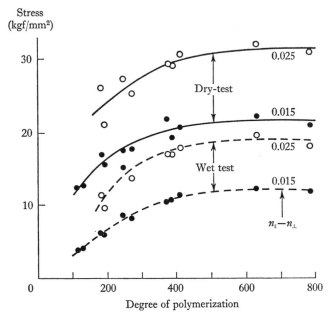

Fig. 10.16. Variation of tenacity of rayon filaments with degree of polymerization at constant orientation ($n_{\parallel} - n_{\perp}$ = birefringence).

As the molecular weight is increased, the strength of the fibre will increase up to a point where molecular slippage is replaced by molecular breakage. In natural cellulosic fibres, the molecular lengths are much greater than the critical value: with regenerated celluloses, the distribution straddles the critical value and hence a regenerated rayon fibre tends to have a lower strength than a natural fibre with a similar degree of orientation. With some of the newer fibres, however, the strength has been increased by improvements in the molecular length distribution. Thus in work on the tenacities of viscose rayon filaments prepared from fractionated samples of cellulose, the strength increases up to a degree of polymerization of about 500 but above this value, the curve levels off presumably because slippage has been eliminated. Similar results have been obtained for breaking extension (Fig. 10.16).

10.8.2 Effect of moisture

The majority of textile fibres absorb water vapour to an extent depending on the number of polar groups in the molecules. Thus the uptake for wool and cotton is very high whereas polyethylene terephthalate absorbs only a very

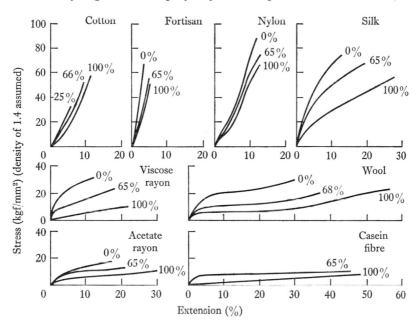

Fig. 10.17. Stress–strain curves at various humidities.

small amount. The changes in stress–strain curves for some fibres which absorb water are shown in Fig. 10.17. It is important therefore in the testing of textiles to specify the humidity conditions, which are normally taken to be 65 % relative humidity.

10.8.3 Effect of temperature on tensile properties

Some results are given in Table 10.3. In general, the stress decreases and strain increases both at yield and break as the temperature is raised. Care must be observed in making the measurements since the quantity of moisture

TABLE 10.3. *Effect of temperature on tensile properties*

Material	Temperature	Tenacity (gf/tex)†	Breaking extension (%)†	Work of rupture (gf/tex)†	Initial modulus (gf/tex)†	Yield stress (gf/tex)†	Yield strain (%)†
Nylon,	−57	79	11	4.2	1010	15	1.9
210 den.	21	65	13	4.0	450	7	1.5
	99	46	14	2.2	226	5	2.0
	177	26	29	4.7	108	—	—
Orlon,	−57	58	14	4.8	1170	35	3.8
100 den.	21	36	14	3.0	665	11	2.1
	99	23	20	2.0	64	—	—
	177	3	28	—	27	—	—
Dacron,	−57	78	8	3.3	1370	27	1.1
70 den.	21	56	8	2.8	1070	17	1.7
	99	41	10	1.4	261	5	2.5
	177	26	19	2.3	45	4	5.8
Fortisan,	−57	71	5	1.9	2830	21	1.0
270 den.	21	57	6	1.8	2010	14	1.0
	99	38	4	0.8	2000	17	1.0
	177	23	3	0.5	1560	13	0.9
Tenasco,	−57	35	12	2.9	1190	14	2.0
300 den.	21	22	20	3.0	667	7	1.9
	99	21	8	1.1	810	11	2.0
	177	17	14	1.5	360	5	1.6
Silk	−57	40	8	2.1	1130	27	3.2
	21	31	19	4.8	613	14	2.8
	99	25	9	1.6	613	13	3.0
	177	20	10	1.3	486	11	2.8
Cotton,	−57	18	5	0.5	567	—	—
20s/4	21	16	9	0.8	270	—	—
combed	99	10	6	0.3	369	—	—
yarn	177	8	7	0.3	252	—	—

† These figures may be converted to kgf/cm² by multiplying by 100 × density.

adsorbed changes with temperature. Unfortunately, the data in the table were not obtained using samples with the same moisture content; the measurements at 21° C were at 65 % relative humidity whereas for the other tests the temperature was changed from this condition, so that the low temperature specimens may contain considerable amounts of moisture whereas the high temperature ones contain only small quantities. In the completely wet state, however, the initial moduli at different temperatures (Fig. 10.18) show considerable differences in behaviour, some decreasing slightly, others markedly.

Increase in temperature increases the mobility of the polymer segments in the non-crystalline regions. In wholly amorphous polymers, the transition through the glass–rubber region is from a hard inextensible, glassy state to the soft extensible rubbery state: in fibres, the change is less severe because the crystalline, or at least fairly well-ordered material, does not take part in the transition. The effects are greatest in poorly crystalline fibres such as acrylics (e.g. Orlon 42).

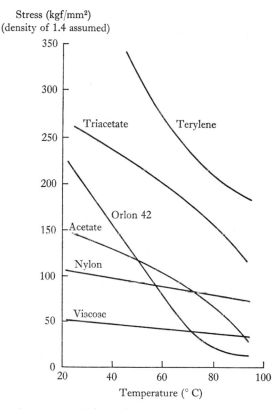

Fig. 10.18. Change in initial tensile modulus of wet fibres with temperature.

10.8.4 Time effects

The determination of a load–extension curve takes time and hence the fibre will yield or creep during the operation to an extent which is determined by the interval during which the fibre is under load (Fig. 10.19). In these circumstances, a difference exists between the experiments under constant rate of extension owing to the different proportions of time spent at the different parts of the curves. For a stress–strain curve which bends towards the strain axis (Fig. 10.20), a greater proportion of the time is spent on the part of the curve at high loads in a constant rate of extension test than is the case with a constant rate of loading. In the example, three-quarters of the time is spent above the point *A* compared to the reverse in the constant rate of loading test. Hence in a constant rate of loading test, whilst there will be slightly less

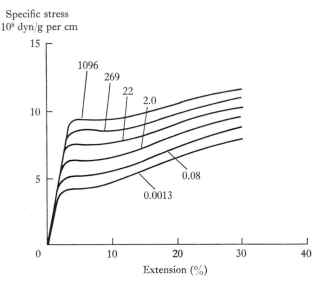

Fig. 10.19. Stress–strain curves at various rates of extension for acetate rayon. The figures against the curves refer to the rates of extension per cent per second.

creep in the earlier stages, there will be a greater total creep at the end of the test since rate of creep is greater at high loads. An example of the effect of different rates of extension is given in Fig. 10.21.

10.8.5 Elastic recovery

The extent to which a fibre is permanently deformed after it is stretched is of great technical importance and may limit its use. When a fibre is subjected to a stress–strain cycle, hysteresis is observed; namely when the load is reduced, the strain is greater than when the loading was increasing. In a cyclic change of stress or strain, the results will not fall on a single line although after the first few cycles, the fibre will become conditioned and the results will tend to fall on a loop. This means that energy is used up by internal friction and

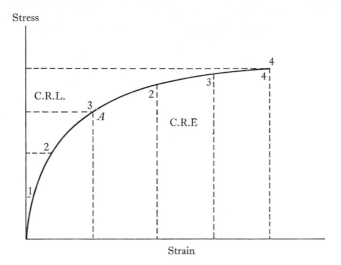

Fig. 10.20. Stress–strain curve showing equal intervals of time at constant rate of loading (C.R.L.) and constant rate of elongation (C.R.E.).

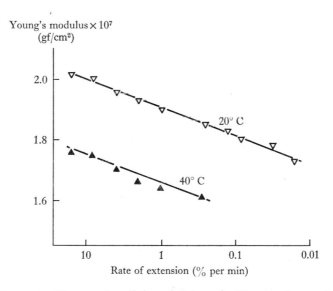

Fig. 10.21. Change of modulus of wet wool with rate of extension.

consequently the material will heat up and if it contains moisture will tend to dry out. This is particularly important for uses in which repeated loading occurs (e.g. tyres) as heating will affect the properties.

On a molecular scale, recoverable or elastic deformation is due to stretching of interatomic and intermolecular bonds whereas the non-recoverable or plastic deformations result from a breaking of bonds and their reforming in new positions. By definition, the former has been referred to as elasticity, the property of a body by virtue of which it tends to recover its original shape after deformation, and the latter as plasticity.

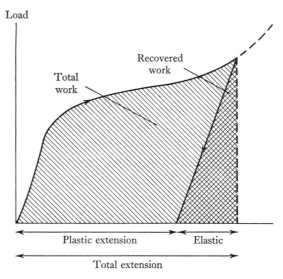

Fig. 10.22. Elastic and plastic extension.

As shown in Fig. 10.22 a deformation may be separated into elastic and plastic portions, the former being recovered when the stress is removed.

Quantitatively, these effects may be calculated as follows.

$$\text{Elastic recovery} = \frac{\text{elastic extension}}{\text{total extension}}.$$

$$\text{Work recovery} = \frac{\text{work returned during recovery}}{\text{total work done in extension}}.$$

It may be noted that (1 − work recovery) gives the proportion of the total work which is dissipated as heat. In measurements of recovery, a particular programme of application and removal of stress is necessary.

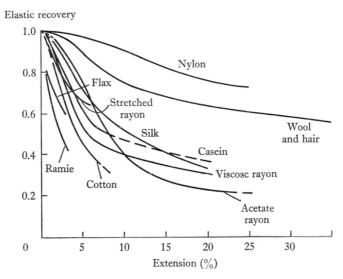

Fig. 10.23. Elastic recovery plotted against strain.

The results for a series of fibres are given in Fig. 10.23 as a function of strain showing the proportion of a given extension which will be recovered.

10.8.6 Mechanical conditioning

Stretching a fibre to an extent which leaves it with permanent set causes other changes in properties. The fibre being subject to loading, unloading cycles gives a stress–strain curve with a higher yield stress and a reduced breaking extension. For example, fibres subjected to 80 % of breaking elongation in 50 cycles are shown in Fig. 10.24. Thus the properties of fibres may be changed by the action of high stresses during processing and if highly strained may not serve their proper function. The rise in yield stress of a fibre after the application of a certain stress for some time usually results in almost perfect recovery from subsequent stresses below that value. The fibre has been mechanically conditioned and if taken repeatedly through a given stress cycle, the loading and unloading curves in successive cycles gradually come together until they form a continuous repeated loop.

In simple terms, the mechanical conditioning effect may be pictured as follows. Once the specimen is stretched to a certain extent, it acquires a new structure with fewer 'cross-links' which can be stretched again with a reversible straining of the links available.

10.8.7   Creep relaxation

The first type is the extension with time under applied load and the second is the reduction of stress with time under a given extension. The continued deformation and possible rupture of the specimen when a load is applied for some time has important consequences when testing the mechanical properties since it means that the results of a test (e.g. a stress–strain curve) will depend on the time taken. If the tests are very rapid, the conditions can become nearly adiabatic whereas tests carried out normally are in general isothermal.

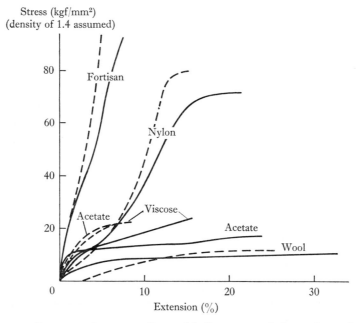

Fig. 10.24. Stress–strain curves after and before repeated elongation to 80 % of breaking extension (— original; - - - conditioned).

10.9   Creep, primary and secondary

When a constant load is applied to a fibre, the instantaneous extension is followed by creep (Fig. 10.25). The removal of the load gives an instantaneous recovery usually equal to the instantaneous extension followed by a partial recovery with time, which still leaves some unrecovered extension. The three features are elastic deformation which is instantaneous and recoverable, the primary creep which is recoverable in time and the secondary creep which is non-recoverable. A similar situation results when the fibre is held at constant extension (Fig. 10.25).

The time effects, creep and relaxation, arise from the thermal vibrations of the segments in the molecules so that the 'cross-links' between molecules are liable to break at random intervals. When the linkage breaks, there is a movement of the segments of the molecules and stress is thrown on to other parts of the structure which in turn may break. Breakages will assist in slippage of the molecules and hence plastic flow. At higher temperatures when

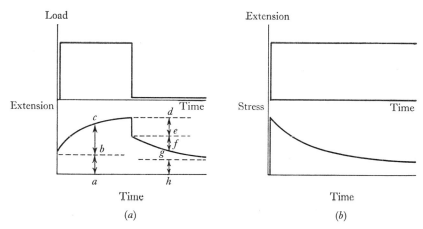

Fig. 10.25. (*a*) Creep under constant load and recovery under zero load, showing instantaneous extension, *a–b* and *d–e*; total creep, *b–c*; primary creep, *e–f*; and secondary creep, *g–h*. (*b*) Relaxation of stress under constant extension.

the system has more energy, breaks occur more frequently and creep is speeded up. On release of stress, the molecules as a result of their thermal vibrations may return over a period of time to their original structure and hence we get the phenomenon of creep recovery.

If after recovery, the same load is re-applied, the rate of creep is less than in the first test on the specimen. The primary creep takes place at its initial rate but the secondary resumes at the rate at which it left off. The mechanical conditioning effect is a special case of this since it means that, if the load had been applied for long enough for the secondary creep to become negligible, there will not be any appreciable secondary creep in later experiments, unless the load is further increased. The secondary creep gives rise to the major-part of the permanent extension of a fibre and is usually negligible below the yield point.

10.10 Dynamic tests

In conventional testing, although time of application of the forces influences the result, the stress and strain could be regarded as being the same at all points in the specimen. However, in some tests, the stress and strain are not continuously increasing or decreasing but are oscillating, that is, a stress–strain loop is directly measured. In some tests, the inertia of the specimen cannot be neglected and the stress and strain will vary along the specimen. When a stress is applied to the end of a specimen, it takes time for this stress to be transmitted and the resulting oscillation to die away. The study of stress pulses is another way of carrying out a dynamic test, or alternatively standing wave patterns may be set up, when, for example, the velocity of longitudinal sound waves in specimens may be deduced. There are a variety of methods used in dynamic testing which cover different frequencies (Table 10.4) and some results are shown in Table 10.5. In general,

TABLE 10.4. *Frequency range of dynamic tests*

Method	Frequency range
Direct observation of stress–strain loop	Up to 5 c/s
Free vibrations	1 to 50 c/s
Forced resonant vibration	1 to 300 c/s
Flexural resonance of specimen	20 c/s to 10 kc/s
Velocity of sound waves—continuous	500 c/s to 30 kc/s
Pulse velocity	10 kc/s to 100 kc/s

TABLE 10.5. *Dynamic parameters at 65 % R.H. and 21° C*

Material	E dyn/cm$^2 \times 10^{10}$			
	Static	1.5 to 100 c/s	10 kc/s	100 kc/s
Viscose	4.2	10.6	17.1	19.5
Tenasco	6.6	—	—	26.2
Acetate	—	5.1	6.5	—
Wool	3.1	—	—	8.1, 9.2
Feather keratin	—	4.4	—	—
Hair	3.6	—	6 to 7	8.1
Nylon	2.9	5.8	7.7, 9.4	7.0
Undrawn nylon	—	—	2.0	—
Raw silk	—	16.8	19.4	—
Degummed silk	—	13.5	—	—
Linen yarn	—	—	46	—
Steel wire	1.93	—	—	1.98

between 1.5 and 100 c/s, the modulus has been shown to be independent of frequency (except in polyethylene) but as the frequency increases, the Young's modulus E increases. Such a result was found even for a small static strain. At 100 kc/s a significant change in dynamic modulus for strains up to 1% was observed in nylon but not for hair and viscose. As larger strains are applied, the modulus for a series of rayons varies after a critical strain is reached. Some results on a variety of fibres are given in Fig. 10.26.

10.11 Strength of glass fibres

Flat glass has a strength between 1.5 and 7 kgf/mm² but commercial fibre strengths are in the order of 140 kgf/mm². This is still considerably smaller than the value of 700 kgf/mm² anticipated on theoretical grounds. The explanation of this lies in the formation of microcracks in the surface and is discussed elsewhere. Furthermore, the strength of glass fibres increases with decrease in diameter, a fact which may be explained by the suggestion that the probability of finding flaws is less in the smaller sample. Although in the original theory of Griffiths, flaws were assumed to occur uniformly throughout the body of the glass, breakage takes place from the surface. Further evidence of the presence of surface flaws in glass comes from the phenomenon of delayed fracture. Glass shows no fatigue under cyclic loading but in the presence of water vapour, the breaking stress for sustained load is less than that for instantaneous loading and delayed fracture can occur. This would seem to arise from the slow diffusion of an adsorbed film in the stressed region at the apex of the crack.

The possibility of surface damage means care must be taken in interpreting the results of strength tests. There are plenty of examples of the damage caused to glass fibres by contact with other materials. Thus in order to see the effect of various variables on their properties, it is necessary to take the fibres straight from the extruder. When this was done, it was found that there was no variation of strength with diameter for fibres drawn at the same temperature and the strength obtained was around 350 kgf/mm².

It is interesting to note that strengths of this magnitude are obtained when the load is applied for very short duration of perhaps a few seconds. If this phenomenon of delayed fracture is due to moisture diffusion in flaws, it must be assumed that fibres contain numerous flaws of the same size and of the order of the atomic spacing; that is, they arise from structural defects.

10.12   Fibre friction

In the processing of fibres to form yarns and fabrics, frictional forces become important in that they govern the tensions operative when the fibre or yarn passes over a rod or guide. The extent to which the frictional forces are operative is described by the coefficient of friction μ defined as the ratio of the force exerted tangentially at the point of contact to that acting along the normal. Several techniques have been used to determine the magnitude of the frictional forces operative, e.g. by sliding one fibre against another, by sliding

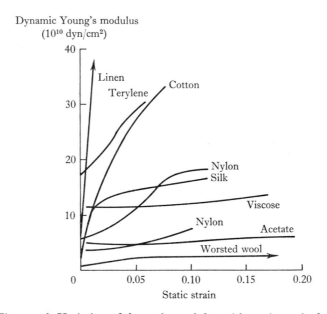

Fig. 10.26. Variation of dynamic modulus with static strain for various textile yarns.

fibres around a rod, by measuring the force required to remove a fibre from a specimen made up by twisting two fibres together or the force required to remove a single fibre from a mass of fibres under pressure, etc.

Measurements show the coefficient of friction to decrease with the normal load and hence empirical relations have been put forward between load and the friction force. One example is $F = aN^n$ where a and n are constants. Some idea of the changes in μ defined as F/N is given in Fig. 10.27. However, though μ is not a constant and varies not only with load but also with speed of movement of the fibre, area of contact etc., the value under particular conditions is useful as a means of comparison. Some typical values are given in Table 10.6.

These variations in μ are undoubtedly related to the physical properties of the fibre. Thus at low loads a narrow track is produced along which the fibre is flattened; at higher loads the deformation is greater and may be accompanied by severe tearing in the centre of the track. With some fibres, for example cellulose acetate, the fibre may be abraided by sticking to the guide over which it is passed and then breaking away. Concomitant with the damage to the fibre is the fact that particles of cellulose acetate were found to adhere to glass after it was rubbed.

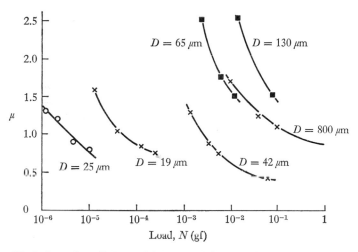

Fig. 10.27. Variation of coefficient of friction of fibres with load (D — fibre diameter). ○, Teflon; × Nylon; ■ Polythene.

The origin of the frictional forces would seem to lie in a welding or union of the two surfaces at the point of contact which break when sliding starts. When examined on a small enough scale, surfaces, with one or two exceptions, are irregular (e.g. cleavage planes of a crystal mica). Contact between two bodies occurs at the tips of the peaks so that when a load is applied, the pressure at these points of contact is very great and they squash until the contact area is sufficient to support the load. Under the intense pressure and with a possible increase in temperature, the junctions weld together and to allow sliding, must be broken by shearing. Frictional forces thus depend on the force needed to shear the junctions. Calculation of the shear strength of plastics from friction measurements have shown some agreement with bulk measurements of shear strength.

TABLE 10.6. *Typical values of μ*

(a) Between fibres

	Crossed fibres	Parallel fibres
Nylon	0.14–0.6	0.47
Silk	0.26	0.52
Viscose rayon	0.19	0.43
Acetate	0.29	0.56
Cotton	0.29, 0.57	0.22
Glass	0.13	—
Jute	—	0.46
Casein	—	0.46
Saran	—	0.55
Terylene	—	0.58
Wool, with scales	0.20–0.25	0.11
Wool, against scales	0.38–0.49	0.14

(b) For yarns passing over guides

	Hard steel	Porcelain	Fibre pulley	Ceramic
Viscose rayon	0.39	0.43	0.36	0.30
Acetate, bright	0.38	0.38	0.19	0.20
Acetate, dull	0.30	0.29	0.20	0.22
Grey cotton	0.29	0.32	0.23	0.24
Nylon	0.32	0.43	0.20	0.19
Linen	0.27	0.29	0.19	—

References to chapter 10

1 Goodman, I., 'Synthetic fibre-forming polymers', Royal Institute of Chemistry Lecture No. 3 (1967).
   I. Goodman & J. A. Rhys, *Polyesters* (London, Iliffe 1965).
   R. H. Peters, *Textile chemistry* 1 (Amsterdam, 1963).
   F. T. Bowden, & D. Tabor, *Friction and lubrication in solids* (Oxford, 1950).
   W. E. Morton & J. W. S. Hearle, *Physical properties of textile fibres* (Butterworth, 1962).
   J. W. S. Hearle & R. H. Peters (Eds.), *Fibre structure* (London, 1963).
   R. Hill (Ed.), *Fibres from synthetic polymers* (Elsevier, 1953).
   H. F. Mark, S. M. Atlas & E. Cerna (Eds.), *Man-made fibres science and technology* 1–3 (New York, 1967).
   R. W. Moncrieff, *Man-made fibres* (London, 1957).
   J. W. S. Hearle, *J. Pol. Sci.* **C20**, 215 (1967).

11 Glasses and glass-ceramic materials

D. R. Uhlmann

11.1 Glasses

The term 'glass' is generally used to describe an amorphous material exhibiting solid-like behaviour, most frequently obtained by cooling from the liquid state. Essential to this description is the matter of structure – in particular the absence of periodicity or long-range order. The most common disordered state is, of course, the liquid state; and the structural similarity between glass and liquid has led to the notion of glass as a frozen-in liquid.

It should be noted, however, that the definition of this freezing-in is somewhat arbitrary, and depends upon the time-scale of the experiment. The standard definitions are based upon relaxation times of minutes to hours (common experimental time-scales); and amorphous materials with viscosities of 10^{13} to 10^{15} poise or more are designated as glasses, while those with lower viscosities are termed liquids.

Glass formation is widely believed to be a universal feature of liquid behaviour, provided the liquid can be cooled to a sufficiently low temperature without the occurrence of crystallization. There is, however, a wide variation from one material to another in glass-forming ability – that is, a wide variation in the rate at which a material can be cooled and still be obtained in the glassy state. Among inorganic oxide materials, the ability to form glasses ranges from B_2O_3 which cannot be crystallized from a dry melt at atmospheric pressure, through materials like SiO_2 which can be obtained as glasses even with quite moderate cooling rates, and materials like $Na_2O.2B_2O_3$ which can be cooled from liquid to glass if relatively high cooling rates are employed, to materials like Al_2O_3 which have been obtained in the amorphous solid state only by condensation from the vapour.

There are, of course, many different types of materials which have been obtained in the glassy state. In addition to the familiar inorganic oxide glasses, these include simple molecular glasses such as toluene and orthoterphenyl, polymer glasses such as polyvinyl acetate and polymethyl methacrylate, simple ionic glasses such as calcium aluminate and calcium–potassium nitrate, metallic glasses such as gold–silicon–germanium alloys, and glassy water. For purposes of space, however, we shall in the present chapter devote most of our attention to inorganic oxide materials.

11.1.1   The structure of inorganic oxide glasses

Several models have been suggested for the structure of inorganic oxide glasses. The first is termed the *crystallite model*. Taking SiO_2 as an example, it was observed that the main broad peak of the diffraction pattern of the glass

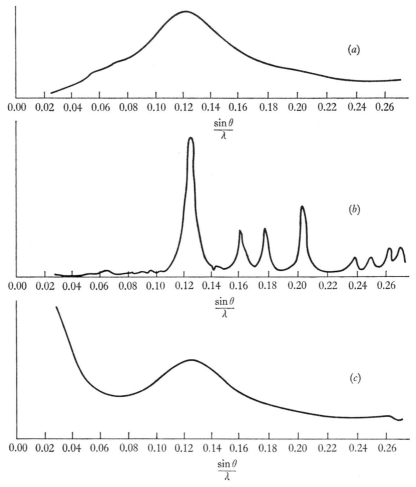

Fig. 11.1. Diffraction patterns from (*a*) fused silica, (*b*) cristobalite and (*c*) silica gel (ref. 1).

(see Fig. 11.1) was located in the same range of angle as the strongest sharp peaks in the diffraction pattern of the corresponding crystal (cristobalite). It was suggested, therefore, that the structure of fused silica was composed of very small crystallites of cristobalite; and the width of the glass diffraction peak was attributed to particle-size broadening, caused by the small size of the cristobalite crystals. In the case of multi-component glasses, the structure was viewed as composed of crystallites of compositions corresponding to compounds in the given system.

A different model of glass structure, termed the *random network model*, views oxide glasses as three-dimensional networks of oxygen polyhedra, lacking symmetry and periodicity – in the case of fused silica, a three-dimen-

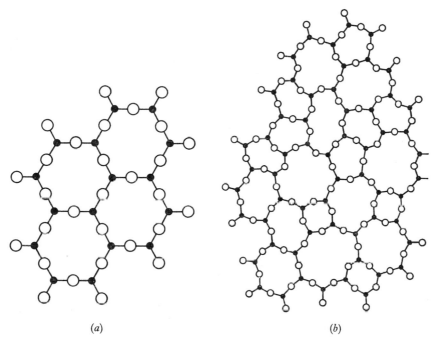

(*a*)   (*b*)

Fig. 11.2. Schematic two-dimensional representation of (*a*) crystal and (*b*) glass (ref. 2).

sional array of SiO_4 tetrahedra in which no unit of the structure is repeated at regular intervals. A schematic two-dimensional representation of such a random network is shown in Fig. 11.2. In the case of multi-component glasses, the cations were pictured as randomly distributed through the Si—O network.

The discussion between proponents of these models was generally decided in favour of the random network model, largely on the strength of the arguments advanced by Warren:

(1) Careful experimental data on fused silica indicated a crystallite size of only 7–8 Å. Since the size of the unit cell of cristobalite is also about 8 Å, any crystallites would be only a single unit cell in extent; and such structures seem at variance with the notion of a regularly repeating crystalline array. This remains a powerful argument even if the estimate of crystallite size were only accurate to within a factor of 2.

(2) In contrast to silica gel, there is no marked small angle scattering from a sample of fused silica (see Fig. 11.1). This indicates that the structure of silica is continuous, and is not composed of discrete particles like the gel. The absence of small angle scattering from silica was confirmed by later work, and indicates that if crystallites are present, there must be a continuous spatial network interconnecting them.

(3) The diffraction patterns from multi-component glasses are also characterized by very broad diffraction peaks. Consequently, any crystallites present in these glasses must also be of the order of a unit cell in size.

Based largely on arguments such as these, the random network model dominated the field of glass structure for nearly two decades. Electron microscopic studies indicated, however, the presence of submicrostructure in the fracture surfaces of a number of optically homogeneous oxide glasses. These submicrostructures were typically in the range from seventy-five to a few hundred ångström units in size, and were interpreted as micelles or paracrystals – that is, domains with enough order to allow their mutual misorientation to be discerned in the electron microscope, but with sufficient disorder that they did not give rise to sharp Bragg reflections.

In recent years, a large number of studies have indicated structural inhomogeneities in many glass systems, but in fused silica and a number of alkali silicates recent work [e.g. 3] has indicated the absence of any such submicrostructure observable in the electron microscope. In all cases studied in detail, the submicrostructures seen in oxide glasses have been associated with the occurrence of miscibility gaps, either stable or metastable, in the particular systems. In these cases, at temperatures above the miscibility gaps, the stable liquid configurations consist of single homogeneous fluid phases. At temperatures within the gaps, however, the stable configurations consist of two or more liquids of different compositions. This is shown schematically in the temperature–composition and free energy–composition diagrams of Fig. 11.3 for the case of a metastable miscibility gap.

The process by which an initially homogeneous liquid separates into two liquids of different composition is of considerable interest at the present time.[4, 5] Depending upon conditions, this phase separation may take place by either a discontinuous nucleation-and-growth process or a continuous process called spinodal decomposition. The scale of the resulting submicrostructures is typically in the range of 50 to 1000 Å. The form of the submicrostructure varies as the composition changes through a miscibility gap, from isolated discrete particles of one phase in a matrix of the second phase, through the

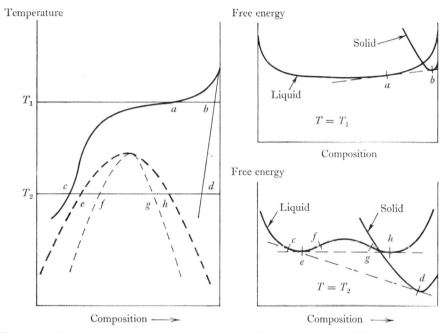

Fig. 11.3. Schematic temperature versus composition and corresponding free energy versus composition diagrams for a metastable miscibility gap (ref. 5).

central region of the gap where both phases are three-dimensionally interconnected, to a region where discrete particles of the second phase in a matrix of the first are observed. Examples of such submicrostructural variations observed[4] in the system BaO–SiO$_2$ are shown in Plate 11.1 *a* to *c* between pages 320 and 321. The miscibility gap for this system is shown in Fig. 11.4.

Miscibility gaps, either stable or metastable, are found in most silicate and borate systems; and the occurrence of diphasal or multiphasal submicrostructures on a scale of 50 to 1000 Å, with their associated diffusion fields, should affect nearly all properties of the glasses.[6–8]

In the case of borate glasses, in addition to the occurrence of phase separation in nearly all systems studied,[9] there is an additional structural feature of

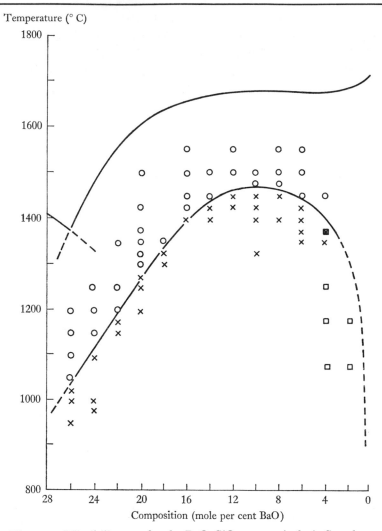

Fig. 11.4. Miscibility gap for the BaO–SiO$_2$ system (ref. 4). See also
Plate 11.1 (facing p. 320).

interest, viz. the change of boron co-ordination from BO$_3$ triangles to BO$_4$
tetrahedra on the addition of modifying cations[10].

Recent X-ray diffraction studies of some simple glasses by Mozzi and
Warren[11, 12] have given new insight into the structure of these glasses.
While they have not uniquely established a particular model as *the* best
representation of glass structure, these studies are useful for providing
information with which any suggested structural model must be consistent.

Along these lines, it might be noted that the structure of glasses may not be amenable to a simple unequivocal description, and will certainly depend upon the heat treatment as well as on the composition of the glasses. While the model of a random array may provide a useful (but not uniquely established) description of the structure of many single-component glasses, it seems inappropriate, because of the occurrence of phase separation, for describing the overall structure of many multi-component glasses. Even in these cases, however, such a model may provide a useful representation of the structures of the individual phases into which the system separates.

At the present time, little is known about the structure of glasses on the scale between about 3 or 4 Å (the size of the basic-co-ordination units) and about 30 to 50 Å (the size of the smallest observed second-phase submicrostructures). Little also is known about the structural state of the common modifying cations, or about the distribution of atomic environments, in even simple glasses.

11.1.2   Elastic properties of glasses

The response of an amorphous material to a stress is commonly divided into three parts: (1) an instantaneous elastic strain, which is instantaneously recoverable on removal of the stress; (2) a delayed elastic strain, which is recoverable with some characteristic relaxation time after removal of the stress; and (3) a viscous flow, which is not recoverable on removal of the stress.

At temperatures well below their glass transition regions, the viscosities of ordinary glasses are of the order of 10^{20} poise or so (see §11.1.3). At such viscosities, even at stress levels as high as 10^{10} dyn/cm^2 one would expect from viscous flow an annual fractional change in length of only one part in 10^3. This kinetic resistance to deformation is reflected in the permanence of shape observed in glass objects, as well as in the generally small magnitude of the delayed elastic strain. Under such conditions, the overall response of the glass to stress may be described as that of an elastic solid.

Since glasses are isotropic materials, their elastic behaviour can be described completely by two independent elastic constants or moduli. For most glasses Poisson's ratio is about 0.25, the shear modulus is about 40% of Young's modulus, and the bulk modulus is approximately two-thirds of Young's modulus.

Values of Young's modulus, E, for most silicate glasses are in the range between about 4 and 12×10^{11} dyn/cm^2. Corresponding values for borate glasses range between about 1 and 8×10^{11} dyn/cm^2.

The moduli of lithium silicate glasses increase slightly with increasing concentration of the modifying cations, while the moduli of sodium silicate and lead silicate glasses decrease somewhat as alkali oxide is added. At a given total concentration of modifier, glasses containing more than one type of cation are often characterized by larger moduli than those containing only a single type of cation.

In alkali borate glasses[8, 13] the elastic moduli increase with increasing concentration of modifying cations, as well as with their field strength (see Fig. 11.5).† The differences in the forms of the modulus versus composition relations for silicates and borates may be related to the structural features of the glasses. Besides the direct effects of the modifying cations, which in both cases should act to increase the density and modulus, there are also the effects on the Si—O and B—O frameworks. With silicates, oxygens bonded to only a single silicon are formed as modifier oxide is added to SiO_2; while with borates, borons bonded to four rather than three oxygens are formed on adding modifier. With silicates, this framework effect might be expected to lower the modulus; while with borates, it should raise the modulus.

The elastic moduli of most glasses, like those of most other materials decrease with increasing temperature. In the case of fused silica, however, the moduli increase with increasing temperature (Young's modulus for silica at 1000° C is approximately 10 % higher than its value at 25° C.[14]) The behaviour of silica has been associated[15] with its anomalously low thermal expansion coefficient, which in turn has been related to the presence of a number of low-lying transverse vibrational modes, perhaps associated with the motion of oxygen atoms normal to the silicon–oxygen–silicon directions. It should be noted, however, that modulus versus temperature relations of positive slope have also been observed in a number of other silicate glasses with appreciably larger expansion coefficients than fused silica.

The elastic moduli of most glasses increase with increasing pressure. Fused silica is again an exception, as its moduli decrease with increasing pressure, at least up to pressures of 30 kilobars or so.[16] This behaviour has also been related to the unusually low thermal expansion coefficient of silica. Consistent with the behaviour at applied pressure, the moduli of fused silica have been observed[17] to increase with increasing tensile strain, Δl. In detail, this increase has been approximated[17] by an expression of the form:

$$\left. \begin{aligned} \Delta l &= 7.3(1 + 5.75\epsilon) \times 10^{11} \text{ dyn/cm}^2 \\ E &= 3.2(1 + 3.06\epsilon) \times 10^{11} \text{ dyn/cm}^2 \end{aligned} \right\} . \qquad (11.1)$$

† The field strength is defined as the ratio of the charge on the cation to the square of its ionic radius, and is often used as a measure of cohesion.

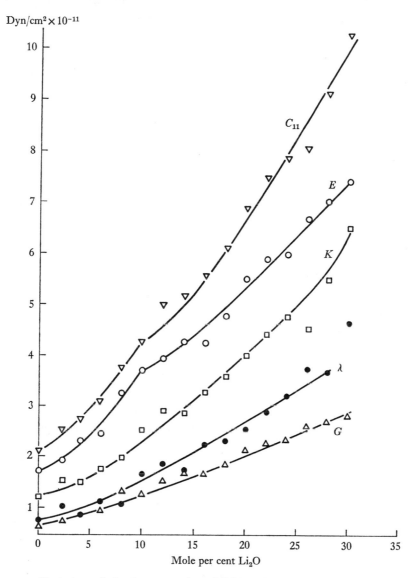

Fig. 11.5. Variation of elastic properties of lithium borate glasses as a function of composition (refs. 8 and 13). C_{11} = stiffness constant, E = Young's modulus, K = bulk modulus, λ = Lamé constant, G = shear modulus.

In contrast, the modulus of a soda glass has shown[17] to decrease with increasing tensile strain. This is consistent with the increase in elastic moduli which most glasses exhibit on application of pressure.

11.1.3   Anelastic properties of glasses

The anelastic (visco-elastic) behaviour of glasses, with emphasis on inorganic oxide materials, has been well treated in a number of recent reviews,[6, 18, 19] and will not be repeated here. In the present discussion we shall not consider anelastic behaviour associated with distortion of the framework or with motions of the modifying cations. Rather, our attention will be directed to the larger scale anelastic response which involves changes in the configuration of the framework.

Surveying various types of materials, it is found that the anelastic response of glasses to stress is quite large in the case of polymeric materials, is significantly smaller in the case of oxides, and is still smaller in the case of simple molecular glasses. The differences among these various classes of materials are undoubtedly related to differences in their structural characteristics; but the detailed relation between these characteristics and the anelastic behaviour has yet to be satisfactorily elucidated.

For a given material, the anelastic contribution to the strain is most significant in a temperature range around the glass transition. At higher temperatures, this contribution reaches its total value in times which are short relative to experimental time-scales, and is generally overshadowed by the process of viscous flow. At temperatures well below the transition region, the anelastic contribution may well be large relative to the steady-state flow, but is itself small if not insignificant in total magnitude.

In many experimental situations, the phenomenon of anelasticity is accompanied by the process termed stabilization. The latter term describes the approach with time of the properties of materials to their equilibrium values during isothermal treatment. This relaxation of the system toward equilibrium is most frequently accompanied by a decrease in volume and an increase in the viscosity of the system with time. In order then to isolate the anelastic response of the system, the stress should be applied to a stabilized (i.e. equilibrium) sample.

The time dependence of the anelastic component of the strain is not simply described by a single relaxation time. Rather than varying simply as $\exp(-t/\tau)$, the variation with time seems describable by a relation of the form: $\exp[-(t/\tau)^n]$, where n approaches unity at high temperatures, decreases with falling temperature, is about one-half near the upper portion of the transformation range,[e.g. 20, 21] and seems to approach one-third at lower temperatures.[22] Such behaviour, for $n \neq 1$, is of course equivalent to a distribution of relaxation times, as: $\sum_i \exp\left(-t/\tau_i\right)$. With such functional dependences, the strain builds up or decays more rapidly at short times than a simple exponential dependence, and more slowly at long times.

Estimates of the magnitudes of the relaxation times are often conveniently given by the Maxwell relaxation time:

$$\tau \equiv \frac{\eta}{G}, \tag{11.2}$$

where η is the viscosity and G is the appropriate modulus. As noted in the preceding chapters, various mechanical models using different elements with specific characteristic times have been used to represent the anelastic behaviour of materials. The relations between these models and elements with their characteristic times and any actual molecular processes have, however, yet to be established.

The forms of the anelastic response curves imply that viscosities higher than about 10^{20} poise cannot be measured to within 1 % of their equilibrium values in a human life time, and in experiments of more reasonable duration, weeks to months, the measurable limit is about 10^{17} to 10^{18} poise.[19, 23] For the same reason, the highest *apparent* viscosity (coefficient relating observed stress to observed strain rate) which is measurable in experiments of weeks to months is of the order of 10^{20} poise or so.[23]

Related to these considerations are a number of observations of the flow of glass at low temperatures, well below the transformation range, on exposure to even small stresses over extended periods of time. It might be noted in this regard that significant differences exist among the reported observations. For example, no observable deformation in a soft glass rod, 4.9 mm diameter × 1 m long, loaded at its centre with 300 gm in a period of 7 years could be noted; in contrast, a 'permanent' deformation of 9 mm at the centre of an alkali-lead silicate tube, 1 cm diameter × 1 mm wall × 1 m long, loaded at its centre with 885 gm for 6 years was observed. In both these cases, the tests were conducted at room temperature, some 400–500° C below the transformation ranges of the glasses. Perhaps the most striking test of this type, however, was reported by Phillips,[24] who loaded a glass rod of unspecified composition to 125 000 lbf/in² for a period of 26 years, and found no evidence of permanent deformation upon removal of the stress. (An anelastic deformation of about 1 % was observed to recover after a period of time under ambient conditions.)

Other forms of deformation at low temperatures, during short times, are exemplified by the formation of hardness indentations, scratches and the like in glasses. Their discussion lies, however, beyond the intent of the present chapter.

11.1.4 Viscous flow

The most familiar response of amorphous materials to stress is viscous flow. Under most conditions, the flow is conveniently described as Newtonian viscous; that is, a linear relation between stress and strain rate is exhibited:

or
$$\left. \begin{array}{l} \tau = \eta\gamma' \text{ (for shear stress)} \\ \sigma = 3\eta\epsilon' \text{ (for normal stress)} \end{array} \right\}. \qquad (11.3)$$

Here τ and σ are the shear and normal stresses (in dyn/cm²); η is the shear viscosity (in poise); and γ' and ϵ' are the shear and normal strain rates (in s⁻¹).

Phenomenologically this relation between stress and strain rate results in the stability against necking which characterizes most liquid state forming processes. Such stability may readily be seen by applying the conservation of mass condition to a volume element of length L and cross-section A of a sample subjected to a normal load F:

$$\epsilon' = \frac{L'}{L} = -\frac{A'}{A}, \qquad (11.4)$$

where the primes indicate the time derivatives of the respective quantities. combining equations (11.3) and (11.4) we have:

$$A' = -A\epsilon' = -\frac{\sigma A}{3\eta} = -\frac{F}{3\eta}. \qquad (11.5)$$

From this, it is apparent that A' does not depend upon the area of the volume element, but merely upon the load and the viscosity.

At high stress levels, the viscosity of a homogeneous silicate liquid has been shown[25] to decrease with increasing stress. The critical stress level for such behaviour was found to be in the range $1-2 \times 10^9$ dyn/cm²; and at lower stresses Newtonian behaviour was observed. In contrast, a phase-separated borosilicate glass failed to exhibit stress-dependent viscosity even at stress levels as high as 2×10^{10} dyn/cm².[26] More generally, phase separation has been shown[26] to affect strongly the viscosity as well as the apparent stress dependence of the viscosity of glasses. This problem is a difficult one to treat analytically, as consideration must be given to the volume fraction, morphology and connectivity of second-phase material as well as to the associated diffusion fields;[6, 26] and even highly approximate treatments are not yet available.

The application of high pressure at a given temperature is observed to increase the viscosity of liquids. Among inorganic oxide materials, this effect has been most extensively studied in B_2O_3,[27] where application of pressures as small as one kilobar has produced increases in viscosity as large as a factor of four. As expected, the effect of pressure is larger at lower temperatures (and higher viscosities under ambient conditions).

In technological practice, there are a number of important points on the viscosity scale, corresponding to different viscosity levels. Among these are the *softening point* ($10^{7.5}$–10^8 poise) at which a body will deform under its own weight; the *annealing point* (10^{13} poise) at which internal stresses are reduced to low values in 15 minutes, and the *strain point* ($10^{14.5}$ poise) at which stresses are reduced to low values in 4 hours. The working range in commercial practice is usually between about 10^4 and 10^7 poise, and a larger temperature interval between these viscosity values is desirable in processing.

In addition to these points and the general region of the working range, it is often important to have information about the viscosity–temperature relation as a whole. The form of such a relation for an inorganic oxide glass-forming material is shown in Fig. 11.6a. The apparent activation energy for flow (the slope of the corresponding log η versus $1/T$ relation) is higher at low temperatures than at high temperatures. For common inorganic glass-formers, the ratio of the low-temperature slope to the high-temperature slope is generally a factor of 2 or 3 or less. In contrast, for simple organic or polymeric glass-formers, this ratio is generally an order of magnitude or more. The low-temperature slopes in the latter case indicate that under such conditions the flow is not a simply-activated process, but rather involves co-operative motion of the molecules.

In describing flow and relaxation phenomena in the vicinity of the glass transition, use is frequently made of the Williams–Landel–Ferry (WLF) relation, which for viscosity takes the form:[29]

$$\eta = \eta_g \exp\left[-\frac{40(T-T_g)}{52+(T-T_g)}\right]. \tag{11.6}$$

Here η_g is the viscosity at the glass transition (generally about 10^{13} poise) and T_g is the glass transition temperature. This expression may be rewritten in the form of the empirical Vogel–Fulcher relation, which has frequently been used to describe log η versus $1/T$ relations in which curvature is observed:

$$\eta = \eta_0 \exp\left(\frac{A}{T - T_0}\right), \qquad (11.6a)$$

where η_0, A and T_0 are constants.

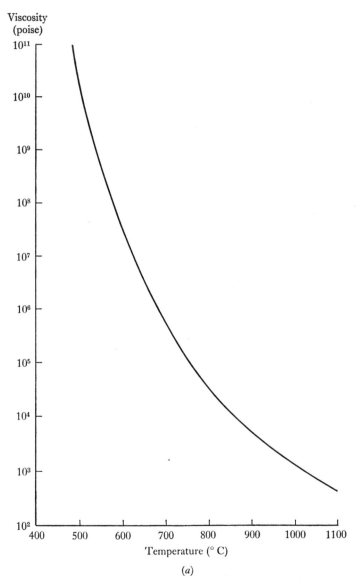

(a)

Fig. 11.6. (a) Viscosity–temperature relation for sodium disilicate (ref. 28).

Expressions equivalent to the WLF relation have been derived on the basis of free volume[30, 31] and excess entropy[32] models for viscous flow and glass formation. Unfortunately, while these models have gained wide acceptance, they have not proved adequate for describing the observed forms of

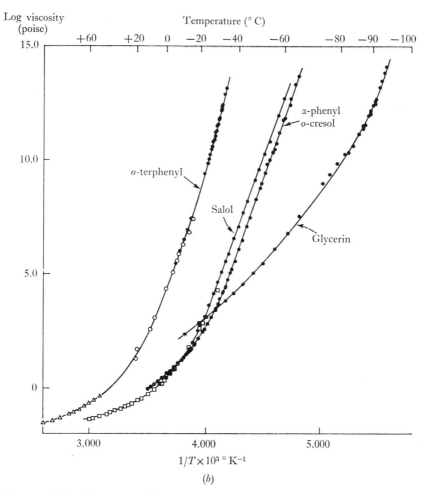

Fig. 11.6. (*b*) Viscosity–temperature relations for glycerin, α-phenyl, *o*-cresol, salol, and *o*-terphenyl, (ref. 35).

viscosity–temperature relations for various materials. For example, some materials, like SiO_2,[33, 34] apparently exhibit an Arrhenius temperature dependence over the full temperature range for which precise data are available (10^6 to 10^{12} poise); some, like salol,[35] exhibit a low-temperature

Arrhenius region (10^5 to 10^{13} poise) and a high-temperature region of curvature (10^{-1} to 10^5 poise); while some, like O-terphenyl,[35, 36] show curved $\log \eta$ versus $1/T$ relations over the full range of measurements (10^{-2} to 10^{13} poise). Examples of the variety of temperature dependences shown by organic glass-formers are illustrated in Fig. 11.6b.

In detail, the manifold forms of the observed viscosity–temperature relations cannot satisfactorily be described by any of the standard models.[30–32] This lack of agreement, together with the observed complexity of form among $\eta - T$ relations for various materials, suggests that the standard models are overly simple; and progress has recently been made toward the development of more complex treatments of the flow process.

Among inorganic oxide materials, the viscosity is often found to be a strong function of the composition as well as temperature. In the case of silicates, the viscosity is almost invariably found to decrease with increasing concentration of modifying cations. In many cases, this variation is quite pronounced. For example, at a temperature of $1700°$ C, the viscosity of fused silica can be decreased by about four orders of magnitude by the addition of as little as 2.5 mole % K_2O; and the effect of larger concentrations of modifiers is illustrated in Fig. 11.7. The temperature dependence of the viscosity is correspondingly decreased by the addition of modifying oxides.

In detail, there is no good picture relating viscosity to molecular structure in these systems. It seems clear, however, that an important effect of adding the modifier is the introduction of singly-bonded oxygens, which serve as weak links in the Si—O network.

In borate glasses, the viscosity at high temperatures is observed to decrease with increasing concentration of alkali oxide at high temperatures. At intermediate temperatures the viscosity decreases with small additions of alkali oxide, then increases to a maximum and then decreases again with increasing modifier concentration, while at low temperatures the viscosity increases with alkali oxide additions. This behaviour is not satisfactorily understood at the present time.

In more complex oxide glasses, the addition of modifying cations generally decreases the viscosity at any given temperature, while the addition of silica or alumina usually increases it. Beyond this, there is a general mixing effect, where the addition of more than one type of alkali or alkaline earth ion results in a higher viscosity than would be obtained with the same total concentration of a single modifier.

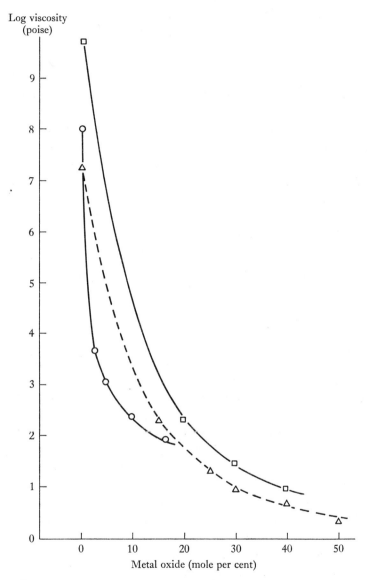

Fig. 11.7. Effect of modifier oxides on viscosity of fused silica (ref. 37). □, Li₂O–SiO₂, 1400° C; ○, K₂O–SiO₂, 1600° C; △, BaO–SiO₂, 1700° C.

11.1.5   Mechanical strength of glasses

Various treatments[e.g. 38] of the theoretical strengths, σ_{th}, of materials yield expressions of the form:

$$\sigma_{th} \approx \left(\frac{E\gamma}{a_0}\right)^{\frac{1}{2}}. \qquad (11.7)$$

Here E is the modulus, γ the surface energy and a_0 the interatomic separation. On this basis, the theoretical strengths of most inorganic oxide glasses should be in the 10^5 kgf/cm^2 range (a few million lbf/in^2); yet the observed strengths of glass bodies are typically only about 0.1 to 1 % of such values.

In accounting for this discrepancy, use is generally made of the now classic suggestion of Griffith that flaws in materials can act as stress concentrators, and that the separation of surfaces in fracture need not take place simultaneously over an entire cross-section. Rather, the separation is viewed as taki.·g place sequentially, with the theoretical strength being exceeded only at one p.'ace at a given time and the mean applied stress being potentially much less than the theoretical strength.

According to the Griffith treatment, a crack of length c in material should result in a lowering of the observed strength, σ_{obs}, from σ_{th} to:

$$\sigma_{obs} \approx \left(\frac{E\gamma}{c}\right)^{\frac{1}{2}}. \qquad (11.8)$$

According to this relation, cracks of only a few microns in size are required for a reduction in strength from the 10^6 to the 10^4 lbf/in^2 range. In cases where plastic or viscous deformation takes place in the vicinity of the crack tip, γ in equation (11.8) should be replaced by $\gamma + p$, where p is the plastic or viscous work.

Before discussing the application of this picture to various experimental situations, let us consider briefly the morphology and limiting velocity of fracture. We shall being by noting that fracture almost invariably begins in a direction normal to the direction of maximum tension. It generally stays in the same direction for part of the way across the diameter of a piece, and then often forks into two more fissures. The part where the fracture stays in a plane is generally smooth and is called the *mirror*, while that where forking takes place is generally rough and is called the *hackle*. In recent observations, roughness on a finer scale, in some cases as fine as a few hundred ångströms, has been observed, and is termed *micro-* or *submicro-hackle*.

Using strobe light techniques, fractures initiated by projectiles striking glass plates at high velocities were observed to proceed by the formation of radiating forked arrays of cracks and it was noted that all cracks in this forking range travel with the same velocity. This velocity varies from about 700 to about 2200 m/s depending on the composition, and is highest for fused silica. It does not depend significantly upon temperature or upon the state of internal stress of the glass. The relation between these limiting crack velocities and the wave velocities in the materials has been explored, but the agreement between experimental results and these simple theories leaves something to be desired.

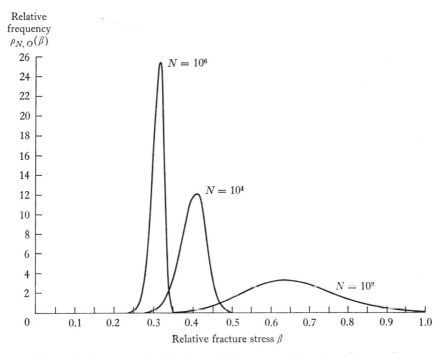

Fig. 11.8. Expected distribution of failures as a function of stress for sample containing N flaws. (ref. 39).

Starting with the concept of initial distributions of flaws (*a priori* probabilities of finding flaws of a given severity in each element of the specimen surface or volume), various statistical theories of fracture have been developed. All such theories predict that larger samples should be weaker, and that the dispersion in failure values would increase as the median strength increases (see Fig. 11.8). Such theories have not directly treated time dependent phenomena, although time effects might be easily included in statistical treatments if the time dependent processes did not change the geometry of the flaw spectrum.

When the surface energy in equation (11.8) is evaluated from studies of fracture with intentionally introduced flaws of known size, interesting results are obtained. For ship steel and glassy polymers[e.g. 40−4] the experimental γ's are generally about three orders of magnitude higher than the theoretical estimates. This difference has been explained by the occurrence of plastic deformation or viscous flow at the tip of the crack, which can absorb a large amount of work during the crack propagation. In the case of glassy polymers, such flow can take place at temperatures well below the glass transition, presumably because the viscosity is stress dependent (see discussion in §11.14.) or because of local heating effects. This flow has been associated with the appearance of colours on fracture surfaces, which have been taken[40, 41] as evidence of molecular orientation. Also seen on many fracture surfaces, particularly at low crack velocities, are crazing and fine cracking, which would also be reflected in a larger surface work of fracture.

The surface work has been observed[42] to decrease as the degree of cross-linking increases. Concomitant with the lowering of the surface work, the colours on fracture surfaces are found to disappear or change to shorter wavelengths. Consistent with this behaviour, the surface work in the case of inorganic oxide glasses is greater than theoretical estimates by a factor of only three or less.[45] In general, some amount of viscous flow might be expected in the vicinity of a moving crack tip, since the stresses there are very likely above the critical stress for non-Newtonian flow even in the case of phase separated glasses. The question of moment, however, is the extent of the flow which may be expected, and this remains an open question.

Turning now to the dependence of glass strength on various parameters, we might note that experimental results may be divided into three ranges:

(1) 140–1400 kgf/cm²: glass products; samples tested without particular care;

(2) 2100–42 000 kgf/cm²: fibres, thin rods, selected areas of plates, etc. tested under careful conditions; glass samples strengthened by surface treatment;

(3) 4.2×10^4—1.4×10^5 kgf/cm²: SiO$_2$ fibres and thin rods tested under careful conditions at low temperatures or under vacuum.

In the first range, the expectations of statistical theories are generally confirmed; and the results are very likely associated with relatively gross flaws introduced by mechanical damage. In the second range, the predictions of the statistical theories are generally not confirmed. The dispersion in fracture values does not increase as the median strength increases; and under appropriate conditions the fracture stress does not decrease with increasing sample size. The behaviour in this range is not satisfactorily understood at the present time; it seems likely, however, that it is associated with some uniform flaw source, perhaps related to the forming conditions, which would be beyond the scope of the statistical theories. In the third range, failure seems associated with flaws only tens of ångströms in size; the origin of these flaws may be chemical inhomogeneities or very small pores or the like.†

For the remainder of this section we shall consider the dependence of glass strength on various material and environmental conditions.

(a) Surface condition

It is by now well established that the flaws which are responsible for the low practical strengths of glass bodies are associated primarily with their surfaces; and to observe strengths in the 10^4 kgf/cm² range, one must ensure that the surfaces of the samples are clean, untouched and undamaged. Even touching the surface with one's fingers can reduce the strength from this range by an order of magnitude or more. Etching the surface of a damaged body with HF often restores the original high strength. As an example of this behaviour, Table 11.1 shows the results[24] on $\frac{1}{4}$ in diameter glass rods tested for one-hour periods.

TABLE 11.1. *Effect of surface condition on strength (ref. 24)*

Treatment	Breaking strength (kgf/cm²)
Severely sandblasted	140
As received from factory	460
Acid etched and lacquered	17 500

Work by Holloway and his co-workers[46] has indicated that microscopic dirt particles, bonded to glass surfaces, may play an important role in limiting the observed strengths. Specifically, fractures were frequently seen to propagate from such particles, which affected the strength in tension, not compression, while original high strengths were found in regions free of such adherent dirt particles. The effect of particles on strength may be related to differences in moduli between particle and glass (as suggested by Holloway) or to differences in their thermal expansion coefficients. Nothing is known, however, about the strength of adhesion of the particles to the glass surfaces; but

† It might be noted that testing under ambient conditions in air results in a mean fracture stress lower by about a factor of 3 than that observed at low temperatures or in vacuum (see discussion under 'Time, temperature and atmosphere' §11.1.5 b).

in any event, the role of the particles may depend significantly on water in effecting local solution and adhesion and providing mobility at relatively low temperatures.

Whether the surface-associated flaws are dirt particles or Griffith cracks or other as yet unidentified imperfections, they are active primarily in tension; and a substantial improvement in strength may be effected by placing the surfaces of the bodies in an initial state of compression. This is generally accomplished in three ways:

(1) Thermal tempering where rapid cooling of the exterior is initiated just below the softening point of the glass. The interior cools more slowly and continues to contract at a relatively high rate after the surface is rigid and contracts more slowly. The practical limit of such treatment is about 2100 kgf/cm² ; and it is not well suited for thin (less than about $\frac{1}{8}$ in) samples.

(2) Chemical tempering[47, 48] in which surface compression is achieved by changing the composition of the surface. This may be accomplished with an overlay or coating of a low expansion material; or, most frequently with aluminosilicate glasses, by ion-exchanging a thin surface layer (typically 100–200 microns in extent), substituting large ions for small ions below the strain point of the glass. With this technique, compressive stresses in the 10⁴ kgf/cm² range may be obtained (albeit in a very thin surface region).

(3) Crystallization[49] in which the surface region, sometimes of a different composition from the bulk, is crystallized to give low-expansion crystals.

(b) Time, temperature and atmosphere

It has long been known that the fracture stress generally decreases with increasing duration of the load. This phenomenon, called *fatigue*, is apparently associated with impurities, most notably water vapour, and is effectively absent in vacuum or at low temperatures. Examples of fatigue behaviour are shown in Fig. 11.9. While the relation proposed by these workers cannot be valid at very short times, it provides a useful description over nearly all the observed range:

$$\log t = A/\sigma + B. \tag{11.9}$$

Here t is the time to fracture, σ is the stress, and A and B are constants depending on the material, temperature and atmosphere. Both cyclic and static fatigue experiments, carried out on soda-lime glass, indicate that the relevant parameter in fatigue is the duration of the tensile stress, and not whether it is static or cyclic.

The essence of the fatigue process has been associated[50-2] with a stress corrosion process, in which a sufficiently high stress† enhances the rate of corrosion at the tip of a crack relative to that at the sides. Under such circumstances, the crack will sharpen and deepen until the stress concentration at its tip builds up to that required for failure. An alternative interpretation would

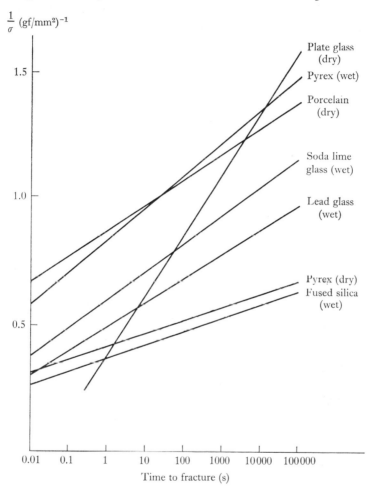

Fig. 11.9. Reciprocal of stress versus log of time to fracture (ref. 50).

be based on the lowering of the surface energy by the adsorption of an active species. At least in the case of fused silica tested at very high stress levels,[53] the former suggestion of surface-controlled stress corrosion seems more

† At low stresses, etching may be expected to increase the strength of the sample by rounding off the flaws.

plausible than the latter; but either mechanism is capable of explaining the observed very long time strength under ambient conditions being about one-third of the very short time strength, and under appropriate circumstances, either may represent the appropriate description.

The strength of glasses seems to increase significantly with decreasing temperature below ambient, as shown in Fig. 11.10. This increase is very

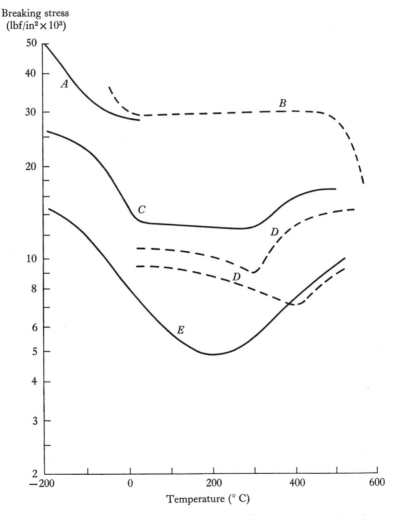

Fig. 11.10. Breaking strength of glass as a function of temperature (ref. 54).
A = fibres, 0.008–0.016 in diameter in tension; B = bars, window glass in bending; C = fibres, 0.052 in diameter in tension; D = strips, soda-lime glass in bending; E = rods, 0.25 in diameter soda-lime glass in bending.

likely associated with decreasing atomic mobility at lower temperatures (and hence a smaller rate of chemical attack or motion of adsorbing species). Below liquid nitrogen temperature, the strength seems to change little with changes in temperatures.[53] At temperatures above about 200° C, the strength seems to increase with increasing temperature, perhaps because of decreasing surface adsorption of atmospheric water or a smaller effect of absorbed dirt particles or increasing deformation work at the crack tip.

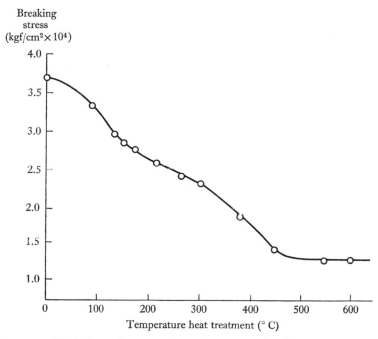

Fig. 11.11. Variation of room-temperature strength with temperature of previous heat-treatment (ref. 55).

Heat treatment at elevated temperatures prior to testing at room temperature seems to have a deletereous effect on the strength of glasses, as shown in Fig. 11.11. The origin of this phenomenon is not clear, but it may be related to the effect of adherent dirt particles or adsorbed corrosive impurities.

In studies of fatigue, the time to failure has been found to increase with decreasing temperature (see Fig. 11.12); and at very low temperatures the times become sufficiently long that samples break upon application of a given load or not at all.

(c) Composition

Little is known in detail about the variation of strength with composition; although it does seem that the sensitivity of glasses to fatigue does increase with increasing alkali concentration, and to date strengths in the 10^5 kgf/cm² range have been found only with SiO_2 fibres. The latter observations may reflect the effect of composition on related phenomena such as solubility and dissolution kinetics.

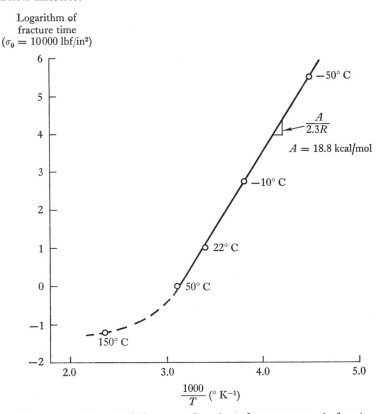

Fig. 11.12. Time to failure as a function of temperature (ref. 50).

Little is known about the possible effect of phase separation on strength. It does seem clear, however, that for a system which separates into two phases having quite different glass transition temperatures, the difference in expansion of the phases could result in local stresses of significant magnitudes. There is, of course, a limitation to the build-up of such stresses, in their relief by flow processes.

Another factor which might influence the strength of glass in the high-stress region is the presence of compositional heterogeneities, including submicroscopic gas bubbles, which are almost invariably present (to a greater or lesser extent) in glass bodies.

(d) Size and forming conditions

Most experimental work in both the high- and low-stress regions has supported the early suggestion that the strength of glass bodies decreases as their size increases. Recent studies have, however, indicated that strengths in the 10^4 kgf/cm² range can be obtained with relatively massive samples. More directly, the results by Otto[56] have indicated that the tensile strength of glass fibres is independent of diameter, provided they are formed at the same forming temperature. The range of diameters covered was 20–60 μm, and the observed strength was about 28 000 kgf/cm². A similar independence of the strength on fibre diameter was found by Thomas[55] provided the drawing temperature was sufficiently high to permit a uniform fibre to be produced.† Again the range of diameters covered was 20—60 μm, and the observed strength was about 37 000 kgf/cm².

Thus in the high-strength range, the strength may or may not be a function of sample size, depending upon the forming conditions (perhaps most importantly the cooling rate). The relation between such conditions and the strength is, however, far from clear. For example, while Otto reported the strength to increase with forming temperature (for a given nozzle and fibre diameter), Thomas found no such temperature effect on strength. It has been suggested by Glicksman[57] that this difference may be related to differences in nozzle tensions and cooling rates for the two sets of experiments, but the suggestion must be regarded as a tentative one. Beyond this, forming conditions seem likely to play an important role in the very small dispersion of strengths which is frequently observed in the high-strength region.

In the low-strength region of nearly all commercial products the strength does indeed seem to decrease with increasing sample size; and the dispersion does apparently increase with increasing strength. In these cases, the conceptual framework of statistical theories seems useful for discussing the experimental results.

† In both studies, lime alumina borosilicate glasses were investigated.

11.2   Glass-ceramic materials

One of the most striking advances in materials technology in recent years has been the development of practical glass-ceramic materials. These materials, which are formed by the controlled crystallization of appropriate glasses, generally consist of a large proportion (typically 90–98 volume per cent) of very small crystals (usually smaller than one micron) with a small amount of residual glass phase making up a pore-free composite. In discussing these materials, we shall first describe the principles involved in their fabrication, then consider specific systems of current interest and the physical basis of the properties have been obtained with these systems, and go on to discuss some recent advances in the area and some likely directions for future development.

11.2.1   Principles of fabrication

The essential features of the production of glass-ceramic materials include melting the constituents and forming the bodies using conventional glass-forming techniques; carrying out controlled nucleation and crystal growth to give the desired conformation of the desired phases; and in some cases carrying out subsequent thermal or chemical treatment, applying an appropriate glaze or the like.

The techniques of glass-melting and glass-forming are well described in the literature[58] and will not be repeated here. Suffice it to say that the techniques are well developed, and can be carried out efficiently with a high degree of automation.

After forming an article in the desired shape, the processes of nucleation and crystal growth are carried out to produce the desired glass-ceramic body. In considering these processes, let us recognize at the outset that whenever a crystal is formed in a liquid, a surface must be created which separates atoms in the new crystalline array from those in the old amorphous configuration. Such a surface can be created only at the expense of energy, and its creation represents a barrier to the change of phase. For this reason, all liquids can be undercooled, at least to some extent, below their respective liquidus temperatures. In the case of most inorganic glass-forming liquids, the magnitude of this surface energy barrier combined with the relatively low atomic mobility results in the effective absence of internal homogeneous nucleation. With such materials, crystallization is almost invariably observed to initiate at the external surfaces, and is observed in the interior only along striae, seeds, or other inhomogeneities.

Because of the large grain size which results from external nucleation, such crystallization would be inappropriate for producing most glass-ceramic materials. For reasons of mechanical strength, to be discussed below, it is desirable to produce bodies having crystal sizes of 1 micron or less. For such small crystals occupying a large volume fraction, one needs numbers of nuclei of the order 10^{12}–10^{15} per cm^3; and to effect nucleation in such numbers throughout the bulk of the glass, nucleating agents are added to the batch during the melting operation. The most commonly used nucleating agents are TiO_2 and ZrO_2; while P_2O_5, the Pt-group and noble metals, and fluorides are also used in some operations. TiO_2 is most frequently used in concentrations of 4 to 12 weight per cent, while ZrO_2 is generally used in concentrations near its solubility limit (4–5 weight per cent in most silicate melts). In some cases, ZnO_2 and TiO_2 are used in combination.

In inquiring into the role of the nucleating agents, a number of possibilities should be considered:

(1) The added nucleant may be important in promoting a phase separation process. The separation in turn can provide a very fine second phase dispersion, which may crystallize directly to one of the major crystalline phases, or may form a nucleant crystalline phase on which crystals of the major phases may later form and grow.

Considered in detail, the role of phase separation in the nucleation process could be related to:

(a) the formation of an amorphous phase of relatively high mobility in the appropriate temperature range (at sufficient motivating potential for crystallization), from which the nucleation of the crystalline phases may proceed at reasonable frequencies;

(b) the introduction of second phase boundaries (between the phase separated regions) on which the nucleation of the first crystalline phase, be it the nucleant phase or the major phase, may take place;

(c) the provision of driving force for the crystallization of the desired phase (prior to the separation process, there may not be a driving force for the formation of the desired crystalline phases).

Among these various suggestions, the first seems likely to be the most important. Because of the small magnitude of the surface energy associated with boundaries between two amorphous phases,[59] the second possibility should be relatively unimportant; and while the third suggestion may be germane in some particular cases, it seems unlikely to have general applicability.

(2) The added nucleant may directly form a crystalline nucleating phase by

a precipitation process, without the precursor step of phase separation. This certainly seems to be the case when Pt-group or noble metals are used as nucleants, and might well be germane in many uses of oxide nucleants, where TiO_2- or ZrO_2- rich crystalline phases may directly form on cooling the melt.

(3) The added nucleant may assist directly in the nucleation of the parent crystalline phases without the occurrence of phase separation or the formation of a crystalline nucleating phase. This might take the form of decreasing the overall melt fluidity in the appropriate temperature interval, or providing the appropriate molecular configurations in the melt to promote nucleation of the crystalline phases, or enhancing the diffusivity of a major species, or the like.

The experimental evidence on the role of the oxide nucleating agents leaves much to be desired. In at least one case,[60] involving the crystallization of a TiO_2-nucleated Li_2O–Al_2O_3–SiO_2 glass ceramic, phase separation on a scale of about 50 Å followed by the formation of a crystalline TiO_2-rich nucleating phase† seems to be the sequence of the nucleation stage. In other similar systems, however, there is no evidence of structural heterogeneities, detectable by either electron microscope or light scattering observations, prior to the appearance of the crystals of the major phases. Whether this merely represents difficulties of detecting small-scale heterogeneities, or whether it reflects a multiplicity of nucleation modes, must await further study.

Despite this uncertainty in the role of the nucleation catalyst, it is clear that the effect of such additions is to produce copious nucleation (of the order of 10^{12} per cm^3 or more) throughout the bulk of the glass; and with such copious nucleation, it is possible to obtain a fine grained, largely crystalline body having desired properties.

Going beyond the role of the nucleant to the general process of crystallizing a glass-ceramic body, we shall refer to the temperature–time cycle of Fig. 11.13. After the material has been melted and formed at elevated temperatures, it is generally cooled to a range of temperature (often ambient) below the temperature where the desired nucleation of the major phases will be carried out. At this stage, the material may be largely homogeneous or may contain phase-separated domains or very small crystals of the nucleant phase. In any case, the sample is then heated at a rate limited by the avoidance of thermal shock (typically about 1–5° C/min) to the temperature where nucleation of the major phases is effected. In accomplishing this, the sample is generally held at the nucleation temperature (where the melt viscosity is usually in the range 10^{11}–10^{12} poise) for 1–2 hours.

† This phase was estimated to contain about 35 weight per cent Ti and about 20 weight per cent Al. The starting material contained 4.75 weight per cent TiO_2.

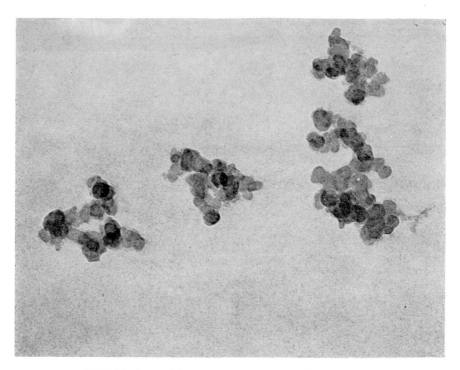

9.1 HAF black particle aggregates at a magnification of 25 000.

10.1 Portion of cotton fibre (photograph by E. Slattery).

(*facing page* 320)

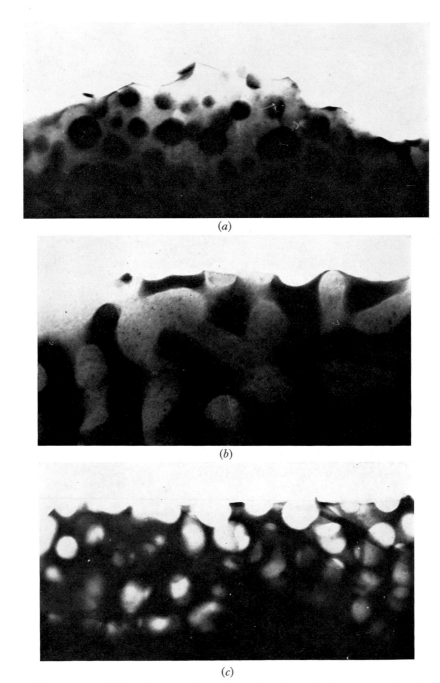

11.1 Direct transmission electron micrographs showing submicrostructural variation as a miscibility gap is traversed: (a) 6 % BaO, (b) 10 % BaO, (c) 16 % BaO.

13.1 Etch pits on the surface of an annealed tungsten crystal ($\times 310$).

13.2 Subgrain boundaries in tungsten single crystal ($\times 310$).

13.3 Slip ellipses in a stretched crystal of solid mercury (ref. 7).

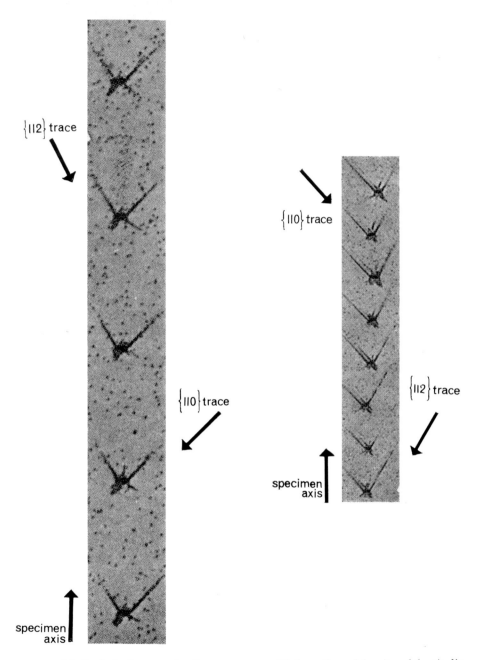

13.4 (*a*) Optical photograph of molybdenum surface etched after indenting (× 125).

13.4 (*b*) Lengthened (110) and (112) slip bands after stress pulsing (× 125).

After nucleation is completed, the material is heated further to carry out the desired growth of the major crystalline phases. This heating is typically done at a rate of about 5° C/min, and continues to the chosen holding temperature for growth. The latter temperature is generally chosen so as to maximize the kinetics of crystal growth, subject to the constraints of obtaining the desired combination of phases and avoiding deformation of the sample or resolution of some of the phases or unwanted phase transformations within the crystalline

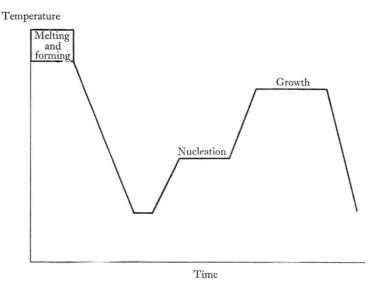

Fig. 11.13. Temperature–time cycle for the crystallization of a glass-ceramic body (schematic).

phases. The holding temperature and the time at which the sample is held at the temperature depend strongly on the particular system and the particular composition selected, as well as upon the properties which are desired in the final body; and under appropriate circumstances, one might want to employ more than a single holding temperature for growth.

In most cases, the crystallization is carried out until the extent of crystallinity reaches about 90–98 %. The typical grain size achieved by this crystallization process is in the range of 0.1–1 micron. After the desired crystallization, the sample bodies are cooled down to ambient conditions, generally at rates of 5–10° C/min or more. In subsequent treatment, a glaze coat and appropriate decoration may be applied as desired.

In most commercially important systems, a great variety of combinations of phases may be obtained, depending upon details of the nucleation and growth procedures selected, as well as upon the composition. For a given composition and heat treatment schedule, the crystalline phases present will change during the course of the crystallization treatment. For details on these matters, the reader is referred to the book by McMillan[61] and other literature sources.[62–4] Suffice it to say that the optimization of the processing of a given glass-ceramic material can represent a formidable task for the crystallization specialist. Also non-trivial are the problems involved in scaling the processing from one sample size to another. These last problems are associated largely with heat flow effects and with the criticality of the crystallization treatment. In overcoming such problems, the crystallization steps are sometimes carried out at relatively low temperatures, where nucleation and growth times may be long relative to the time for changes in sample temperature.

11.2.2   Important glass-ceramic systems

Among the systems in which technologically interesting glass-ceramic materials have been produced, the following seem most worthy of note:

(a) $Li_2O–A_2lO_3–SiO_2$

This system is well-known for glass-ceramic materials having very low thermal expansion coefficients, and hence very high resistance to thermal shock. Among the trade names for materials in this system might be noted Corning's Corning Ware, Owens-Illinois' Cer-Vit, and PPG's Hercuvit. The very low expansion coefficients in this system, which can be appreciably lower than that of fused silica, are associated with the presence in the crystallized materials of crystalline β-spodumene ($Li_2O.Al_2O_3.4SiO_2$) which has a low expansion coefficient and β-eucryptite ($Li_2O.Al_2O_3.2SiO_2$) which has an expansion coefficient that is larger in magnitude and negative. The compositions of typical glass-ceramic materials in this system cover a range about (in weight per cent): Li_2O (5), Al_2O_3 (25), SiO_2 (65), and TiO_2 (5).

(b) $MgO–Al_2O_3–SiO_2$

This system is well-known for glass-ceramic materials having high electrical resistivity and high mechanical strength. The high strength has been associated with the presence in the crystallized materials of crystalline α-cordierite ($2MgO.2Al_2O_3.5SiO_2$). The compositions of typical glass-ceramic materials in this system cover a range about MgO (13), Al_2O_3 (30), SiO_2 (47), and TiO_2 (10).

(c) $Li_2O–MgO–SiO_2$

Glass-ceramics in this system are noted for their high expansion coefficients. They have been nucleated, in practice, with P_2O_5.

(d) $Li_2O–ZnO–SiO_2$

Glass-ceramics in this system are noted for high mechanical strengths with a wide range of expansion coefficients. They have been nucleated in practice with P_2O_5 or the noble metals.

(e) $Na_2O–Al_2O_3–SiO_2$

Glass-ceramics in this system are noted for the case with which their surfaces can be chemically strengthened. The important crystalline phase in these materials, nepheline, occurs over a range of composition.

(f) $LiO_2–MgO–Al_2O_3–SiO_2$

Glass-ceramics in this system are noted for their variable (in some cases low or negative) thermal expansion coefficients, transparency (in some cases), and the ease with which they can be chemically strengthened. The important crystalline phase is a stuffed β-quartz solid solution.

11.2.3 Properties of glass-ceramic materials

Many advantages of glass-ceramic materials over conventional glasses and ceramics are related to the absence of porosity, and occurrence of well-dispersed, very small crystals, and the particular material properties of the crystalline and glass phases present in the glass-ceramics. The absence of porosity in turn is related to the small volume changes involved in crystallizing these systems† and to the fact that changes in volume can be accommodated by flow. The small crystal size is, as noted above, a result of copious nucleation achieved by adding the nucleant to the melt; and the particular material properties can to a significant extent be programmed by selection of the compositon and crystallization treatment.

Under ambient conditions, glass-ceramic materials behave as brittle, elastic solids. They exhibit little ductility, but rather deform elastically up to

† One of the criteria used in selecting systems for commercial application is that the volume change on crystallization be relatively small, so that the dimensions of formed pieces can be reasonably maintained. In contrast, in the powder processing of most ceramic materials, volume changes on firing can often be substantial, and the final products frequently have closed porosities of the order of 2–10 %.

the load which causes fracture. Like the inorganic glasses from which they are derived, glass-ceramics are characterized by large elastic moduli and fail by the propagation of flaws.

(a) *Elastic properties*

Most glass-ceramic materials have Young's moduli in the range $8-14 \times 10^5$ kgf/cm². The relation of the elastic moduli of the crystallized bodies to those of the constituent phases may be viewed as a problem in the elasticity of composite materials.[65, 66]

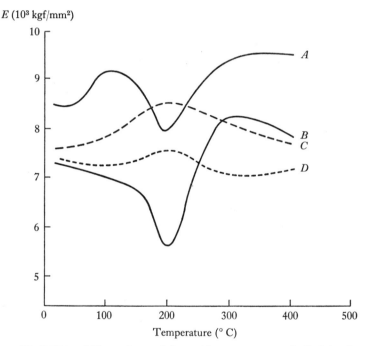

Fig. 11.14. Variation of Young's modulus with temperature (ref. 61). $A = Li_2O-ZnO-SiO_2$ (low ZnO); $B = Li_2O-ZnO-SiO_2$ (high ZnO); C, $D = Li_2O-ZnO-Al_2O_3-SiO_2$ (different compositions).

The variation with temperature of the elastic moduli of glass-ceramic materials depends strongly on the particular crystalline phases which are present. This is illustrated in Fig. 11.14, where compositions A and B (which exhibit pronounced minima) contain cristobalite as a major phase, while compositions C and D do not. Cristobalite has a well-known $\alpha-\beta$ inversion which generally occurs between 200 and 275° C.

(b) Mechanical strength

The practical strengths, without special surface treatments, of glass-ceramic materials are generally higher than those of most glasses. In some systems, for example, rupture strengths as high as 1400–2800 kgf/cm² have been obtained on abraded samples under ambient conditions. These high mechanical strengths are very likely related to the absence of porosity and the presence of many very small crystals, often of a mechanically hard phase, as well as to their relatively high chemical durability. The chemical durability affects their fatigue behaviour and also their susceptibility to surface damage; the absence of porosity avoids this frequent source of weakness; and the presence of many small crystals affects the susceptibility to mechanical damage as well as the inherent strength. Cracks propagating through the glass phase may be arrested or diverted or caused to fork at the interface with a mechanically stronger crystalline phase. This may be associated with the crack, initially propagating in a direction normal to the direction of maximum tensile stress and unable to propagate through the harder crystal, being required to move in a direction having a smaller normal tensile stress in order to propagate further through the glass phase. Alternatively, in some systems, residual compressive stresses may be found in regions of the glass adjacent to crystals. Further, the small size and large volume fraction of the crystals limit the size of incipient flaws in the glass (or crystal) as well as the boundary stresses between crystals and surroundings.

The effect of composition on the strength of glass-ceramics is illustrated in Table 11.2. As indicated above, the high strengths of glass-ceramics in the $MgO–Al_2O_3–SiO_2$ system seems associated, at least in part, with the presence of cordierite as a major crystalline phase. The significantly lower strengths in the $Li_2O–Al_2O_3–SiO_2$ system have been associated[61] with internal stresses resulting from the presence of crystals having low and (more importantly) negative expansion coefficients. The existence of additional other factors, such as durability and susceptibility to surface damage, seems to be suggested by the results obtained on the other systems, as well as by observations of surface strengthening.†

With reference to this last, it has been found that the mechanical strengths of glass-ceramic bodies may be materially increased by surface treatments similar to those regularly employed with glasses (discussed in §11.1.5). The effectiveness of such strengthening, which will be considered in some detail in

† Internal stresses may, indeed, play their most significant role in promoting atmospheric attack and mechanical damage in the surface regions.

TABLE 11.2. *Effect of composition on the strength of glass-ceramics (ref. 67)*

System	Rupture strengths (kgf/cm^2)
$Li_2O–Al_2O_3–SiO_2$	1100–1300
$MgO–Al_2O_3–SiO_2$	1200–2700
$CaO–Al_2O_3–SiO_2$	1200
$SrO–Al_2O_3–SiO_2$	1100
$BaO–Al_2O_3–SiO_2$	550–630
$ZnO–Al_2O_3–SiO_2$	420–1300
$CdO–Al_2O_3–SiO_2$	350–1300

the following section, suggests that the strength-limiting flaws are associated primarily with the surfaces of the bodies (just as in the case of glasses).

Because of the presence of the crystalline phases, glass-ceramic materials invariably are more refractory than corresponding glasses; and softening temperatures in excess of 1000° C have been reported.[61] The limitation in this regard is undoubtedly imposed by the residual glass phase, and refractoriness is increased by increasing the percentage crystallinity and by achieving appropriate residual glass compositions.

In discussing the effect of temperature on the strength of glass-ceramic materials, it might be noted that most systems exhibit strengths which decrease with increasing temperature. An example of this behaviour for a $MgO–Al_2O_3–SiO_2$ glass-ceramic is shown in Fig. 11.15. For some systems, however, such as $Li_2O–Al_2O_3–ZnO–SiO_2$, the strength decreases as temperature is raised above ambient, goes through a minimum, and then increases with further increases in temperature.

Glass-ceramic materials having thermal expansion coefficients higher than 160×10^{-7}° C^{-1}, lower than 0.5×10^{-7}° C^{-1}, and as strongly negative as -35×10^{-7}° C^{-1} have been produced.[61] These are illustrated in Fig. 11.16, together with data on other familiar materials. The variety of observed thermal expansion coefficients for glass-ceramics are directly related to a substantial variation in the thermal expansion of the crystalline phases obtained in the various systems. The greatest technological interest has been directed to low-expansion glass-ceramics, which can have remarkable dimensional stability and resistance to thermal shock. These are being used in a variety of applications from telescope mirror blanks to home cookware.

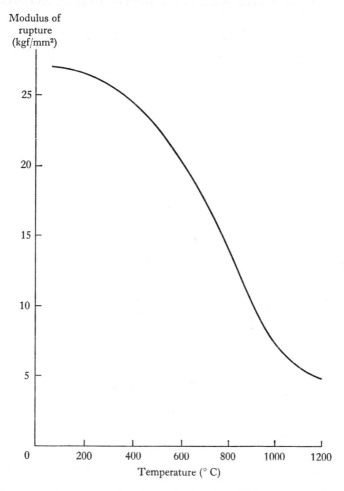

Fig. 11.15. Variation of mechanical strength with temperature for
MgO–Al$_2$O$_3$–SiO$_2$ glass-ceramic (ref. 61).

11.2.4 Recent advances and future developments

Some of the more striking advances in recent years have been concerned with
the systematic development of glass-ceramic materials with particular
desired properties, and with the strengthening of the materials by appropriate
surface treatment. The former category may be illustrated by the development
of Cer-Vit materials having expansion coefficients less than $0.5 \times 10^{-7}\,°$ C^{-1}.
This was accomplished by the systematic variation of the final grouping of

phases through systematic variation of the composition and crystallization treatment.

In the development of glass-ceramic materials having substantially higher engineering strengths, use has been made of the techniques of surface treat-

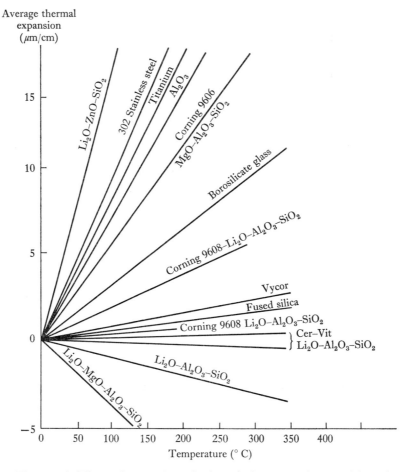

Fig. 11.16. Thermal expansion of selected glass-ceramic materials and other familiar materials (ref. 68).

ment employed to strengthen glasses. This has taken the form of care in the handling of the glass body prior to carrying out the crystallization, relatively rapid cooling of the surfaces of the crystallized body over a range of temperatures below the softening point of the residual glass phase, overlaying a relatively high expansion crystallized body with a lower expansion glaze[69] and most dramatically by ion-exchanging the surface layers of the crystallized bodies.[63, 69—72]

In the last case, this treatment includes not only the direct strengthening of the glass phase as discussed in §11.1.5, but also more important effects which involve the crystalline phases as well: phase transformations to phases of different volumes and expansion coefficients, and solid solution in the already-existing crystalline phases, changing their volumes and expansion coefficients. This development has led to the exploration of glass-ceramic systems having desirable ion-exchange characteristics.

With such surface treatments, abraded rupture strengths in the 10^4 kgf/cm^2 range have been obtained. While such chemical tempering of glass-ceramics seems promising for future development, it has inherent problems associated with the small thickness of the surface compressive layers; should a pointed object penetrate through this layer, failure is immediate.

In addition to the further development of such techniques of strengthening, their application in new situations, and the concomitant development of systems with desirable surface-treatment characteristics, one can look forward to the development of systems with inherently higher mechanical strengths. Indeed, the central feature of the glass-ceramic concept may well be the possibility which it opens for developing materials with given desired properties or combinations of properties.

Within the latter category will likely come materials having low expansion coefficients together with high mechanical strength. Beyond this, one can anticipate the development of glass-ceramic materials which combine particular desired properties (as chemical inertness and high mechanical strength) with low cost. One might also look forward to the application of the glass-ceramic concept to other classes of materials.

In such developments, greater attention will be directed to questions such as the relation between the properties of the composite glass-ceramic and the properties of the individual phases, the effect of various constituents on phase equilibria and on the kinetics and morphology of crystal growth, the effect of the morphology and configuration of the crystalline phases on the material properties, the dependence of the surface strengthening processes on time, temperature, nature of the crystalline phases, and material properties, etc.

Because of the substantial potential rewards, one can be confident that such materials will be developed and such questions answered, and in this process, our understanding of the elasticity, plasticity and structure of materials should be refined and enlarged.

References to chapter 11

1   B. E. Warren & J. Biscoe, *J. Amer. Ceram. Soc.* **21**, 49 (1938).
2   W. H. Zachariasen, *J. Amer. Chem. Soc.* **54**, 3841 (1932).
3   T. P. Seward, D. R. Uhlmann, D. Turnbull & G. R. Pierce, *J. Amer. Ceram. Soc.* **50**, 25 (1967).
4   T. P. Seward, D. R. Uhlmann & D. Turnbull, *J. Am. Ceram. Soc.* **51**, 634 (1968): **51**, 278 (1968).
5   J. W. Cahn & R. J. Charles, *Physics Chem. Glasses* **6**, 181 (1965).
6   M. Goldstein, *Trans. Soc. Rheol.* **12**, 69 (1968).
7   R. R. Shaw & D. R. Uhlmann, *J. Non-Cryst. Solids* **1**, 474 (1969).
8   R. R. Shaw & D. R. Uhlmann, *J. Non-Cryst. Solids* **5**, 237 (1971).
9   R. R. Shaw & D. R. Uhlmann, *J. Amer. Ceram. Soc.* **51**, 377 (1968).
10   P. J. Bray in *Interaction of radiation with solids* (New York, 1967).
11   R. L. Mozzi & B. E. Warren, *J. Appl. Cryst.* **2**, 164 (1969).
12   R. L. Mozzi, 'An X-ray diffraction study of the structure of glass', Sc.D. thesis, Massachusetts Institute of Technology, Cambridge, Mass., 1967.
13   R. R. Shaw & D. R. Uhlmann, 'Physical, optical, and elastic properties of alkali borate glasses', to be published.
14   S. Spinner, *J. Amer. Ceram. Soc.* **45**, 394 (1962).
15   O. L. Anderson & G. J. Dienes, in *Non-crystalline solids* (New York, 1960).
16   L. Peselnick, R. Meister & W. H. Wilson, *J. Phys. Chem. Solids* **28**, 635 (1967).
17   F. P. Mallinder & B. A. Proctor, *Physics Chem. Glasses* **5**, 91 (1964).
18   R. W. Douglas, *Brit. J. App. Phys.* **17**, 435 (1966).
19   I. L. Hopkins & C. R. Kurkjian in *Physical acoustics* **2B** (New York, 1965).
20   C. R. Kurkjian, *Physics Chem. Glasses* **4**, 128 (1963).
21   J. DeBast & P. Gilard, *Physics Chem. Glasses* **4**, 117, (1963).
22   R. W. Douglas, P. H. Duke & O. V. Mazurin, *Physics Chem. Glasses* **9**, 169 (1968).
23   J. H. Li & D. R. Uhlmann, *J. Non-Cryst. Solids* **1**, 339 (1969).
24   C. J. Phillips, *Amer. Scientist* **53**, 20 (1965).
25   J. H. Li & D. R. Uhlmann, *J. Non-Cryst. Solids* **3**, 127 (1970).
26   J. H. Li & D. R. Uhlmann, *J. Non-Cryst. Solids* **3**, 205 (1970).
27   L. L. Sperry & J. D. Mackenzie, *Physics Chem. Glasses* **9**, 91 (1968).
28   G. S. Meiling & D. R. Uhlmann, *Physics Chem. Glasses* **8**, 62 (1967).
29   M. L. Williams, R. F. Landel & J. D. Ferry, *J. Amer. Chem. Soc.* **77**, 3701 (1955).
30   D. Turnbull & M. H. Cohen, *J. Chem. Phys.* **34**, 120 (1961).
31   F. Bueche, *Physical Properties of Polymers* (New York 1962).
32   G. Adam & J. H. Gibbs, *J. Chem. Phys.* **43**, 139 (1965).
33   E. H. Fontana & W. A. Plummer, *Physics Chem. Glasses* **7**, 139 (1966).
34   R. Bruckner, *Glastech. Ber.* **37**, 413 (1964).
35   W. T. Laughlin, 'Viscous flow and volume relaxation in simple glass-forming liquids', Sc.D. thesis, Massachusetts Institute of Technology, Cambridge, Mass. 1969.
36   R. J. Greet & D. Turnbull, *J. Chem. Phys.* **46**, 1243 (1967).
37   J. O'M. Bockris, J. D. Mackenzie J. A. Kitchener, *Trans. Faraday Soc.* **51**, 1734 (1955).
38   A. H. Cottrell, *The mechanical properties of matter* (New York, 1964).
39   J. C. Fisher & J. H. Hollomon, *Trans. Met. Soc. AIME* **171**, 546 (1947).

40 J. P. Berry, *J. Polymer Sci.* **50**, 107 (1961).
41 L. J. Broutman & F. J. McGarry, *J. App. Polymer Sci.* **9**, 589 (1965).
42 L. J. Broutman & F. J. McGarry, *J. App. Polymer Sci.* **9**, 609 (1965).
43 E. Orowan, *Welding J. Res. Suppl.* **34**, 157 (1955).
44 G. R. Irwin & J. A. Kies, *Welding J. Res. Suppl.* **33**, 193 (1954).
45 S. M. Wiederhorn, *J. Amer. Ceram. Soc.* **52**, 99 (1969).
46 W. Brearley & D. G. Holloway, *Physics Chem. Glasses* **4**, 69 (1963).
47 S. S. Kistler, *J. Amer. Ceram. Soc.* **45**, 59 (1962).
48 M. E. Nordberg, E. L. Mochel, H. M. Garfinkel & J. S. Olcott, *J. Amer. Ceram. Soc.* **47**, 215 (1964).
49 S. D. Stookey, J. S. Olcott, H. M. Garfinkel & D. L. Rothermal in *Advances in glass technology* (New York, 1962).
50 R. J. Charles in *Progress in ceramic science* **1** (New York, 1961).
51 R. J. Charles & W. B. Hillig in *Symposium on the mechanical strength of glass and ways of improving it* (Union Scientifique Continentale du Verre, Charleroi, Belgium, 1962).
52 W. B. Hillig in *Modern aspects of the vitreous state* **2** (New York, 1962).
53 W. B. Hillig in ref. 51
54 E. B. Shand, *Glass Engineering Handbook* (New York, 1958).
55 W. F. Thomas, *Physics Chem. Glasses* **1**, 4 (1960).
56 W. H. Otto, *J. Amer. Ceram. Soc.* **38**, 122 (1955).
57 L. R. Glicksman, *Glass Technol.* **9**, 131 (1968).
58 F. V. Tooley (Ed.), *Handbook of Glass Manufacture* (2 vols; New York, 1961).
59 J. J. Hammel, *J. Chem. Phys.* **46**, 2234 (1967).
60 P. E. Doherty, D. W. Lee & R. S. Davis, *J. Amer. Ceram. Soc.* **50**, 77 (1967).
61 P. W. McMillan, *Glass-ceramics* (New York, 1964).
62 E. A. Porai-Koshits (Ed.), *The structure of glass* **3** (New York, 1964).
63 G. H. Beall, B. R. Karstetter & H. L. Rittler, *J. Amer. Ceram. Soc.* **50**, 181 (1967).
64 R. A. Eppler, *J. Amer. Ceram. Soc.* **46**, 97 (1963).
65 Z. Hashin, *J. Appl. Mech.* **29**, 143 (1962).
66 Z. Hashin & S. Shtrikman, *J. Mech. Phys. Solids* **11**, 127 (1963).
67 S. D. Stookey, British Patent 829,447 (1960).
68 L. L. Hench, Notes from Summer Program on Materials Fabrication Processes, M.I.T., July, 1968.
69 S. D. Stookey in *High strength materials* (New York, 1965).
70 D. A. Duke, J. F. MacDowell & B. R. Karstetter, *J. Amer. Ceram. Soc.* **50**, 67 (1967).
71 B. R. Karstetter & R. O. Voss, *J. Amer. Ceram. Soc.* **50**, 133 (1967).
72 M. E. Nordberg, E. L. Mochel, H. M. Garfinkel & J. S. Olcott, *J. Amer. Ceram. Soc.* **47**, 215 (1964).

12 Thermohardening resins

A. Heslinga

12.1 Internal structure

From a technological viewpoint the hardening or two-stage resins are a very important class of polymers. Their chief characteristic is that of hardening or setting by a condensation reaction at increased temperature (thermosetting) or by a catalytically induced addition polymerization.

Before the final manufacturing step they consist of molecules with a low degree of polymerization containing reactive groups. At this so-called *A*-stage the resins can melt and flow under the influence of heat and pressure. With further polymerization they pass via the so-called *B*-stage into the *C*-stage, becoming infusible and insoluble. Highly complex three-dimensional molecular structures are formed this way by cross-linking the original low molecular weight molecules. The physical and mechanical properties of the resins are profoundly and irreversible altered in the course of moulding or casting processes. This cross-linking or cure is a very complex phenomenon dependent on many variables and difficult to unravel because the cured insoluble resins are not amenable to most of the common methods of examination. It is understandable in this connection that a complete theory relating structure and mechanical properties is still lacking, although the technical and commercial development of the hardening resins has been quite successful.

In practice the resins are in most cases mixed with various fillers and compounding agents which have considerable effect on the properties of the end product.

The main representatives of the thermosetting resins are phenolformaldehyde and the amino (urea or melamine) formaldehyde types. The phenolic resins will serve as a model to give some insight into the structurally important aspects of this group.

As shown in Fig. 12.1 formaldehyde reacts with phenols to form hydroxymethyl–phenol compounds, which react further in a secondary process with other phenol molecules to give a polymer linked by methylene bridges. This reaction is an example of a mechanism resulting in three-dimensional growth and structure.

This reaction is accompanied by the formation of water which, depending

332

on reaction conditions, is sometimes present in the cross-linked product as a separate phase.

The condensation reaction from small molecules to larger aggregates up to a molecular weight of about 1000, comprising some 10 phenolic nuclei, shows itself clearly when the gel point is reached. At this stage the molecular complexes are joined together by primary bonds and it is customary to say

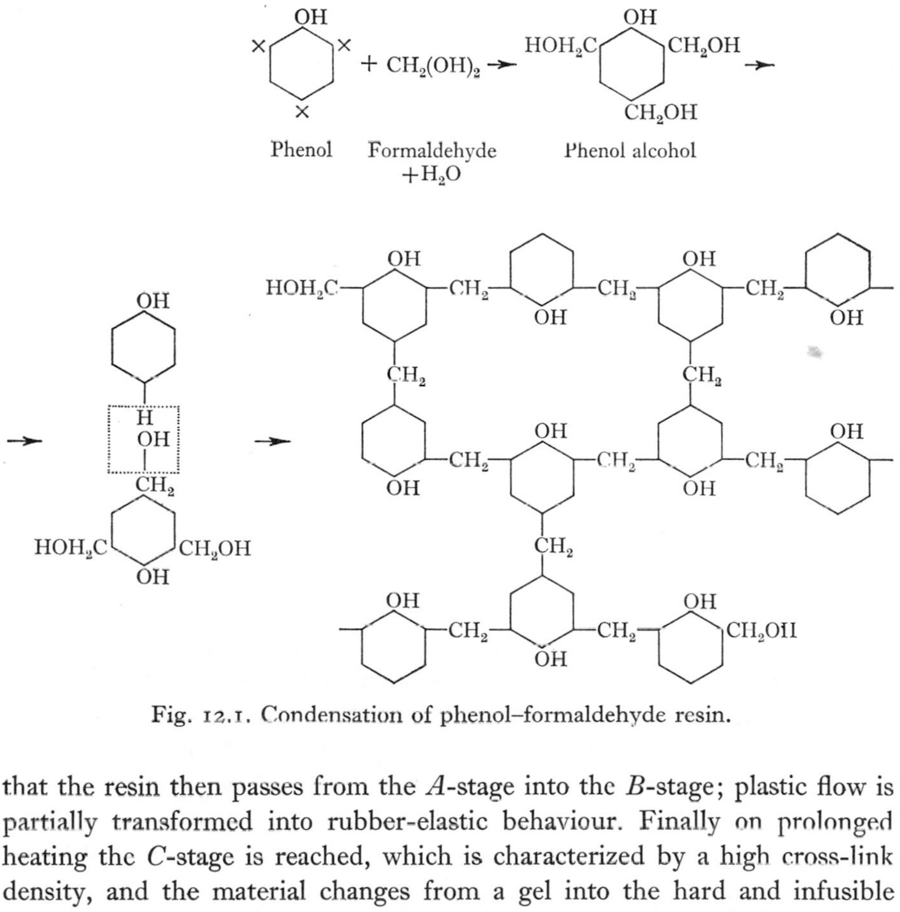

Phenol   Formaldehyde   Phenol alcohol
          +H₂O

Fig. 12.1. Condensation of phenol–formaldehyde resin.

that the resin then passes from the *A*-stage into the *B*-stage; plastic flow is partially transformed into rubber-elastic behaviour. Finally on prolonged heating the *C*-stage is reached, which is characterized by a high cross-link density, and the material changes from a gel into the hard and infusible condition of the end product. The main representatives of the addition polymerization resins are polyesters, polyisocyanates and epoxy resins.

As with the older thermosetting resins, the first pre-condensates carry functional groups such as vinyl, isocyanate, hydroxyl and epoxy groups. In a secondary process, these groups are reacted catalytically or by heat, cross-linking the resin molecules to a more or less homogeneous network. The greater flexibility of these resins compared with the phenol-formaldehyde polymers is partly due to the greater distance between the points where cross-linking occurs and hence a higher degree of free rotation of the molecular parts.

As an example, Fig. 12.2 shows the structure of the starting material for epoxy resins. These resins are characterized by a reactive epoxy group or oxirane ring $CH_2\!\!-\!\!CH-$, which can react with curing agents possessing active H-atoms, such as polyamines, or acid anhydrides.

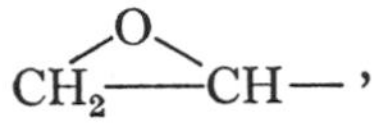

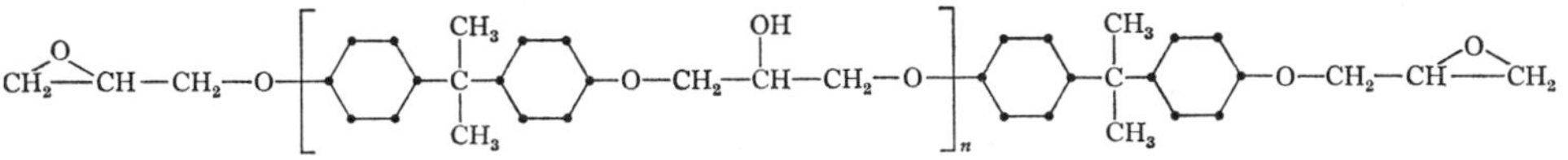

Fig. 12.2. Starting material for epoxy resins. Cross-linking occurs at epoxy or/and hydroxyl groups.

The final addition reactions are mostly very fast and very exothermic, and care must be taken to control the release of excessive heat. On the other hand this reactivity makes it possible, in contrast to phenolic resins, to harden the resins at room temperature.

12.2   Elastic properties

Assuming a homogeneous network consisting of all possible primary links De Boer[1] has calculated the strength of a phenolic resin in the final *C*-stage. In Table 12.1 several values are given in comparison with other materials. The theoretical value of 4000 kgf/mm^2 in no way compares to the value actually reached in practice of about 8 kgf/mm^2.

In an effort to explain this discrepancy Houwink[2] put forward his *Lockerstellen* theory for resin macromolecules. The weak points in the internal structure are a consequence of the steric nature of the condensation reaction. The chance that a single molecule will be added at each active centre of such a growing molecule becomes smaller as the molecule to be added begins itself to increase in size (see Fig. 12.3).

If one assumes that at a given moment the macromolecules *A* and *B* (whose chemically active points are marked *x*) are grown together at points 1, 2 and 3, no further attachment of *A* and *B* is possible since the freedom of movement necessary for such an addition is lacking. The structural features of the molecular entities also make the chance of additions at points 4, 5, 6

TABLE 12.1. *Comparison of experimental and theoral deticata for strength of certain amorphous materials*

Substance	σ_{max} (observed) kgf mm^{-2}	σ_{max} (calculated) kgf mm^{-2}
Phenol-formaldehyde resin (*C*-stage) at $-195°$ C	7.8	4300
Cresol-formaldehyde resin (*C*-stage) at $-195°$ C	3.8	3800
Glass, normally	3.5–8.5	1100†
Glass, under special conditions	to 630	
NaCl crystal, normally	0.6	200–400
NaCl crystal, under special conditions	to 160	

† This value was obtained by extrapolation from experimental data. Calculations based on atomic considerations would lead to much higher values.

and 7 extremely small. From this picture it follows that weak spots (*Lockerstellen*) may be expected to be formed in a resin conglomerate (Fig. 12.3*b*). On p. 23 another theory was presented to explain the low strength.

In accordance with this picture it follows that the polymer structure is entirely amorphous; no evidence of any crystallization phenomena, not even very small crystallites can be detected.[2]

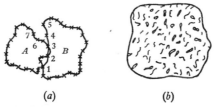

(*a*)   (*b*)

Fig. 12.3. (*a*) Further formation of *Lockerstellen* in a hardening resin. × represents chemically active sites; 1, 2, 3, points of juncture; 4, 5, 6 and 7, sites where further growth is extremely improbable. (*b*) A macromolecule with *Lockerstellen*.

The structure of phenol–formaldehyde resins may therefore be likened to a sponge having a framework of more or less cross-linked macromolecules. Between these substrates there will be sections which, depending on the degree of cure, contain material of a more thermoplastic character. It is evident that a concentration of stress at the faults will occur during deformation in a way similar to that postulated in the theory of Smekal for metals. The strength is determined only by the concentration of stresses at the most 'dangerous' spots.

Table 12.2 shows the moduli and tensile strengths of some resins with increasing degrees of polymerization.[2] Taking the phenol-formaldehyde resin as an example, the increase of E upon polymerization is from 290 to 595 and that of $\sigma_{\max}$ is from 0.01 to 6.7 kgf/mm². The replacement of secondary by primary bonds was practically complete within 10 min at 150° C.

TABLE 12.2. *Modulus of elasticity and tensile strength of some hardening resins*

Resin	Modulus of elasticity E (kgf mm⁻²)		Tensile strength σ max (kgf mm⁻²)		Elongation at proportional limit in per cent	
	At 20° C	At −195° C	At 20° C	At −195° C	At 20° C	At −195° C
Cresol-formaldehyde (A-stage)	152	—	$\pm 1 \times 10^{-2}$	—	—	—
Moulded 10 min at 150° C (C-stage)	513	—	2.5	—	—	—
Moulded 60 min at 150° C (C-stage)	500	803	3.3	3.8	0.6	0.5
Phenol-formaldehyde (A-stage)	290	—	$\pm 1 \times 10^{-2}$	—	—	—
Moulded 10 min at 150° C (C-stage)	591	—	6.0	—	—	—
Moulded 60 min at 150° C (C-stage)	595	1050	6.7	7.8	1.1	0.7
Urea-formaldehyde (cast) (C-stage)	310–380	—	3	—	—	—
Shellac Originally	135	—	1.3	—	—	—
After 20 hours at 110° C	260	—	—	—	—	—

The deformation of such resins in the C-stage is truly elastic. Some stress–strain curves have been reproduced in Fig. 12.4. Polymerization causes a regular increase of tensile strength until a maximum which in moulding practice is reached within 1 min per mm of wall thickness at 150° C.

The epoxy resins, containing only aromatic ring structures, like the phenolics in general, also give hard, rigid compositions with rather low impact strength. A high degree of flexibility is often desirable for such applications as adhesives, pottings and encapsulating. This can be achieved by mixing the aromatic-bisphenol-A epoxy with long chain aliphatic compounds bearing epoxy rings.

The most widely used available modifiers are polyglycol di-epoxides with the following structure:

$$H_2C \overset{O}{\diagdown} CH-CH_2-O-\left[CH_2-\overset{\overset{R}{|}}{CH}-O\right]_n CH_2-CH \overset{O}{\diagdown} CH_2$$

Commercial products are available where *n* varies from 2 to 7. They are normally used in blends with the aromatic resins. After curing, the cross-linking points are further from each other and free rotation is increased by the long aliphatic chains giving end-products with greater flexibility.

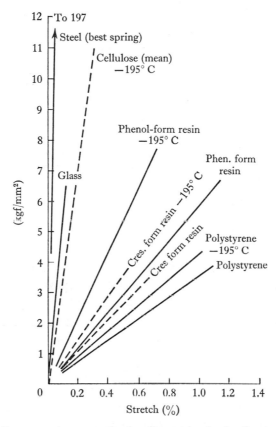

Fig. 12.4. Stress–strain curves for harding resins in the *C*-stage compared with other materials.

Table 11.5 shows the effect of a polyglycol di-epoxide on the physical properties of a cured epoxy-resin system. The addition of flexible resins increases the tensile strength, ultimate elongation and impact strength;

TABLE 12.3. *Flexible resin/bisphenol-A resin blend cured with methylene di-aniline (ref. 3)*

	70 % Bisphenol 30 % Flexible resin	100 % Bisphenol resin
Viscosity (c/s) at 70° C	58	100
Heat deflection temp (° C)	84	157
Flexural strength (lbf/in²)	14.060	16.970
Flexural modulus	2.7×10^5	2.3×10^5
Compress. strength (lbf/in²)	24.320	32.000
Tensile strength (lbf/in²)	9.160	8.150
Ultimate elongation (%)	8.1	3.8
Izod impact strength (ft/lbf/in notch)	1.16	0.44
Hardness (Rockwell M)	97	106

hardness and heat deflection are lower. The amount and type of the modifying resin to be used is determined by the end-use and properties required.

12.3 Plastic properties

In the *A*-stage the hardening resins show plastic and thermoplastic behaviour similar to the non-hardening resins (see §8.2). True flow can easily be shown to exist at high temperatures but when polymerization and cross-linking have proceeded to a certain degree there is only the possibility of quasi-flow; thermo-recovery becomes then clearly noticeable (*B*-stage). This behaviour is of great importance in practice. In order to obtain products of good appearance and reasonable strength a good balance between flow and degree of thermo-elasticity must be met.

Many technical methods are known for testing the flow properties of resin mixtures. Fig. 12.5 demonstrates[4] to what extent the flow is influenced by the resin content of the mixtures. From Fig. 12.6 it appears that the maximum flow is strongly dependent upon the temperature. The viscosity decrease due to heating is counteracted by a viscosity increase as a consequence of the cross-linking reaction. Consequently the moulding process must be per-formed between narrow limits of temperature (about 150° C) and at a considerable pressure (250 kgf/mm²). These high pressures are specially important for the thermosetting resins to act as a counteracting force against

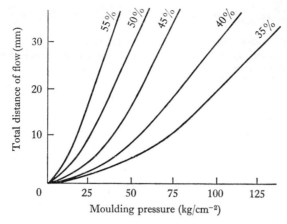

Fig. 12.5. Flow of some moulding mixtures, differing in resin content, plotted against pressure (temperature = 150° C).

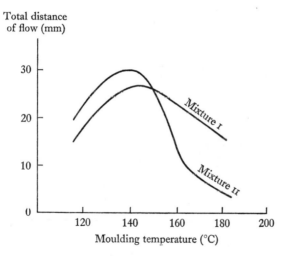

Fig. 12.6. Flow of some moulding mixtures plotted against temperature (constant pressure).

the vapour pressure of small molecules which are formed during the condensation polymerization at high temperature. If the moulding pressure is too low the mechanical properties are seriously affected by the formation of bubbles or voids.

In this connection it is clear that polyester and epoxy resins can often be cured at very low contact pressures as no gaseous by-products are produced.

References to chapter 12

1   J. H. der Boer, *Trans. Faraday Soc.* **32**, 10 (1936).
2   R. Houwink, *Elasticity, plasticity and structure of matter* (Cambridge, 1954).
3   P. Bruins, *Epoxy resin technology* (New York, 1968).
4   J. L. Peakes, *Brit. Plastics* **6**, 421, 475, 516 (1934).

13 Deformation of crystalline solids

D. Robert Hay and George E. Dieter, Jr

13.1 Crystal structure

Crystalline solids comprise a large group of materials which are characterized structurally by a highly regular, three-dimensional arrangement of their atomic constituents. An extensive record in the literature of structural features and deformation behaviour of metallic, ionic, covalent, and molecular crystals has shown the mechanical properties to be characteristic of the type and strength of bonding as well as the crystal structure.

It is possible to consider the formation of a crystal by placing atoms or groups of atoms at *space lattice* sites. In such an array, each site has identical surroundings and the lattice points occur at the intersection of three sets of parallel planes. The basic lattice building block is the elementary parallelepiped or *unit cell* shown in Fig. 13.1. Crystal symmetry is characterized by the angles α, β, and γ between the unit cell axes and the relative lengths of the lattice constants a, b, and c. A unit cell may contain one or more atoms varying in complexity of arrangement from highly symmetrical cubic and hexagonal structures of many elemental metals to the more complex, multi-atomic, lower symmetry inorganic chemical compounds. Most familiar materials, however, crystallize in high symmetry lattices often of cubic or hexagonal symmetry. Several examples are shown in Fig. 13.2. In cubic crystals $\alpha = \beta = \gamma = 90°$ and $a = b = c$ while in hexagonal crystals, $a = b \neq c$ and $\alpha = \beta = 90°$ and $\gamma = 120°$.

Crystallographic planes and directions are represented in the Miller index notation. Indices of a crystallographic plane are specified by the reciprocals of its intercepts on the three cell axes. The intercepts are measured from the origin in multiples of their respective lattice constants and are reduced by a common denominator to the set of lowest prime integers. As an example, the plane *ABCD* in Fig. 13.3 is parallel to the x and y axes and intersects the z axis at a distance of one lattice constant c. Thus the intercepts of the plane are $(\infty, \infty, 1)$ and the indices (001). Indices in parentheses denote the specific plane *ABCD*. Similarly the plane *ABFE* would be given the representation ($\bar{1}$00) since the origin can be moved to *H*, an equivalent space lattice point, and an overbar implies a negative index. Indices in braces designate a family

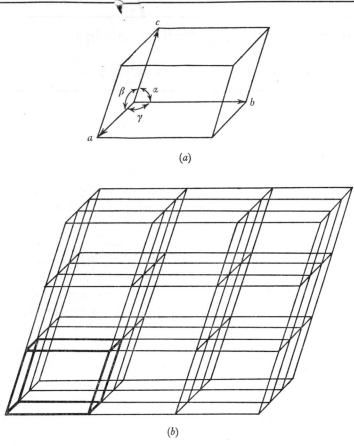

(a)

(b)

Fig. 13.1. (a) A unit cell. (b) Construction of a lattice by translation of the unit cell.

of crystallographically equivalent planes {100}, {010}, and {001}. Fig. 13.3 shows several examples of crystallographic planes and their representation in Miller indices. The Miller indices of a crystallographic direction are specified by the co-ordinates of a point on a line in the direction of interest through the origin. These co-ordinates are expressed in multiples of lattice constants and are reduced to the set of lowest prime integers. For example, the coordinates of the point G on the line EG are a, b, and 0 and the Miller indices of the direction EG are [110]. The family of equivalent directions is denoted by indices in carets ⟨110⟩. Examples using this notation are given in Fig. 13.3. In the cubic system only, the normal to a plane has the same Miller indices as the plane.

Conceptually it is often convenient to consider the atoms in a crystal as

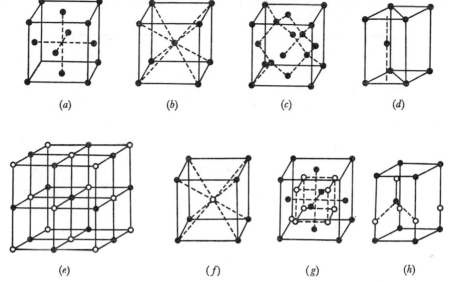

Fig. 13.2. (*a*) Face-centred cubic: Al, Ni, Cu, Au, Ag. (*b*) Body-centred cubic: Fe, W, Cr, Mo, Ta. (*c*) Diamond cubic: C (diamond), Si, Ge. (*d*) Close-packed hexagonal: Zn, Mg, Be, Ti, Zr. (*e*) Sodium chloride: MgO, NaCl, LiF. (*f*) Caesium chloride: CsCl, NiAl (ordered). (*g*) Fluorite: CaF_2, Cu_2S, Mg_2Si. (*h*) Wurtzite: Zns, AgI, MgTe.

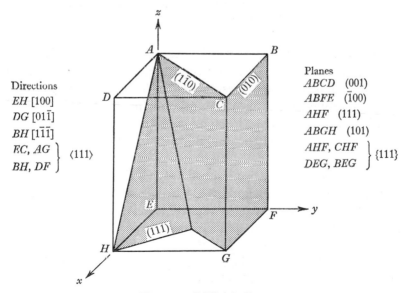

Directions

EH	[100]
DG	[01$\bar{1}$]
BH	[1$\bar{1}\bar{1}$]
EC, AG	$\}\ \langle 111 \rangle$
BH, DF	

Planes

$ABCD$	(001)
$ABFE$	($\bar{1}$00)
AHF	(111)
$ABGH$	(101)
AHF, CHF	$\}\ \{111\}$
DEG, BEG	

Fig. 13.3. Miller indices.

hard spheres. The arrangement of highest-density planar packing (close-packing) of spheres is shown in Fig. 13.4a. Both the face-centred cubic (fcc) and close-packed hexagonal (cph) structures may be constructed by stacking such planes. In the cph lattice the close-packed planes are the {001} or basal planes while in fcc crystals the close-packed planes are the {111}. A second close-packed plane may be set into the depressions in the layer of atoms labelled A in Fig. 13.4a, locating in either sites labelled B or C. Assume the second layer fills the B sites. The third layer may then be set in depressions

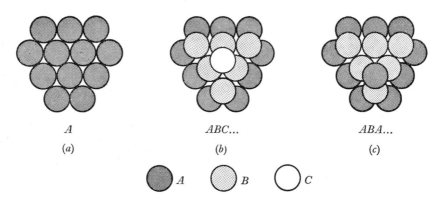

A $ABC...$ $ABA...$

(a)   (b)   (c)

A B C

Fig. 13.4. (a) Close-packing of spheres. (b) Cubic close-packing. (c) Hexagonal close-packing.

in the second layer over A sites or in C sites. Should this third layer enter A sites giving an $ABABA...$ stacking sequence, a cph lattice is produced (Fig. 13.4c). If it is put into C sites, the $ABCABCA...$ stacking sequence of an fcc lattice results (Fig. 13.4b). Seventy-four per cent of the volume of a close-packed lattice is occupied by atoms compared to sixty-eight per cent for a body-centred cubic (bcc) lattice and thirty-four per cent for diamond cubic crystals.

Although the inherent mechanical behaviour is determined by the type of bonding and the crystal structure, the actual responses of the material are influenced to such a great extent by crystalline defects that predictions based on a perfect crystal are often different from experimental observations by several orders of magnitude.

Any deviation from the periodic arrangement of the lattice is termed a defect. The defects are generally classified with respect to their extent in the crystal. When the lattice perturbation is localized to the vicinity of only a few atoms it is termed a *point defect*. Three types of point defects are shown in Fig. 13.5. The vacant lattice site or *vacancy* may be created by thermal excitation according to

$$\frac{n}{N} = \exp\left[-\frac{E_v}{kT}\right],$$

where n is the number of vacant sites in a total of N sites and E_v is the energy required to move an atom from the interior of a crystal to its surface. A

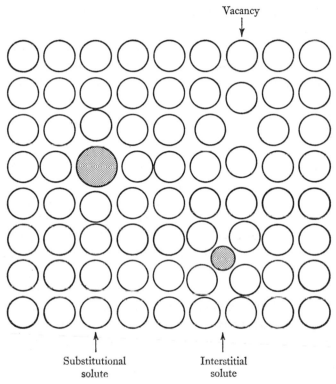

Vacancy

Substitutional solute

Interstitial solute

Fig. 13.5. Point defects.

typical value of E_v for a metal is 1 eV and, at the melting point, the fraction of vacant lattice sites might be 10^{-5} to 10^{-4}. Higher than equilibrium concentrations of vacancies at room temperature are produced by quenching from elevated temperatures, extensive plastic deformation, or bombardment with high energy nuclear particles. *Interstitial* point defects are atoms on non-lattice sites. These may be atoms of the metal composing the lattice (solvent atoms) or impurity (solute) atoms. *Substitutional* point defects are solute atoms on solvent lattice sites. In each point defect, the disturbance will extend beyond the immediate vicinity of the defect due to relaxation of neighbouring solute atoms around the vacant site or as a result of lattice distortion produced by size or bonding differences of the interstitial atoms. Thus a stress field extending to near neighbours is associated with each point defect.

The most significant defect influencing plastic deformation of crystals is the line defect or dislocation. A dislocation moves through the lattice as a deformation front producing plastic flow in the crystal. It is thus a dividing line between deformed and undeformed regions of the crystal and has associated with it, lattice disregistry and higher energy than the adjacent perfect material. One of the common methods of observing dislocations takes advantage of the higher chemical reactivity of a crystal at a dislocation. When a crystal surface is chemically etched, pits are developed at the points where dislocations emerge. Plate 13.1 (between pages 320 and 321) shows the etch pits on the surface of an annealed tungsten crystal. The dislocation density, defined as the total length of dislocation line in a unit volume of the specimen, is often expressed in terms of the number of dislocation lines intersected by a plane of unit area. The density shown in Plate 13.1 of 10^6 lines/cm^2 is typical of an annealed or as-solidified metal. For many solutes the dislocation provides a preferential site for them to precipitate, resulting in segregation of solute to the dislocation. In crystals which are normally transparent to visible or infrared radiation, the precipitated solute atoms scatter the light and the dislocation lines have been observed in NaCl and Si[1] where the dislocations were arrayed in the form of a network. In well-annealed crystals, dislocations are generally arranged in this kind of array, sometimes less well-defined, called a *Frank net*. The distribution of dislocations in a crystal depends upon a number of factors including crystal structure, the temperature, and the rate and amount of straining. The roles of dislocations in producing plastic deformation are discussed in a subsequent section.

Real crystalline objects are not generally single crystals with point and line defects, but are an aggregate of microscopic *grains* of crystalline material. Such a material is termed *polycrystalline*. The relative lattice orientation of each crystallite is, except in special cases, very different from its adjacent grains leading to an interface of lattice disregistry at the *grain boundary*. This constitutes a surface defect as do interfaces between phases in a multiphase material as well as the external surfaces of a crystal. Within the grains of a polycrystal there may exist a substructure of subgrains whose angle of relative misorientation is small ($< 1°$). This may be observed in Plate 13.2, where the sub-boundary assumes the structure of a row of dislocations.

Another type of surface defect is the *stacking fault*, an error in the stacking sequence. The intrinsic stacking fault is equivalent to omission of a layer from the sequence and the extrinsic stacking fault equivalent to insertion of an extra layer. A closely related defect is the *twin*, a surface across which the stacking sequence inverts to the mirror-image configuration. These are illustrated in Fig. 13.6 for an fcc crystal. Stacking faults and twins are produced during growth of a crystal or through deformation. They are commonly observed in fcc and cph metals and in mineral crystals such as wurtzite.

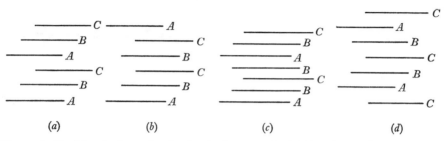

Fig. 13.6. Stacking faults and twinning in fcc crystals. (*a*) Regular stacking sequence. (*b*) Intrinsic fault. (*c*) Extrinsic fault. (*d*) Twin.

13.2 Elasticity of crystals

Recoverable elastic deformation occurs in crystals at only small strains, typically $< 10^{-5}$. At an atomistic level, this deformation is accompanied by both bond extension and shear distortion of the bond angle as shown in Fig. 13.7. When a randomly oriented shear or tensile stress is applied to a crystal the resultant elastic deformation will be a complex combination of extension and shear distortion. In general, the application of a tensile stress does not result only in an extension in the direction of the stress and a lateral Poisson contraction, as [described in Chapter 1] for isotropic materials, but produces other geometric changes as shown in Fig. 13.8. Anisotropy, rather than isotropy, is the general case, even in high symmetry crystals.

To describe elastic response of anisotropic materials it is necessary to extend the definitions of stress and strain of Chapter 1. A body to which forces are applied is said to be in a state of stress. If F is a force applied to the surface of a body and ΔF is the amount of F distributed over the element of area ΔA, then the stress σ is

$$\sigma = \lim_{\Delta A \to 0} \frac{\Delta F}{\Delta A}.$$

Any general stress may be decomposed into three components, a normal stress perpendicular to the surface and two shear stresses in the tangent plane. The complete state of stress on an elementary cube is shown in Fig. 13.9. The state of stress is described by specifying six components, three normal stresses σ_{11}, σ_{22}, and σ_{33} which produce volume changes, and three shear stresses

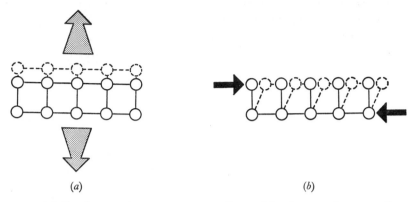

(a)   (b)

Fig. 13.7. (*a*) Bond extension across a crystallographic plane under a tensile stress normal to the plane. (*b*) Shear distortion across a crystallographic plane under a shear stress in the plane.

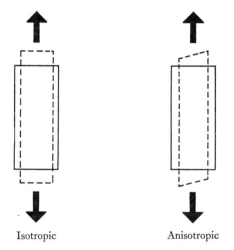

Isotropic   Anisotropic

Fig. 13.8. Effect of anisotropy on elastic deformation.

σ_{23}, σ_{13}, and σ_{12} which produce distortions with no volume change. We note that $\sigma_{12} = \sigma_{21}$, etc., for a body in static equilibrium. Similarly six strain components are defined; the normal strains ϵ_{11}, ϵ_{22}, ϵ_{33} and the shear strains ϵ_{12}, ϵ_{13}, and ϵ_{23} where $\epsilon_{13} = \epsilon_{31}$, etc., where there is no rotational displacement.

The normal components correspond to extensions of the cube, and shear components to changes in angles between the cube axes.

When the strains are small and time-dependent phenomena contributions negligible, the stress and strain components are linearly related through constants called *compliances*. For example, all the stress components contribute to the strain ϵ_{11} in the most general case through

$$\epsilon_{11} = S_{11}\sigma_{11} + S_{12}\sigma_{22} + S_{13}\sigma_{33} + S_{14}\sigma_{23} + S_{15}\sigma_{13} + S_{16}\sigma_{12},$$

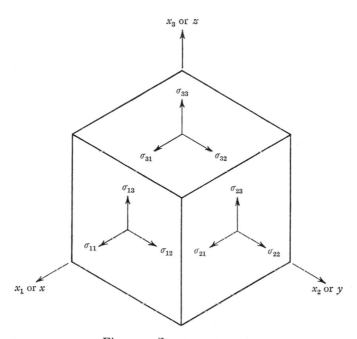

Fig. 13.9. Stress components.

where S_{ij} are the elastic compliances. Thermodynamic considerations reduce the number of independent compliances from 36 to 21 for the most anisotropic crystals and crystal symmetry may further reduce the number of independent constants. Hexagonal crystals require five constants while three suffice for the cubic system and two describe the response of isotropic materials.

Examples from the cubic system will illustrate the effects of crystalline elastic anisotropy. While the physical significance of the compliances and stiffnesses may be difficult to interpret, the three constants are conveniently represented as the bulk modulus B and two shear moduli C and C'. The bulk modulus describes the resistance to volume change under a hydrostatic pressure P:

$$B = -V \frac{dP}{dV}.$$

C describes the resistance to deformation by a shear stress applied in the (001) plane in the [010] direction and C' the resistance to a shear stress in the (001) plane in the [1$\bar{1}$0] direction. These three constants C, C' and B are related to the elastic compliances through

$$B = \frac{1}{3(S_{11}+2S_{12})},$$

$$C = \frac{1}{S_{44}},$$

$$C' = \frac{1}{2(S_{11}-S_{12})}.$$

As the direction of measuring the Young's and shear moduli is changed, their magnitudes vary according to

$$\frac{1}{E} = S_{11}-2[S_{11}-S_{12}-\tfrac{1}{2}S_{44}] \quad (l^2m^2+n^2m^2+n^2l^2),$$

$$\frac{1}{G} = S_{44}+4[S_{11}-S_{12}-\tfrac{1}{2}S_{44}] \quad (l^2m^2+n^2m^2+n^2l^2),$$

where l, m, and n are the direction cosines of the direction of observation relative to the cubic 100 directions. The orientation dependence is described by the orientation factor $\Omega = (l^2m^2+m^2n^2+n^2l^2)$. The anisotropy factor $A = [(S_{11}-S_{12})-\tfrac{1}{2}S_{44}]$ determines the magnitude of the orientation effect. Note that when $C = C'$, $A = 0$ and E and G are independent of orientation.

Examples of the elastic properties of several cubic crystals are recorded in Tables 13.1 and 13.2. The elastic constants of copper lead to a large anisotropy factor and a strong orientation dependence of Young's modulus in which it varies from 0.68×10^{12} dyn/cm^2 when measured in the $\langle 100 \rangle$ to 1.98×10^{12} dyn/cm^2 in the $\langle 111 \rangle$. Tungsten on the other hand, has an anisotropy factor of zero and exhibits isotropic elastic behaviour. The examples in Table 13.2 illustrate a spectrum of elastic behaviours, both with regard to magnitudes of the constants and the degree of anisotropy they exhibit even in high symmetry crystals.

A prediction of the elastic moduli of a polycrystal from its single crystal elastic constant poses a difficult problem in determining the correct compro-

TABLE 13.1. *Elastic constants of some cubic crystals*[2]

Material	Elastic compliances (10^{-12} cm²/dyn)			Elastic moduli (10^{-12} dyn/cm²)			Anisotropy factor
	S_{11}	S_{12}	S_{44}	C	C'	B	A
Copper	1.50	−0.63	1.33	0.75	0.24	1.40	2.84
Tungsten	0.24	0.07	0.63	1.60	1.60	3.08	0.00
Diamond	0.105	−0.01	0.17	10.76	4.75	8.69	0.02

TABLE 13.2. *Orientation dependence of Young's modulus in some cubic crystals*

Material	Young's modulus (10^{12} dyn/cm²)		
	$\langle 100 \rangle$	$\langle 110 \rangle$	$\langle 111 \rangle$
Copper	0.68	1.35	1.98
Tungsten	3.89	3.89	3.89
Diamond	9.43	10.20	11.36

mise between the conditions of uniform local stress or local strain which approximates the physical situation. Calculations by Voigt[3], based on uniform local strain, show the situation for fibrous grains aligned parallel to the stress axis and lead to the following constants for polycrystals with cubic lattice symmetry:

$$B_V = B,$$

$$G_V = \tfrac{1}{5}[2C' + 3C].$$

Similar calculations by Reuss[5] for the assumption of uniform local stress, the condition in a specimen of layered grains give, for cubic materials,

$$B_R = B,$$

$$G_R = \frac{5CC'}{2C + 3C'}.$$

Both assumptions lead to the same values of B and Hill[6] has shown that, based upon energy considerations, these estimates provide bounds upon the Young's and shear moduli. In the Voigt–Reuss–Hill approximation, it is assumed that the arithmetic mean of the Voigt and Reuss moduli represents the values of the polycrystalline material. That this is a good approximation is shown in Table 13.3 for Young's modulus of polycrystals, calculated from

$$E = \frac{9BG}{3B+G}$$

for the highly anisotropic materials Cu, MgO, and CaF_2.

TABLE 13.3. *Young's modulus of polycrystals, experimental and predicted*

Material	V–R–H Young's modulus (10^{12} dyn/cm^2)	Measured Young's modulus (10^{12} dyn/cm^2)
Copper	1.23	1.23
Magnesium oxide[4]	3.09	3.07
Calcium fluoride	1.09	1.08

Although structural changes such as presence of defects can induce variations in the elastic properties by amounts up to a few per cent, this effect is much less pronounced than for many other properties such as plastic flow parameters. In view of this, the elastic behaviour is usually considered to be structure-insensitive.

13.3   Role of dislocations in plastic deformation

Many solids if subjected to a constant load of suitable magnitude for a sufficiently long time show a continuously increasing deformation. This phenomenon of 'flow' may be considered as resulting from the tendency of the atoms or molecules to readjust themselves in such a way that, if the deformation were kept constant, a release of stress would take place. The mechanism of relaxation is different in amorphous and crystalline substances, notwithstanding a certain similarity in the macroscopic appearance of the phenomenon of flow. While plastic deformation of amorphous substances must be considered as a process in which the constituent atoms or molecules readjust *individually*, in crystals a large number of them, coupled together elastically, move co-operatively so that a very definite change in shape is brought about. This may be described as a process of *slip* or shear along definite crystallographic planes.

Deformation is purely elastic until a critical stress, the elastic limit, is surpassed and processes characteristic of slip become active. Plate 13.3 (between pages 320 and 321) shows a system of elliptical slip lines on the surface of a stretched metal crystal indicative of the discrete nature of plastic deformation. Furthermore, slip occurs along definite crystallographic planes in well-defined directions. For the simultaneous slip of two portions of a crystal (Fig. 13.10) various estimates of the theoretical maximum shear strength of a crystal have been placed between $G/50$ and $G/2\pi$. Shear strengths approaching this theoretical maximum have been observed in extremely small, highly perfect

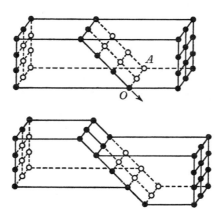

Fig. 13.10. Schematic picture of simultaneous slip along a crystallographic plane.

TABLE 13.4. *Shear strengths of whiskers*[9]

Whisker	Shear strength/G
Iron	0.060
Copper	0.022
Silver	0.031

crystals or *whiskers* (Table 13.4). However, in less perfect single crystals and polycrystals, plastic flow begins at stresses as low as $G/10^5$, a phenomenon associated with the defect structure.[8] In the ranges of intermediate deformation rates and moderate temperatures, plastic deformation occurs by dislocation motion. Under other more extreme conditions of high and low temperatures and deformation rates other defects may make significant contributions.

In a crystal, dislocations assume the edge or screw configurations or a combination of both termed the *mixed dislocation*. A dislocation was previously described as the boundary between slipped and unslipped regions of a crystal or as a deformation front. The atomic arrangement in the vicinity of edge and screw dislocations is shown in Fig. 13.11 and the manner in which they and the mixed produce slip illustrated in Fig. 13.12.

The presence of a large number of dislocations has prompted some speculation as to their origin. Possibilities include formation during solidification

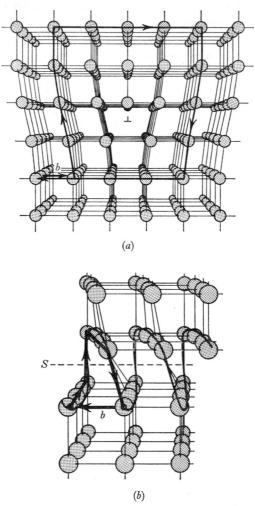

(a)

(b)

Fig. 13.11 (a) Edge dislocation. (b) Screw dislocation (ref. 11).

and nucleation at stress concentrations arising from the defect structure. During solidification, bending of dendrite tips by mechanical and convective disturbances may produce dislocations. Also, low-angle boundaries may be produced when individual solidifying nuclei grow and come into contact.

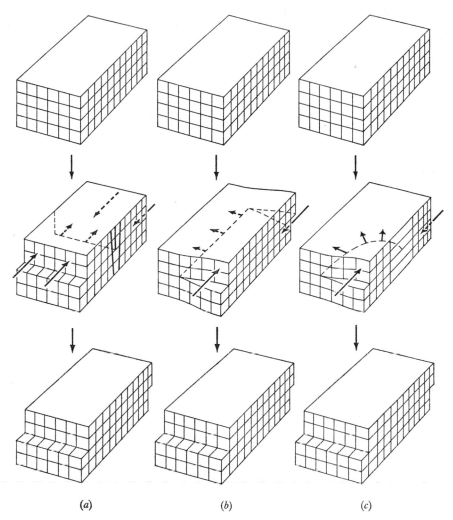

Fig. 13.12. Mode of propagation of dislocations. (*a*) Edge. (*b*) Screw. (*c*) Mixed.

Differential thermal contraction at crystal defects may induce internal strains of sufficient magnitude to generate dislocations. Such stresses must exceed the theoretical strength $\approx G/10$. Of these possibilities, nucleation at impurity particles and grain boundaries are the most likely[10] as is surface damage caused by handling. Once some dislocations have been nucleated, their number can be increased rapidly by the operation of dislocation sources.

Each dislocation is characterized by its Burgers vector b which defines the magnitude and direction of the slip it produces. The Burgers vector is determined by the closure failure of a Burgers circuit, an atom-to-atom path around the line defects, which, in the absence of the defect, would return to its starting point. The sense of the Burgers vector is established by encircling the defect with a clockwise Burgers circuit when looking in the positive direction of a vector t tangent on the line. Fig. 13.11 shows the Burgers circuits and vectors of edge and screw dislocations. For an edge dislocation, the Burgers vector is perpendicular to the dislocation line and the slip plane is uniquely defined by the tangent and Burgers vectors. A screw dislocation has parallel Burgers and tangent vectors and is thus only confined to slip on the set of planes in which these vectors lie. Those dislocations with Burgers vectors equal in magnitude to an interatomic distance are termed *perfect*. Burgers vectors of *partial dislocations* do not correspond to a complete lattice translation distance.

Reference to the diagrams of edge and screw dislocations shows an elastic strain field and, thus, a self energy associated with each dislocation. At the dislocation core where strains are large, closed-form analytical expression of the strain field is difficult. However, at distances beyond a core of radius r_0 an estimate of the strain and corresponding stress field can be obtained from isotropic continuum elasticity theory. A straight screw dislocation in an isotropic material has cylindrical symmetry and the stress field contains only a shear component of magnitude equal to $(Gb)/(2\pi r)$ where r is the distance from the dislocation line. The strain energy per unit length of dislocation associated with the field is

$$E_\mathrm{s} = \frac{Gb^2}{4\pi} \ln\left(\frac{r}{r_0}\right).$$

Edge dislocations have radial and tangential normal components as well as shear components in their stress fields resulting in a region of tension in the area below the edge dislocation in Fig. 13.14 and a region of compression above.

The corresponding self energy of an edge dislocation is

$$E_\mathrm{E} = \frac{Gb^2}{4\pi(1-\mu)} \ln\left(\frac{r}{r_0}\right).$$

The total energy of a dislocation is actually the sum of the elastic and core contributions. Strain energy per unit length of dislocation is equal to αGb^2, where α is a constant whose value lies between 0.5 and 1.5 and thus amounts to about 10 eV for each atomic plane threaded by the dislocation. The total energy of a crystal containing a large density of dislocations is the sum of their individual contributions plus terms describing the mutual interactions of dislocation and interactions with other defects. The most stable (lowest energy) perfect dislocations are those with the shortest Burgers vectors, and thus lie in the closest-packed lattice direction.

When an external stress σ of sufficient magnitude is applied to a crystal, the dislocations move and produce slip. Thus, a force acts upon the dislocation which, per unit length of dislocation line, is

$$F = \sigma b.$$

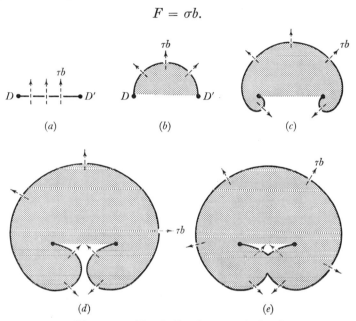

Fig. 13.13. Frank–Read source (ref. 12).

The force is normal to the dislocation line at every point along its length and is directed toward the unslipped portion of the slip plane.

Since the strain energy of a dislocation is proportional to its length, work is required to increase its length. This is manifested as a line tension of a dislocation. A dislocation pinned at two points along its length tends to bow out (Fig. 13.13) under an applied shear; the stress σ required to activate this dislocation segment and produce an arc with radius R is

$$\sigma_0 = \frac{Gb}{R}.$$

This stress reaches a maximum when the dislocation assumes a semicircular configuration between the pinning points. Beyond this point, the dislocation segment will continue to propagate without further increase in stress as shown in Fig. 13.13. Upon completion of one cycle, a dislocation loop is generated and a new segment is in the original position ready to repeat the process. This mechanism whereby a single dislocation segment can continually generate new loops is called a *Frank–Read source*. An accumulation of dislocations emanating from a source produces a *slip band*. Since screw dislocations are not confined to move on a particular slip plane, they may change slip planes by *cross-slip*. In Fig. 13.14. the *double cross-slip* mechanism is

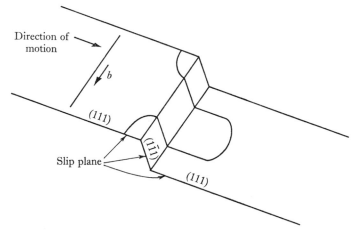

Fig. 13.14. Double cross-slip.

illustrated for an fcc metal where the screw dislocation in the (111) plane cross-slips on the (1$\bar{1}$1) plane to another (111) plane. If the leading dislocation in a slip band becomes blocked, a pile-up is produced and the resulting back stress will ultimately render the source inoperative. When the dislocation segment on the lower (111) plane in Fig. 13.14 is free to expand, the double cross-slip process also is a mechanism of multiplication akin to the Frank–Read source. Other dislocation sources include grain boundaries and single-ended Frank–Read sources (Fig. 13.15).

A fundamental theorem in dislocation theory, based upon topological considerations, holds that a dislocation cannot end within a crystal but must either terminate at a surface or form a closed loop within the crystal. Loops may consist of edge, mixed, and screw components or, in the case of prismatic loops, of only edge components. Both types are shown in Fig. 13.16 with the deformations they produce. The components of a mixed loop move in the plane of the loop whereas the prismatic loop moves normal to the plane.

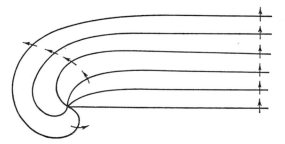

Fig. 13.15. Single-ended Frank–Read source.

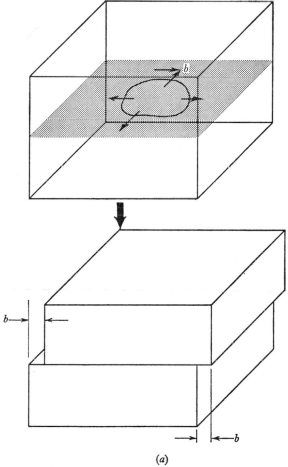

(a)

Fig. 13.16. (a) Deformation produced by expansion of a mixed dislocation loop.

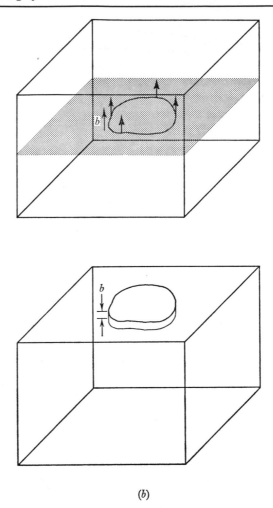

(b)

Fig. 13.16. (b) Deformation produced by motion of a prismatic loop.

The total energy of a crystal can often be reduced when a perfect dislocation b splits into two or more partials, b_i by the reaction.

$$b = \sum_i b_i.$$

Since the strain energy is proportional to the square of the Burgers vector, the reaction will proceed if $b^2 > \sum b_i^2.$

The product of a dislocation reaction may be mobile dislocations. However, several types of reactions produce dislocations whose Burgers vectors do not lie in the glide plane and cannot move by glide. These dislocations have been termed *sessile*.

13.4 Motion of dislocations

Calculations of the inherent lattice friction stress opposing the motion of a dislocation in an otherwise perfect lattice require a knowledge of the structure and properties of the dislocation core. In the case of an edge dislocation, for example, two types of forces exist at the core, those which tend to spread the atoms due to the presence of the extra half plane of atoms, and those attractive forces resulting from their displacement from equilibrium. When a dislocation is displaced from its equilibrium position by a stress, the attractive and repulsive forces do not act in the same way giving rise to a small friction stress. Estimates of the frictional stress using force laws of limited validity, are very sensitive to the material properties and crystallography. One of these estimates[13] of the friction or *Peierls* stress σ_P for edge dislocations is

$$\sigma_P = G \exp \left(\frac{-2\pi a}{(1-\mu)b} \right)$$

where b is the Burgers vector and a the spacing between slip planes. This expression provides two useful indications; the Peierls stress $\sim G/10^4$ is several orders of magnitude lower than the theoretical strength and the Peierls stress is lowest on the planes where a is the largest (the closest-packed planes).

In the previous section, elastic strain energy considerations dictated that the most stable dislocations are those with the shortest Burgers vector. This observation, combined with the low friction stress on close packed planes generally define the most active slip systems. These *slip systems*, comprising a slip direction and a slip plane, thus consist of dislocations lying in the closest-packed planes and moving in the closest-packed directions. Table 13.5 summarizes the slip systems active in several materials. These systems are the predominant slip systems while others may become active at high temperatures and strain-rates. In fcc and cph metals such as copper and zinc, slip is usually only observed on the systems recorded in Table 13.5. However, in bcc metals, the tendency is for slip in the $\langle 111 \rangle$ direction on planes containing this direction with an increasing preference for only $\{110\}$ slip with decreasing temperature and increasing strain rate. In ionic crystals, the additional factor of the polarizability of ions must be considered; hence the difference between NaCl and PbS, both of which have the NaCl structure.

TABLE 13.5. *Slip systems*

	Slip plane	Slip direction
Copper	{111}	⟨110⟩
Iron	{110}	
	{112}	⟨111⟩
Sodium chloride	{110}	⟨110⟩
Lead sulphide	{100}	⟨110⟩
Zinc	{0001}	⟨11$\bar{2}$0⟩

The fcc lattice may be constructed by stacking {110} planes in an *ABAB*...
sequence. Creation of a perfect edge dislocation requires that an *AB* pair of
extra half planes be inserted. If the two planes in the *AB* pair become separated,
each corresponds to a partial dislocation of the ⟨211⟩ type and the region
between the two dislocations in a stacking fault. The two ⟨211⟩ partials repel
each other until an equilibrium distance *d* is reached which depends upon the
energy E_F associated with the stacking fault:

$$d = \frac{Gb^2}{2E_F}.$$

Thus separation of a perfect dislocation into partials will only occur when the
stacking fault energy is relatively low. Table 13.6 lists the stacking fault energy
of a number of fcc metals. Transmission electron microscopy studies show
that partials are prevalent in stainless steel while the tendency to dissociate is
small in aluminium where the stacking fault energy is high.

Shear deformation is determined by the stress component of the force *F*
on a crystal resolved in the slip plane into the slip direction. This resolved
shear stress σ_s in Fig. 13.17 is

$$\sigma_s = \frac{F}{A} \cos \lambda \cos \phi,$$

where λ is the angle between the load axis and the slip direction, ϕ is the angle

TABLE 13.6. *Stacking fault energies* (*erg/cm²*)

Austenitic stainless steel	13
Gold	30
Copper	40
Silver	80
Aluminium	200

between the slip plane normal and the load axis, A is the cross-sectional area and F is the applied load. In fcc and cph metals with well-defined slip planes the onset of slip is characterized by a unique value of σ_s.

When a crystal is tested in a conventional tensile test the ends are constrained and simple shear strain of type shown in Fig. 13.18a is not possible. In fact, near the centre of the specimen the slip planes must rotate (Fig.

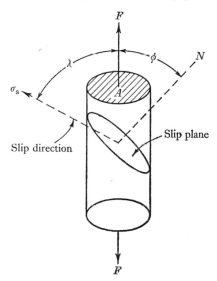

Fig. 13.17. Resolution of force into a shear stress in the slip direction.

13.18b). Thus, a slip system which initially had a high σ_s and exhibited slip, rotates away from this favourable orientation and, simultaneously, new systems become more favourable. The appearance of more than one slip system during deformation is usually described as *multiple slip*.

When dislocation motion has been initiated in a crystal and a density ρ of mobile dislocations contributes to plastic deformation, the plastic strain ϵ_p induced in the specimen as a result of their moving an average distance x is

$$\epsilon_p = \rho b x$$

and the rate of plastic straining $\dot{\epsilon}_p$ is

$$\dot{\epsilon}_p = \rho b v$$

where v is the average dislocation velocity. Dislocation velocities have been measured by an etch pit method devised by Johnston and Gilman.[14] Mobile dislocations are introduced to a crystal by scratching a surface and their positions located by etching. Their positions are then redetermined after imposing a stress pulse of known magnitude and duration. Plate 13.4 (facing p. 321) shows the growth of a slip band emanating from an indentation during application of a stress pulse.[15] Experiments in LiF and several other materials show a general stress dependence of the dislocation velocity given by

$$v = v_0 \exp\left(-\frac{A}{\sigma}\right),$$

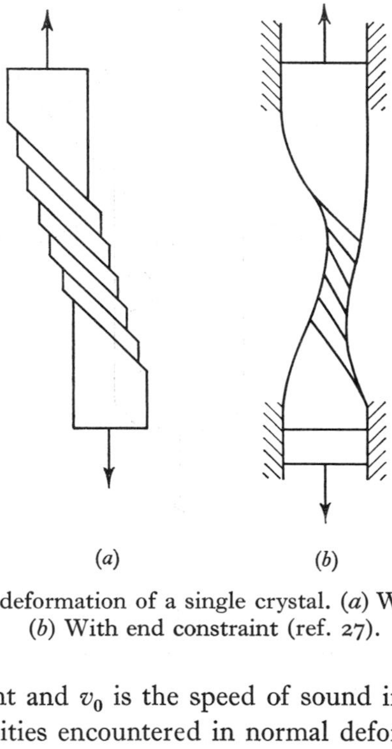

(a)   (b)

Fig. 13.18. Tensile deformation of a single crystal. (a) Without end constraint. (b) With end constraint (ref. 27).

where A is a constant and v_0 is the speed of sound in the material. At the low dislocation velocities encountered in normal deformation processes this relation is well described for several materials by

$$v = \left(\frac{\sigma}{\sigma_0}\right)^n,$$

where σ_0 and n are material constants; n is termed the *dislocation velocity constant* and describes the sensitivity of dislocation velocity to stress. These equations provide the fundamentals of a description of stress–strain behaviour called *dislocation dynamics*. Noting that the strain rate may be expressed as

$$\dot{\epsilon} = \rho b \left(\frac{\sigma}{\sigma_0}\right)^n,$$

measurements of the strain-rate sensitivity of the flow stress provides measure of n. Values of n for several materials are shown in Table 13.7 obtained by etch pit and other indirect techniques. Those with a high dislocation velocity constant, notably the fcc metals silver and copper, are more ductile than, for instance, the bcc metals molybdenum, tungsten, and iron–silicon and ionic crystals for which n is lower by about an order of magnitude.

TABLE 13.7. *Dislocation velocity constant n at room temperature*

Material	n
Tungsten	7
Molybdenum	13
Titanium	15
Iron–2.35 % Silicon	35
Copper	170–250
Silver	300

Superimposed upon the stress activated deformation described above may be a thermal assist provided by thermal energy in the crystal. Thermally-assisted deformation was originally suggested by Becker[16] in 1925. A contemporary model developed by Conrad[17] suggests the dislocations are influenced by long-range energy barriers such as the elastic stress fields of other dislocations as well as short-range barriers of the extent of about a Burgers vector. These latter barriers result from frictional stress such as the Peierls stress. At temperatures above 0° K the thermal energy of a lattice is manifested as atomic vibration. Those atoms in the vicinity of a dislocation vibrate co-operatively such that the dislocation line vibrates as a unit about the equilibrium position. The dislocation thus vibrates as a string between nodal points at intervals characteristic of the material, its internal defect structure, the temperature, and the type of barrier. Using the theory of activated processes it is reasonable to conceive of some of these fluctuations being successful in moving a segment of the dislocation to an adjacent equilibrium position producing a bulge and two kinks in the line as illustrated in Fig. 13.19. This double kink method originally proposed by Seeger[18] then suggests that the rate of formation of bulges ν is

$$\nu = \nu_0 \exp\left(-\frac{E}{kT}\right),$$

where E is the activation energy and ν_0 is the frequency of vibration of the dislocation segment. If the dislocation velocity is given by

$$v = \nu s,$$

where s is the average distance moved on each successful fluctuation, the strain rate is then

$$\dot{\epsilon} = \rho b s \nu_0 \left[-\frac{E(\sigma)}{kT} \right].$$

Upon the application of a stress σ, strain energy is provided to the dislocation and a smaller contribution from thermal energy is required to overcome the barrier. Thus E is a function of stress. In those crystals with a high frictional barrier, exceeding the long-range barriers in magnitude, a strong temperature and strain rate dependence of the stress required for plastic flow is observed.

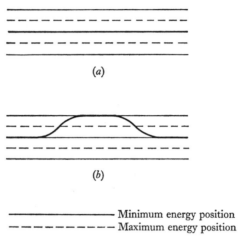

(a)

(b)

———————— Minimum energy position
– – – – – – – – Maximum energy position

Fig. 13.19. (a) Dislocation in minimum energy position. (b) Dislocation with bulge.

A strong temperature dependence of flow is characteristic of bcc metals. Ionic and covalently bonded crystals also exhibit a strong temperature and strain-rate sensitivity whereas fcc metals show a much weaker dependence on these parameters. In fcc metals the deformation is controlled to a large extent by the long-range barriers whereas frictional stresses are larger in the other crystals. Highly directional bonds in covalent crystals result in a high Peierls stress while in ionic crystals electrostatic forces encountered by like atoms during the slip process lead to the high friction stress. A large segment of the literature suggests that the friction stress in bcc metals results from the Peierls stress. However, hardening arising from the presence of residual interstitial impurities may be more significant.[19]

A requirement of a comprehensive theory of dislocation dynamics is a satisfactory interpretation of the yield drop generally observed in those crystals which have a strong temperature and strain-rate sensitivity. These materials have a much lower dislocation velocity constant n than do the fcc metals. Writing the expression for strain rate $\dot{\epsilon}$ in the form

$$\dot{\epsilon} = \rho b \left(\frac{\sigma}{\sigma_0}\right)^n,$$

it is apparent that, in a constant strain-rate test, if at the yield point where dislocation motion begins, rapid multiplication of dislocations should occur, the increase in dislocation density must be compensated by a decrease in σ, that is, a drop in the flow stress. This effect is most pronounced when n is small. Generally, the bcc metals with small n in Table 13.7 show a yield point while it is absent in the fcc metals with high n.

13.5 Strain hardening

A dislocation may encounter several types of lattice irregularities during its motion. Interaction with these irregularities produces resistance to dislocation motion and hardening of the crystal. If the irregularities are other dislocations, grain boundaries, or defects produced as a result of deformation, the effect is called *strain hardening*. These interactions are long-range and essentially athermal and the number of these long-range, athermal barriers increases continuously during the course of deformation.

An arbitrary slip plane in a crystal will intersect a large number of the as-grown dislocations These dislocations which thread through the slip plane are called *forest dislocations*. The most significant strain-hardening mechanisms arise from interactions of mobile dislocations with those of the forest and other mobile dislocations. Mechanisms of interaction include long-range forces exerted by the dislocation stress field, intersection of the dislocations, bowing out between nodal points in dislocation tangles, and immobilization by dislocation reaction.

When intersection occurs, a step is formed in the dislocation as shown in Fig. 13.20 If the step lies in the slip plane it is called a *kink* and may be removed by simple slip of the dislocation. A step running between parallel slip planes is a *jog*. Generally the Burgers vector of a jog in an edge dislocation lies in a slip plane and the jog may move with its parent dislocation. The Burgers vector of a jog in a screw dislocation will not generally lie in a slip plane and the jog cannot glide with its parent. Thus screw dislocations are more difficult to move through the forest. The increased length of dislocation line formed upon intersection requires expenditure of energy and produces hardening. When a screw dislocation containing a sessile jog moves, drawing the jog with it, point defects, vacancies or interstitials, are created in the wake of the jog. The additional energy required to create the point defects creates a drag on the dislocation.

In instances where the moving dislocation does not intersect the forest dislocations but is repelled by their stress fields, immobile nodal points are formed along the mobile dislocation. It can then bow out under an increased shear stress and operate as a Frank–Read source as in Fig. 13.21.

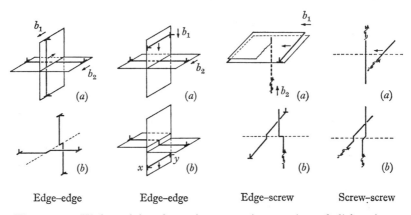

| Edge–edge | Edge–edge | Edge–screw | Screw–screw |

Fig. 13.20. Kink and jog formation upon intersection of dislocations. (*a*) Before intersection. (*b*) After intersection (ref. 12).

Immobilization of dislocations by dislocation reaction occurs when the product of reaction is a sessile dislocation. Such an interaction product is a dislocation with a Burgers vector which does not lie in a slip plane and cannot glide. In addition to immobilizing the reacting dislocations, a barrier to further dislocation motion on these planes is created.

Investigations of a number of bcc, fcc, and cph metals as well as ionic and organic crystals and intermetallic compounds have revealed that, under special conditions, a stress–strain curve exhibiting three-stage hardening may be observed. Five distinct regions can be distinguished on such a curve as shown in Fig. 13.22.

Stage 0, a region of elastic deformation in which the onset of plastic flow also begins. A yield point may be present in this stage.

Stage I, a region of *easy glide* characterized by a low, linear strain hardening rate θ_{I}.

a *Transition Stage,*

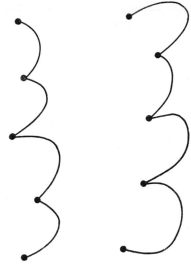

Fig. 13.21. Bowing of dislocations between nodes of intersections with forest dislocations.

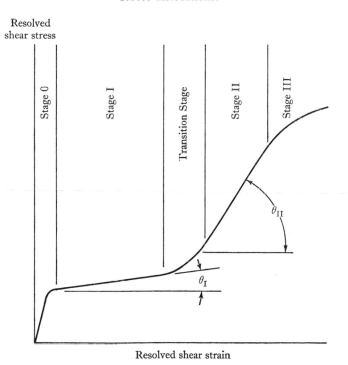

Fig. 13.22. Stress–strain curve with three-stage hardening.

Stage II, a region of high linear work hardening rate θ_{II}.

Stage III, a region of continuously decreasing work hardening rate.

Stage I is observed only in single crystals oriented to produce slip on only a single system. In some materials, the bcc metals for example, Stage I is observed only over a restricted range of temperature and strain rate. Also, three-stage hardening is promoted by high purity.

Transmission electron microscope studies throughout the three-stage hardening curve have been useful in providing an interpretation of this behaviour. Observations of dislocation structure[20] in tantalum are typical of those recorded for other materials. Following Stage 0 in which dislocation motion was initiated and multiplication and interaction of dislocations has begun, tangles of dislocations, surrounded by dislocation-free regions form parallel to the primary slip plane. In these tangles dislocation interaction and immobilization mechanisms outlined previously are presumed to occur. The distance between dislocation tangles is large enough to keep the strain hardening rate low in Stage I.

In the Transition Stage, increased secondary slip (slip on less favourably oriented systems) occurs between the tangles and the density of secondary dislocations approaches that of the primary dislocations. The tangles become joined by secondary dislocations forming a network, the walls of which are low angle boundaries. In Stage II the cells are elongated and proceed toward an equiaxed configuration in Stage III. The processes of *dynamical recovery* which may occur at the high stresses in Stage III enable the dislocations to take part in processes suppressed at lower stresses. It is thought that cross-slip is the main process whereby dislocations can escape barriers which had previously impeded their motion making hardening less effective.

In crystals oriented for multiple slip (several slip systems equally favoured) and in polycrystals where strain compatibility imposes multiple slip, rapid hardening occurs much earlier and Stage I is absent. Fig. 13.23 shows the effects of orientation and polycrystallinity on the stress–strain curve.

The stress–strain curves of several types of crystals are shown in Fig. 13.24. Stage I is promoted in cph metals where slip is confined to only a few slip systems and secondary slip postponed to large strains. The curve for bcc metals reflects the higher inherent lattice strength and a shorter Stage I due to an earlier onset of secondary slip.

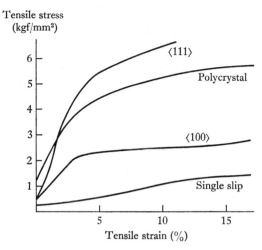

Fig. 13.23. Effect of orientation of single crystals and polycrystallinity on the stress–strain curve of aluminium (ref. 21).

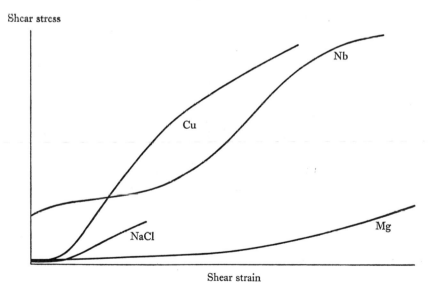

Fig. 13.24. Three-stage stress–strain curves of several types of crystals.

13.6 Strengthening by chemical additions

Hardening resulting from lattice disregistry due to such point defects as vacancies, interstitials, impurities, minor alloying elements and from agglomerations of these defects in solution is called *solid solution strengthening*. If the clusters extend to only a few atomic spacings the effect is termed ordering and is distinguished from larger aggregates such as inclusions, intermetallic second phases, and inert dispersions. Defect hardening due to point defects results from two effects, the local distortion or *size effect* and local shear modulus changes or the *modulus effect*. The latter comprises a number of effects such as local electronic structure changes and local bonding energy changes. In each case there is an interaction energy between a defect and a dislocation resulting in attraction or repulsion and extra work must be done to pass the defect. For example, when a number of solvent atoms are removed from the stress field of a dislocation and replaced by defects which produce distortions, work must be done against the dislocation stress field. The amount of work depends on the position of the defect in the stress field. Thus the energy of interaction is a function of the relative positions of the dislocations and defects, and forces are exerted between them. Similarly the local modulus differences produced by lattice defects make the displacements required by the dislocation stress fields either more difficult or easier, resulting in attraction or repulsion depending upon whether the local modulus is lower or greater than the matrix. However, in both the size and modulus effects, dislocation motion is impeded regardless of whether the forces are attractive or repulsive. In the former case the dislocation is difficult to free from the defects and in the latter it is difficult to move it toward them. Both effects provide major contributions in most metal systems.

In a finely dispersed array of defects the dislocation must have some flexibility in order to sense the above effects. If it remains straight, the average interaction energy is independent of position and the net force on the dislocation is zero. A flexible dislocation can, however, assume a lower energy configuration by lying in the low energy valleys between defects and solution hardening results.

A distinction between two types of defect hardening, rapid or slow, on the basis of the concentration dependence of the strengthening effect, that is, whether $d\tau/dc$ is large or small, has been noted.[22] Slow hardening $\partial\tau/\partial c \approx G/10$ results from defects which are pure dilatation or contraction centres such as substitutional impurities in metals and vacancies in alkali halide crystals. Rapid hardening $\partial\tau/\partial c \approx 1\text{--}10\,G$ results from non-spherical distortion such as interstitials in bcc metals, a divalent ion-vacancy pair in NaCl and vacancy discs. Since screw dislocations only have a shear component in their strain field, they cannot interact with dilatation centres. However, they do interact with the non-spherical defects and are impeded in their motion as are edges which interact with both types of defects. Thus rapid hardening is associated with the extra impediment to screw dislocations.

Some solid solutions do not have a random distribution of solute. When a solvent atom tends to surround itself preferentially by solute atoms the effect is termed *short-range order*. When atoms of the same type exhibit a preference for one another *clustering* results. A dislocation passing through the above regions will increase the randomization in these areas, replacing the preferred types of bonds by the unpreferred and increasing the energy of the lattice. Since a large amount of deformation is required to remove short-range order or clustering, this effect should be large even at high deformations.

In crystals with *long-range order*, each atom occupies a preferred type of sublattice, as for example the two types of sites in the NaCl structure. When a dislocation moves through the lattice, atoms across the slip plane are not in the correct relative positions and the slip plane becomes an *antiphase* boundary. Two dislocations bounding an antiphase interface are similar to two partial dislocations enclosing a stacking fault. In a perfectly ordered crystal, if the dislocations move in pairs, a large retarding force is not expected since the order destroyed by the first is restored by the second.

Only a few alloy systems exhibit extensive solid solubility between two or more elements. Most commercial alloys, therefore, have a heterogeneous microstructure comprising two or more metallurgical phases. Strengthening produced by second phase particles usually is supplementary to solid-solution strengthening. In addition to second phases produced by supersaturation of the continuous phase with alloying elements, called *precipitation hardening*, fine distributions of inert particles may be added deliberately for strengthening. The latter effect is called *dispersion hardening*.

Many factors govern the degree of strengthening, including size, shape, number, and distribution of the second phase particles, the strength, ductility, and strain hardening behaviour of the matrix and second phase, the crystallographic fit between the phases, and the interfacial energy and bond between the phases. Experimental control and evaluation of these parameters is difficult and existing knowledge of these effects is empirical and incomplete.

Effective strengthening is achieved in a number of ductile metallic matrices by the presence of a hard, brittle second phase. If the brittle phase is present as a grain-boundary film, the alloy is brittle. When the second phase is present as a fine uniform dispersion in the metallic matrix, optimum strength and ductility are obtained. Dislocation models of second phase hardening consider that the particles act as obstacles to dislocation motion. Orowan[23] has proposed a model in which dislocations bow out between particles, leaving loops around the particles as they pass (Fig. 13.25). If the average inter-particle spacing is λ, then from the stress required to operate a Frank–Read source an estimate of the stress required to force a dislocation between the obstacles is

$$\sigma = \frac{2Gb}{\lambda}.$$

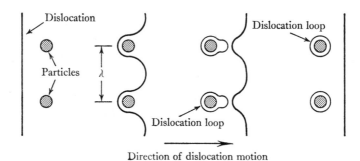

Direction of dislocation motion

Fig. 13.25. Bowing of a dislocation around second phase particles, leaving loops around the particles (ref. 11).

The loops increase the effective size of the particles providing an additional strain hardening mechanism. As loops built up around the particles the shear stress developed by the loops increases to the point where it is sufficient to shear the particles. This latter mechanism, proposed by Fisher, Hart and Pry,[24] has proved to be significant when the dispersion is fine and the interparticle distance small. Orowan's mechanism is more applicable to systems with larger particles and interparticle spacings.

13.7 Roles of grain boundaries

Although single crystals are a convenient simplification for discussing and describing dislocation motions, most real materials are polycrystalline. The grain boundaries may exert a great influence on deformation behaviour. Grain boundaries may either strengthen or weaken a material depending upon the type of crystal, its purity, the temperature, and rate of straining. At low temperatures, less than half the melting temperature, and at high strain rates, the grain boundaries usually strengthen the material. Part of the increase in strength results from the barriers to dislocation motion presented by the boundaries. Pile-ups of dislocations in a slip plane at the boundary exert a back stress upon a source rendering it inoperative. As the stress level is continually increased, the stress concentration ahead of the pile-up results in the initiation of slip in adjacent grains.

Generally a greater increase in strength of polycrystals compared to single crystals is observed than can be accounted for simply on the basis of the grain boundary as a barrier. Taylor[25] has shown that slip on at least five independent slip systems is required to maintain continuity between grains in a polycrystal. Thus in crystals with a low number of active slip systems such as cph metals the grain boundary strengthening may be a large effect. In order to maximize the grain boundary strengthening effect, the grain size should be small so that large amount of grain boundary area is present. Both theoretically and experimentally, the grain size dependence of the yield strength σ_y is given by

$$\sigma_y = \sigma_0 + k_y d^{-\frac{1}{2}},$$

where σ_0, a friction stress including contributions from the Peierls stress, solid solution strengthening, second phase strengthening, and subgrains, is a function of purity and temperature, k_y is a measure of the tendency to pile-up and is not a function of temperature, and d is the average grain diameter. Generally, those properties associated with early stages of deformation are those most greatly affected by grain boundaries. At large deformations, dislocation interactions within grains tend to control deformation.

The role of grain boundaries changes with temperature and strain rate to one of weakening at elevated temperatures (usually greater than half the melting point) and at low strain rates. At elevated temperatures edge dislocations may move in a direction normal to their slit planes by *climb*. Point defects, particularly vacancies, will move to jogs in the dislocation as shown in Fig. 13.26a, and when combined with the dislocation at this point, effect a motion of a dislocation segment normal to the slip plane (Fig. 13.26b).

Grain boundaries and free surfaces provide low energy sources of vacancies which at elevated temperatures diffuse more rapidly to dislocation jogs enhancing deformation by climb. Another prominent mode of deformation at high temperatures is grain boundary sliding which may contribute as much as 30 % to the total deformation.[26] However, relative motion of the grains at their boundaries must be accompanied by a second mode of deformation, possibly grain boundary migration by glide and diffusion, or by opening of cracks at the boundary. A number of metals when deformed under tension at elevated temperatures are capable of exhibiting homogeneous plastic strains in excess of 1000 %. For this phenomenon of *superplasticity* to occur, the rate of recovery from strain hardening effects must exceed the strain rate. Superplasticity is promoted by an extremely fine grain size suggesting that grain boundary sliding may be an important mechanism. Also, the grain boundaries may act as dislocation sinks preventing strain hardening.

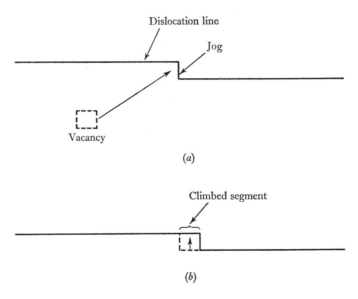

Fig. 13.26. Dislocation climb. (*a*) Before vacancy migration to jog. (*b*) After vacancy migration to jog.

Both high- and low-temperature grain boundary effects may be influenced by segregation of impurities or alloying elements at the boundaries. The usual result is additional strengthening by solid solution strengthening at low temperatures and by impeding boundary glide at elevated temperatures.

Many fabrication processes tend to produce in crystals a preferred relative crystallographic orientation of the individual grains. This type of material often exhibits behaviour characteristic of single crystals rather than the isotropic properties expected from randomly oriented aggregate of grains.

13.8 Twinning

Another atomic scale mechanism of shear deformation wherein the shear deformation is distributed over a large volume of material instead of being confined to a single discrete slip plane is mechanical twinning. In this instance, shear translation across each atomic plane is less than the identity distance. Atomistically, the mechanism is shown in Fig. 13.27 where the atomic configurations are mirror images of one another on each side of the twin plane.

Twinning is rarely observed in those crystals with a number of active slip

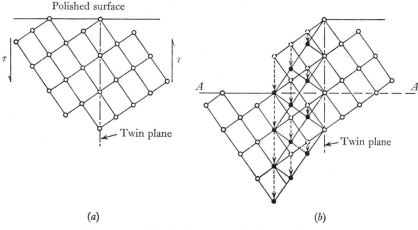

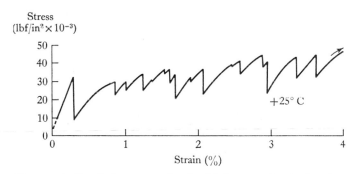

Fig. 13.27. Classical picture of twinning (ref. 27).

Fig. 13.28. Load drops associated with bursts of twinning in a Mo-39% Re alloy (ref. 28).

systems but in crystals where slip is difficult twinning is more common. Generally the stress required for twinning is less than that required for slip especially at low temperature and high strain-rate in materials where plastic flow is sensitive to these parameters. Twinning occurs as discrete events in the stress–strain curves and is accompanied by a drop in the stress level as shown in Fig. 13.28 for a Mo–39% Re alloy.

13.9   Strengthened metals

In addition to providing an interpretation of the mechanical behaviour of solids, dislocation theory and its associated mechanistic theories have led to the development of materials with superior mechanical properties both at room and elevated temperatures. These high strength materials have exploited strain hardening and solid solution, precipitation, and grain boundary strengthening.

High strength aluminium alloys have been produced using solid solution, precipitation, and dispersion hardening. A common alloy, duralumin, contains 4.5 % Cu, 0.6 % Mn, and 1.5 % Mg. At elevated temperatures these elements are in solid solution. At room temperatures, their concentrations are in excess of the solubility limit and they tend to form precipitated crystalline aggregates in an aluminium-rich matrix. During the initial stages of precipitation, the crystal lattices of the precipitates and the aluminium matrix are coherent: no distinct misorientation at interface exists between the precipitate and matrix phases. In order to accommodate the differences in lattice parameter of the two phases, large strains are established in the matrix in the vicinity of the precipitates. Motion of dislocations through these strain fields is more difficult than through the unstrained matrix and the material is strengthened. The process by which the precipitation is effected is termed *ageing* and is a nucleation and growth process aided by thermal energy. Should the precipitate grow excessively, an interface between the precipitate and matrix is produced and the strength lowered in this annealed or *averaged* condition. Particles more stable than the precipitate particles in duralumin are required when the metal is used at elevated temperatures. In these cases, a fine dispersion of inert Al_2O_3 particles is used for strengthening. Table 13.8 compares the strength of several aluminium-base materials.

TABLE 13.8. *Strengths of some Al-base materials*

Material	0.2 % Yield strength kgf/cm^2	Tensile strength kgf/cm^2
Commercially pure aluminium	350	910
6061 Alloy (1.0 % Mg, 0.6 % Si, 25 % Cu, 0.25 % Cr)		
Annealed	560	1260
Room temperature ageing	1470	2450
Elevated temperature (artificial) ageing	2800	2870
SAP (dispersion of $Al_2O_3. \sim 12$ % oxide)	2310	2520

The most versatile metal with respect to mechanical properties is, of course, steel, an iron–carbon alloy. Mechanical properties range from the hard, brittle, metastable martensitic solid solution to the ductile, machinable, spheroidized steel in which the carbon is present as spheroidal carbide particles. The structure is manipulated by heat treatments which involve heating the metal to an elevated temperature where γ-iron, an fcc phase in which the carbon is maintained in solution, is produced. In a very rapid quench from this temperature the γ-iron tends to revert to the bcc structure while the carbon remains in solution, severely distorting the lattice. This structure, with the supersaturation of carbon in solid solution is called martensite and is a hard, brittle material. With slower cooling rates structures containing *pearlite* are produced. The carbon precipitates as a carbide, *cementite*, in platelets alternating with platelets of bcc iron. The result is a material strengthened by a non-coherent precipitate but still having reasonable ductility. Rapid cooling rates tend to promote formation of finer platelets and produce stronger metal. During prolonged heating at temperatures lower than that at which the structure changes to fcc iron, the platelets tend to form spheroids (a structure called *spheroidite*). This structure is the most ductile and workable for a given steel. A very attractive combination of strength, toughness, and ductility is produced when a steel is mechanically worked while undergoing the fcc → bcc transformation. The result is a fine carbide dispersion and fine grain size.

The most satisfactory metals for use at elevated temperatures are the nickel and cobalt superalloys which experience service up to 1100° C. These are natural matrix materials for their oxidation resistance, but are actually complex alloys with a critically controlled microstructure. A nickel-base alloy, Udimet 700, will illustrate some of the principles of strengthening. Fig. 13.29 shows an idealized structure of this alloy. The matrix is the fcc solid solution phase strengthened by the molybdenum and chromium. Uniformly dispersed in the matrix is a γ' phase, nickel–aluminium–titanium intermetallic precipitates. The predominant carbide phase is the $M_{23}C_6$, for example $Cr_{21}Mo_2C_6$, at the grain boundaries. Other MC carbides such as TiC are also extremely stable. The carbides enhance the grain boundary strength and prevent grain growth. Grain boundary vacancies are annihilated by boron additions, reducing the possibility of fracture initiation. The grain boundary γ' produces a ductile grain boundary layer which prevents initiation of fracture. This ductile layer may be the γ' itself or the γ' depletion near the grain boundaries.

13.10   Fracture

With continued strain, plastic deformation eventually terminates in the initiation of a crack and its propagation until the body fractures into two or more pieces. The physics of fracture has been an active area of research for the past thirty years (especially since the development of dislocation theory), while engineers have been coping with fracture in its various aspects for the past several hundred years. If fracture occurs with little gross deformation it is called *brittle fracture*, while if it occurs after large plastic strain (as in the usual tensile test) it is termed *ductile fracture*. For static loads at high temperature, increased atomic and dislocation mobility result in special fracture

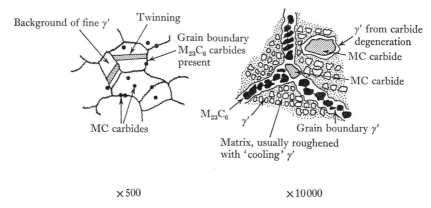

Fig. 13.29. Idealized microstructure of wrought Udimet-700 (Ni base alloy with 4.3 % Al, 3.4 % Ti, 15 % Cr, 53 % Mo, 17 % Co, 0.006 B, 0.05 % C). (Ref. 29.)

mechanisms involved with *creep rupture*. Fracture resulting from many repeated cycles of stress is called *fatigue fracture*. Environmental factors can produce special forms of fracture. Thus a stressed solid in a corrosive environment specific to that material (e.g. chloride ions in contact with austenitic stainless steel) will result in a time-dependent *stress corrosion* fracture, or hydrogen in certain metals will lead to *hydrogen embrittlement*. Repeated temperature cycling will result in fracture from thermal fatigue, while if the material is a brittle ceramic only one cycle may be needed to cause fracture from *thermal shock*. These categories of special fracture conditions could be extended at considerable length, but space will permit only a discussion of some of the more fundamental facts about fracture.

The theoretical fracture strength or cohesive strength of a brittle ideal elastic solid can be developed simply by assuming a sinusoidal relationship for the force–displacement curve between two atoms. The area under the curve is the work done per unit area in creating a fracture. Since two new surfaces, with surface energy γ_s, are created by the process of fracture, the work of fracture is equated to the surface energy. Thus, the theoretical *cohesive strength* of the solid is given by

$$\sigma_c = \left(\frac{E\gamma_s}{a_0}\right)^{\frac{1}{2}},$$

where E is Young's modulus, γ_s is the surface energy, and a_0 is the interatomic spacing of the crystal. Using typical values

$$(E = 10^{12} \text{ dyn/cm}^2; \quad \gamma_s = 10^3 \text{ erg/cm}^2; \quad a_0 = 3 \times 10^{-8} \text{ cm})$$

the cohesive strength is $\sigma_c = 1.82 \times 10^{11}$ dyn/cm^2, or $\sigma_c \approx E/6$. More precise expressions for the force–displacement curve lead to $\sigma_c \approx E/5$ to $\sigma_c \approx E/15$. In general, we can say that $\sigma_c \approx E/10$. The experimental measurement of the fracture stress of solids reveals that for most materials the fracture stress is 10 to 100 times less than the theoretical cohesive strength. Notable exceptions are freshly drawn silica fibres and small metallic whiskers.

The low fracture stress of crystalline solids led to the postulate that these materials contained small flaws or cracks which decreased their strength below the theoretical value. Originally it was proposed that the cracks were present initially before testing as a result of the preparation of the material, but more recently it has been realized that the flaws can be generated by plastic deformation prior to fracture.

Consider an elliptical crack of length $2c$ at the centre of a thin plate loaded in tension, Fig. 13.30. It is well-known from the theory of elasticity that such a discontinuity results in a non-uniform stress distribution with the maximum stress at the ends of the crack appreciably greater than the average stress in the plate. Inglis[30] showed that the maximum stress for a crack of length $2c$ which had a tip radius ρ was given by

$$\sigma_{\max} = \sigma\left(1 + 2\sqrt{\frac{c}{\rho}}\right) \approx 2\sigma\sqrt{\frac{c}{\rho}} \quad \text{for} \quad c \gg \rho.$$

If fracture occurs when the maximum stress at the crack tip reaches the cohesive strength of the solid, then

$$\sigma_{\max} = 2\sigma\sqrt{\left(\frac{c}{\rho}\right)} = \sigma_c = \sqrt{\left(\frac{E\gamma_s}{a_0}\right)}.$$

The nominal stress σ at which $\sigma_{\max} = \sigma_c$ is thus the fracture stress of the solid,

$$\sigma_f = \sqrt{\left(\frac{E\gamma_s}{4c}\frac{\rho}{a_0}\right)}.$$

Since the sharpest possible crack is one where $\rho = a_0$,

$$\sigma_f = \sqrt{\frac{E\gamma_s}{4c}}.$$

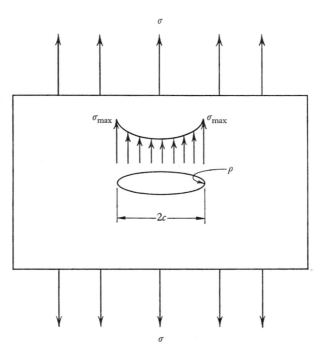

Fig. 13.30. Stress concentration at flaw.

The above basic relationship, that the fracture strength of a solid is inversely proportional to the square root of the length of the largest flaw, can also be developed from energy considerations. Griffith[31] suggested that a crack will propagate into a catastrophic brittle fracture when the decrease in elastic strain energy due to an increase in crack length is at least equal to the energy required to create the new crack surfaces. This leads to an equivalent expression for the fracture stress

$$\sigma_f = \left(\frac{2E\gamma_s}{\pi c}\right)^{\frac{1}{2}}.$$

Griffith's equation satisfactorily predicts the fracture stress of a completely brittle material such as glass, but for metals it leads to the prediction of flaw sizes of the order of several millimeters, which is a physical impossibility.

Orowan[32] suggested that the Griffith equation could be made compatible with brittle fracture in metals by including a term γ_p for the plastic work required to extend a crack in a metal with some ductility. He estimated $\gamma_p \approx 10^6$ erg/cm^2,

$$\sigma_f = \left[\frac{2E(\gamma_s + \gamma_p)}{\pi c}\right]^{\frac{1}{2}} \approx \left(\frac{E\gamma_p}{c}\right)^{\frac{1}{2}}.$$

Although this is a reasonable approach, γ_p is difficult to measure with precision. so that the Orowan–Griffith equation did not provide a practical method for calculating the fracture stress of a metal.

Irwin[33] proposed that the fracture stress of a metal can be determined by measuring the *crack extension force* $\mathscr{G}$, where

$$\mathscr{G} = \tfrac{1}{2}P^2 \frac{\partial(1/m)}{\partial c}.$$

In the above equation, m is the slope of a plot of load P versus extension for a cracked plate with a crack of length $2c$. $\mathscr{G}$ may also be interpreted as the *strain energy release rate*, the rate of loss of energy from the elastic stress field around the crack of the inelastic process of crack propagation:

$$\mathscr{G} = \frac{\sigma^2}{E}(\pi c).$$

A measured value of $\mathscr{G}$ includes the energy of plastic deformation required to slowly propagate the crack. For a given material, temperature, strain rate, and triaxiality of stress, $\mathscr{G}$ increases with crack length, until at some critical value $\mathscr{G}_c$, the crack becomes unstable and propagates in a rapid brittle fashion. Therefore, the Orowan–Griffith equation can be written

$$\sigma_f = \left(\frac{E\mathscr{G}_c}{\pi c}\right)^{\frac{1}{2}}.$$

If tests are carried out under proper conditions of plane strain (maximum triaxial constraint) the measured value, $\mathscr{G}_{Ic}$, is a valid material property which is independent of specimen thickness and design. This is the underlying basis for the active area of *fracture mechanics*. Fracture mechanics[34] has as its aim the measurement of fracture toughness and the development of analytical expressions for describing the fracture stress of structures with different crack geometries and stress systems. It turns out that it is easier to describe these problems in terms of the stress distribution instead of energy. Thus, the stress around a flaw can be written as

$$\sigma = \frac{K}{\alpha\sqrt{(\pi c)}},$$

where K is the *stress intensity factor*, and α is a geometrical correction factor ($\alpha = 1$ for a central crack in a thin, infinite plate). There is complete equivalence between the stress and energy approaches, since $K^2 = GE$. In the stress approach, the critical material parameter is the *plane strain fracture toughness*, K_{Ic}.

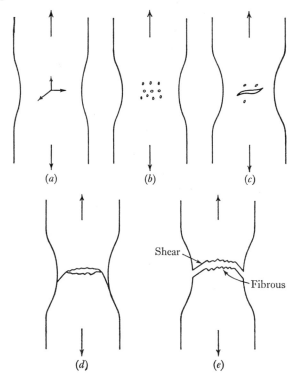

Fig. 13.31. Stages in the formation of a cup-and-cone fracture (ref. 27).

13.11   Mechanisms of fracture

If environmental and material conditions are favourable for unhindered dislocation motion then fracture will be a high energy ductile fracture. The classic case is the cup-and-cone fracture that is found in the tensile test of most ductile metals, Fig. 13.31. Fracture begins internally at the centre of the specimen where the hydrostatic tension is the greatest. It takes the form of small voids or cavities which start at fine second phase particles or inclusions. The voids elongate and grow as a 'void sheet' along planes of shear stress, giving rise to the zig-zag fibrous fracture surface as the crack propagates outward toward the surface. Final separation occurs as the specimen shears at roughly 45° to the tensile axis to form the cone region of the fracture surface. Truly ductile fracture is basically a ductile tearing process in which extensive dislocation motion occurs.

As the yield strength of a material is increased by various strengthening mechanisms, or as the yield stress is increased by environmental factors such as the introduction of a triaxial tensile stress state (usually by the creation of a notch), a decrease in test temperature, or an increase in the strain rate, the fracture process becomes less ductile. It appears that all crystalline solids in which dislocations can move fast only at high stresses tend toward brittle fracture, since the dislocations cannot run fast enough to relax the stresses at the tip of a high speed brittle crack.

A key point in understanding brittle fracture was the realization that pre-existing flaws are not required, but that critical size cracks can be generated by the restricted plastic deformation which precedes brittle fracture in metals. One experimental point in support of this is the fractographic observation of a 'river-pattern' structure consisting of small steps on the surface of brittle fractures.

Dislocation models of brittle fracture are based on the early recognized fact that when dislocations of one sign are piled up against a barrier due to an applied shear stress, a high tensile stress is created at the head of the pile-up. The high tensile stresses at the head of the pile-up are either relieved by local slip or the dislocations coalesce into a wedge-shaped crack. Commonly observed fracture mechanisms for dislocation pile-ups at barriers are illustrated in Fig. 13.32.

The most general mechanism for crack initiation by a dislocation pile-up[35] is at a grain boundary, Fig. 13.32a. The tensile stress at the head of a pile-up of n dislocations will be of the order of $n\tau$, where τ is the effective shear stress on the slip plane. The number of dislocations that can occupy a slipped length L between the source and the barrier is

$$n \approx \frac{L\tau}{Gb},$$

so that the stresses at the head of the pile-up will be lower in fine grain size material than in coarse grain size material. This is consistent with the well-established fact that coarse grain size material is more brittle than the fine

grain size material. Crack initiation may be made easy at the grain boundary
by the presence of a brittle film or phase. In certain materials a fine dispersion
of second phase particles within the grains may serve to limit the length of the
dislocation pile-ups and minimize brittle behaviour. In single crystals, where
grain boundaries are absent, twin bands may serve as barriers to initiate
cracks. However, a more common mechanism involves glide dislocations on

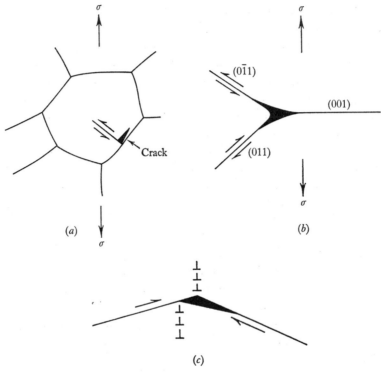

Fig. 13.32. Fracture initiation by dislocation pile-ups at barriers. (*a*) Pile-up at grain
boundary. (*b*) Slip on intersecting planes. (*c*) Crack formation at tilt boundary.

intersecting slip planes which interact to form a cavity dislocation. Fig.
13.32*b* illustrates the model proposed for bcc iron in which dislocations on
intersecting {110} planes interact to produce a dislocation whose Burgers
vector lies in the (001) cleavage plane. This fracture mechanism has been
verified in iron and MgO.

Hexagonal metals fracture on the basal plane, which is also the slip plane.
The mechanism by which fracture occurs is illustrated in Fig. 13.32*c* where a
tilt boundary consisting of a wall of edge dislocations originally lay normal to

the basal plane. Slip on the basal plane causes a displacement of the tilt boundary which leads to the formation of a crack along the slip plane.

Thus, there are a variety of mechanisms by which local plastic deformation can initiate microcracks in a brittle or semi-brittle material. However, frequent observation that cracks have initiated at grain boundaries and not propagated to complete separation suggests that crack growth is more difficult than crack nucleation in most engineering materials. The strong influence of a notch in raising the temperature at which brittle fracture occurs lends support to this position. The chief effect of a notch is to introduce hydrostatic tensile stresses. A hydrostatic state of stress will not increase the number of dislocations in a pile-up or squeeze them more closely together, but it will assist in the growth of an existing crack.

13.12 Ductile-brittle transition

Body-centred cubic and cph metals show a transition from high-energy ductile fracture to low-energy brittle fractures as the temperature of testing is decreased. The phenomenon is particularly important in structural steels where it has resulted in the catastrophic brittle fracture of ships, pressure vessels, storage tanks and other engineering structures. A very considerable effort has resulted in better engineering procedures for the control of brittle fracture and in greater understanding of the phenomenon in terms of dislocation theory.

In simple phenomenological terms, brittle fracture occurs when the yield stress of the material reaches the fracture stress. The yield stress of bcc metals rises steeply with decreasing temperature, and thus at some low temperature a transition occurs from ductile to brittle fracture. Increasing the strain rate (as in impact loading) also raises the yield stress. More important is the influence of the state of stress. The presence of triaxial or biaxial tensile stresses will increase the level at which yielding occurs. The latter effects are all present at a notch or crack in a structure. As shown in Fig. 13.33 a crack or notch will produce a high local stress at the crack tip which will be accompanied by steep strain gradients. The local strain concentration leads to high local strain rates, and the triaxiality of stress at the notch raises the yield stress. Even when the stress level is relieved by local yielding at the root of the notch the yield stress is raised to about three times what it would be for uniaxial tension. These various effects are shown schematically in Fig. 13.33.

The ductile-to-brittle transition is measured routinely in structural steels by a notched-impact test such as the Charpy test. Fig. 13.34 shows how the energy required to fracture a standard Charpy specimen falls off at the transition temperature. For more critical design work the *nil ductility temperature*, the temperature below which the material is completely brittle, may be

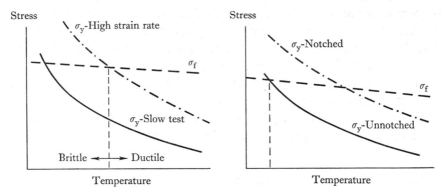

Fig. 13.33. Schematic influence of strain rate and plastic constraint of a notch on ductile-to-brittle transition temperature.

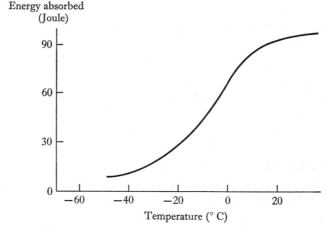

Fig. 13.34. Charpy transition temperature curve for a steel specimen.

determined with a drop weight test or an explosive bulge test. It should be emphasized that the transition temperature is not a fixed temperature for a given material, but depends strongly on defect structure and metallurgical structure as well as the rate of deformation.

The variables that control the transition temperature of a material can be interrelated in a realistic way by a dislocation model based on crack initiation by a pile-up of dislocations. Cottrell[36] assumed that the propagation of the crack is the critical step, and thus occurs when the yield stress exceeds the stress for the growth of the crack. If b dislocations form the crack nucleus, then nb is the displacement between the faces of the dislocation cavity, and the Griffith equation can be written in the form $bn\tau \approx 2\gamma_s$. The left side of this equation is an energy term, which is equated to the surface energy term. The shear strain ϵ in the crack is

$$\frac{nb}{L} = \frac{\tau - \tau_1}{G} = \epsilon,$$

where $\tau - \tau_1$ is the effective shear stress in the pile-up of length L. For a brittle material $\tau_t = \tau_y - \tau_i + kd^{-\frac{1}{2}}$, and, if $L = d$,

$$nb \approx \frac{kd^{\frac{1}{2}}}{G},$$

which when substituted into $nb \approx \dfrac{2\gamma_s}{\tau}$ results in

$$\tau_y kd^{\frac{1}{2}} = \beta G \gamma_s.$$

This equation expresses the limiting condition for the formation of a propagating crack from a pile-up of dislocations. If conditions are such that the left side of the equation is less than the right, a crack can form but it cannot propagate. When the left side is greater than the right a propagating brittle fracture can be produced by a shear stress equal to the yield stress. Note that small grain sizes lead to difficulty in propagating the crack. β is a term expressing the state of stress. For torsion $\beta = 2$, while for a notch in tension $\beta = 1/3$. Material purity and structure is expressed through k and γ_s, the surface energy. The term k expresses the number of dislocations that are released into the pile-up.

13.13 Summary

The considerations of crystalline deformation described in this chapter emphasize the important role of defects as well as the inherent crystal structure in governing deformation behaviour. While classical continuum descriptions continue to evolve slowly, the most rapid advances in theoretical interpretation have been based upon microscopic deformation mechanisms and their extrapolation to the macroscopic deformation level.

The continuum descriptions of solid deformations discussed in earlier chapters will show greater applicability to solids as approximations based upon microscopic observations are incorporated into these phenomenological theories.

References to chapter 13

1   In NaCl: S. Amelinckx in *Dislocations and mechanical properties of crystals* (New York, 1957).
    In Si: W. C. Dash, *J. Appl. Phys.* **27**, 1193 (1956).
2   H. B. Huntington, *Solid State Physics* **1** (1958).
3   W. Voigt, *Lehrbuch der Kristallphysik* (Leipzig), p. 716
4   D. H. Chung & W. R. Buessem, *J. Appl. Phys.* **38**, 2535–40 (1967) and *Proc. Int. Symp. on Anisotropy in single crystal refractory compounds* (New York, 1968).
5   A. Reuss, *Z. Angew. Math. Mech.* **9**, 55 (1929).
6   R. Hill, *Proc. Phys.* **65**, 349 (London, 1952).
7   E. N. daC Andrade & P. J. Hutchings, *Proc. Roy. Soc. A* **148**, 120 (1935).
8   J. D. Meakin, Proc. Int. Conf. on Deformation of Crystalline Solids, *Can. J. Phys.* **45**, 2, 1121–34 (1967).
9   S. S. Brenner, *J. Appl. Phys.* **27**, 1484–91 (1956).
10 G. Thomas, *J. Metals*, 365–9 (April 1964).
11 A. G. Guy, *Elements of physical metallurgy* (Reading, Mass, 1959).
12 W. T. Read, Jr, *Dislocations in crystals* (New York, 1953).
13 R. Peierls, *Proc. Phys. Soc.* **52**, 34 (1940).
14 J. J. Gilman & V. G. Johnston in *Dislocations and mechanical properties of crystals* (John Wiley & Sons, 1957).
15 H. L. Prekel, A. Lawley & H. Conrad, *Acta Met.* **16**, 337–45 (1968).
16 R. Becker, *Z. Physik* **26**, 919 (1925).
17 H. Conrad, *J. Metals* 582–88 (1964).
18 A. Seeger, *Phil. May* **1**, 651 (1956).
19 R. L. Fleischer, *Acta Met.* **15**, 1513–19 (1967).
20 T. E. Mitchell & W. A. Spitzig, *Acta Met.* **13**, 1169–79 (1965).
21 O. F. Kocks, *Acta Met.* **8**, 345–52 (1960).
22 R. L. Fleischer and W. R. Hibbard, *The relation between structure and mechanical properties of metals* **1**, 261 (Her Majesty's Stationary Office, 1963).
23 E. Orowan discussion in *Symposium on internal stresses*, 451 (London: Institute of Metals, 1947).
24 J. C. Fisher, E. W. Hart & R. H. Pry, *Acta Met.* **1**, 336 (1953).
25 G. I. Taylor, J. *Inst. Metals* **62**, 307 (1938).
26 Y. I. Shida, A. W. Mullendore & N. J. Grant, *Trans. Met. Soc. AIME* **233** (1967).
27 G. E. Dieter, Jr., *Mechanical metallurgy* (New York, 1961).
28 A. Lawley & R. Maddin, *Trans. Met. Soc. AIME* **224**, 573–83 (1962).
29 C. T. Sims, *J. Met.*, 1119–30 (October 1966).
30 C. E. Inglis, *Trans. Inst. Naval Architects* **56**, 219 (1913).
31 A. A. Griffith, *Phil. Trans. Roy. Soc. London* **A221**, 163 (1920).
32 E. Orowan, *Rep. Prog. Phys.* **12**, 185 (1948).
33 G. R. Irwin & J. A. Kies, *Welding J.* **31**, 450 (1952).
34 A. S. Tetelman & A. J. McEvily, *Fracture of structural materials* (New York, 1967). 'Fracture touchness testing', ASTM Spec. Tech. Rep. 381 (Philadelphia), *Amer. Soc. Testing & Materials* (1965).
35 E. Smith, *Acta Met.* **16**, 313 (1968).
36 A. H. Cottrell, *Trans. AIME* **212**, 192 (1958).

14 Bakers' dough[1]

A. H. Bloksma

14.1 The role of dough properties in the breadmaking process[2]

The breadmaking process consists of three stages: mixing the ingredients into a dough, some fermentation periods with punches in between, and finally the baking. (Punches are a variety of mechanical treatments of the dough intended to drive out occluded gas.) Within this framework large variations are found in the duration and rate of mixing, and in the number and duration of the fermentation periods. These variations, among others, are related to the type of bread produced and to the extent of mechanization of the process.

The main ingredients of bread doughs are flour and water; water contents of doughs are usually about 45 %, including the flour water. Other ingredients are yeast, or other leavening agents, and common salt. The effect of oxidizing agents, added in quantities at about 10–100 mg per kg of dough, will be discussed in §14.5.

The purpose of the process is to produce a loaf of bread with a light crumb. The crumb structure is fixed in the baking stage at 55–65° C when the starch swells or gelatinizes. Up to that point the dough must retain the gas in its cells, mainly carbon dioxide, produced by the yeast, and water vapour. Mixing and fermentation are carried out to produce sufficient gas retention in the dough. Another condition for a light crumb is that the dough should enter the baking stage with a large amount of gas in it, about 70 % of its total volume. This is attained during the last fermentation period or final proving of about one hour, after which the dough is transferred to the oven without punching.

Mechanical work is required to achieve 'dough development', that is to bring about sufficient gas retention in the dough. In conventional breadmaking processes, the mixing energy is of the order of 3.5 kcal per kg of dough. This is not enough; the dough is satisfactorily developed only after further work has been exerted on it during the first periods of fermentation and punching. Alternatively, one can mix faster, applying about 12 kcal per kg of dough over a few minutes. After this 'mechanical dough development', fermentation and punching are no longer necessary to obtain gas retention; the fermentation process can be reduced to no more than a final proving.

This review of the breadmaking process shows that for a light crumb a sufficient production of carbon dioxide, as well as gas retention, is required. Insufficient gas production is easily remedied by the addition of sugars, amylase, or of more yeast; gas retention is the more important intrinsic property of dough.

Gas retention is related to the rheological properties of dough. A slow dough development, and a combination of a high resistance against deformation and a high extensibility, were empirically found to be desirable in breadmaking. Doughs from a high quality breadmaking flour are characterized by a long relaxation time. These requirements are far from being understood. The advantage of a high extensibility only is obvious; the dough membranes between gas cells can then sustain large extensions without rupture. Although a low resistance facilititates the expansion of gas cells, the resistance must be sufficiently high to prevent their ascent and the spreading of the dough under the influence of gravity.

14.2   Dough structure

A complete dough contains four phases, that is, a continuous phase of highly swollen protein, in which starch kernels, gas cells, and yeast cells are dispersed. For the sake of simplicity, most rheological studies have been made with unleavened doughs. The volume fraction of gas cells in these test-pieces is rarely over 10%, and will be neglected in this discussion.

By kneading a dough in a stream of a dilute salt solution, one removes most of the starch; in addition, the more soluble proteins are lost. The residue, which is called gluten, contains about 80% of the flour proteins; its water content is about 65% and its visco-elasticity is similar to that of dough. Isolated gluten is an approximate model for the continuous phase of dough.

The properties of a composite material like dough depend primarily on its continuous phase. Accordingly, quality differences between flours are explained to a great extent by differences in the amount and in the nature of their proteins. Generally, a high protein content, that is, more than 11% of dry matter, makes a flour suitable for breadmaking. There are, however, also differences in protein quality. The effect of oxidizing and reducing agents on dough properties is due to a modification of its proteins.

The proteins in gluten form a network of linear macromolecules; the nature of the cross-links in this network will be discussed in the next section. Their number and reactivity determine the rheological properties of dough and gluten. Dough development involves the formation of this coherent network from the constituents of separate flour particles. This requires the unfolding of protein molecules, and the formation of cross-links between molecules originating from different flour particles.

Although the gluten phase is most important for dough properties, the starch fraction cannot be neglected. As a first approximation, starch can be considered as a rigid filler in a visco-elastic matrix. Its effect may be more complicated, since the volume fraction of starch in dough, about 60 %, is high enough for interactions between starch kernels to affect dough properties. This effect can be demonstrated in starch pastes in the absence of gluten. Addition of glyceryl monostearate to a dilatant starch paste converts it into a pseudo-plastic system; this is explained by the change of a stable colloid into a flocculated system by the adsorption of glyceryl monostearate. We do not know how important the interaction between starch kernels is during fermentation of normal doughs, in which gluten is present.

14.3   Wheat proteins[3]

The wheat proteins can be divided into three groups. (1) About 20 % of the proteins are soluble in dilute salt solutions: albumins and globulins; most of them are lost in the isolation of gluten. Gluten can be divided into: (2) gliadin, soluble in a mixture of ethanol and water (70:30), and (3) glutenin, soluble in dilute acid or alkali. Gliadin and glutenin make up about equal parts of gluten. Zone electrophoresis in a starch gel and chromatography make possible a further fractionation into about 30 components. Molecular weights of gliadin components are about 30000–40000; glutenin has a molecular weight distribution from about 50000 to some millions.

The peculiar amino acid composition of the gluten proteins explains why they are insoluble in water. They contain about 30 % glutamine, the amide group of which can act both as hydrogen donor and as an acceptor in hydrogen bond formation. The presence of about 10 % proline with restricted rotation hinders the formation of a helix. This and the high protein concentration in dough favour the formation of intermolecular rather than intramolecular hydrogen bonds. The insolubility of gluten is due to these hydrogen bonds.

Gluten proteins contain about 100 moles of disulphide groups (cystine residues; see Fig. 14.1) per 10^6 g protein, or about 25 half-cystines per 1000 amino acid residues. If the disulphide groups in glutenin are broken by thiol compounds or by oxidation, one obtains a number of components with molecular weights of about 20000. With gliadin, a change in conformation, but not in molecular weight, is observed. Apparently, glutenin is composed of units that are bound one to another by disulphide groups. Contrary to this, the disulphide groups of gliadin are intrachain groups.

The native flour proteins also contain a small number of thiol groups (cysteine residues; see Fig. 14.1), that is 5 to 10 moles per 10^6 g protein or about 5–10 % of the disulphide content. Thiol and disulphide groups can be converted into another by oxidation and reduction according to the scheme in Fig. 14.1. In §14.5 we shall discuss the way in which dough properties depend upon thiol and disulphide groups.

Thiol groups
Cysteine

Disulphide cross-links
Cystine

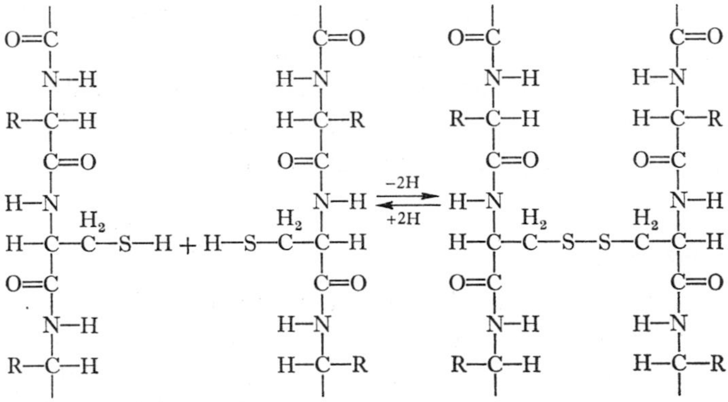

Fig. 14.1. Conversion of thiol groups into disulphide cross-links by oxidation and the reverse by reduction.

Besides hydrogen bonds and disulphide groups, two other types of cross-links contribute to the stability of the structure of proteins; they are electrostatic forces between basic and acidic side chains, and hydrophobic bonds between non-polar amino acid residues. It is improbable that electrostatic bonds are essential for gluten structure. About the importance of hydrophobic bonds little information is available.

14.4   General description of the rheological properties of dough

Dough behaviour can be roughly described by the mechanical model in Fig. 14.2, which allows for instantaneous and retarded elasticity, a yield value, and plastic flow. The modulus of dough, represented by springs B and F in series, is of the order of 10^{-4} to 10^{-5} kgf mm^{-2}. Results for the viscosity vary more widely. Most of the published data fall within the range from 10 to

$10^3 \, \text{kg mm}^{-1} \, \text{s}^{-1}$. Measurements at high stress produced values as low as $1 \, \text{kg mm}^{-1} \, \text{s}^{-1}$. Data on the yield value are scarce and vary widely: from 1.5×10^{-6} to $3 \times 10^{-5} \, \text{kgf mm}^{-2}$. Most reports on the relaxation time mention values between 5 and 100 s, but longer times, up to 2500 s, are also reported; stress relaxation in dough cannot be described by a single relaxation time. The retardation time is of the order of 10 s.[4]

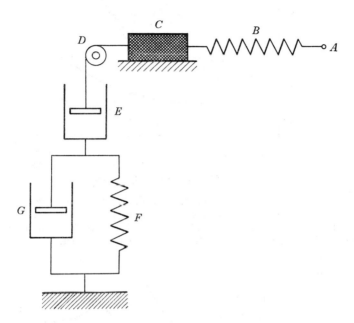

Fig. 14.2. Mechanical model for the description of dough properties.

The model in Fig. 14.2 is a simplification of dough behaviour. Some deviations from this model are discussed in the following paragraphs.

Perhaps the most striking deviation is that dough behaves non-linearly. The compliance, which is the ratio of deformation and stress, increases with increasing stress. Fig. 14.3 demonstrates that this increase is mainly due to the viscous part of the compliance, and hardly to the elastic part. Linear behaviour was found with dynamic experiments in which shear did not exceed 0.002.[5] The observation that the viscous compliance depends heavily upon the shear stress explains why various workers, using different experimental conditions, arrived at widely varying results for the viscosity of dough.

In addition, dough properties depend on the amount of strain. The creep curves in Fig. 14.4 show a rapid start and, after that, a decreasing slope.

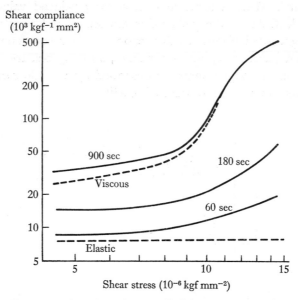

Fig. 14.3. Compliance as a function of stress. Full lines represent the total compliance after three time intervals; dashed lines correspond to the elastic and viscous components of the compliance after 900 s. This figure is a compilation of a series of creep experiments of which the experiments in Fig. 14.4 form a part.

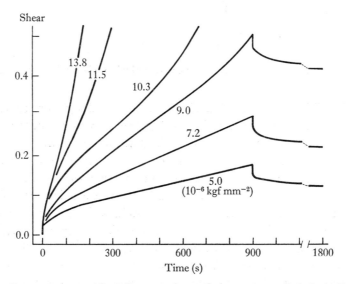

Fig. 14.4. Creep curves with different values of shear stress. Labels indicate shear stress in 10^{-6} kgf mm^{-2}. After 900 s the stress was removed.

Apparently, the viscosity increases with increasing strain. When the shear passes beyond a critical value, which is of the order of 0.5, the slope of the curve increases again.

Another complication is the effect of time. This phenomenon can, among other means, be demonstrated with the Brabender extensigraph. In this instrument, a test-piece is shaped by rounding and rolling it. After a certain rest period it is subjected to a load–extension test. If a series of identical test-pieces are stretched after various rest periods, the resistance to extension decreases and the extensibility increases with increasing rest period. An example of these changes can be found in Fig. 14.6, in which measurements of resistance and extensibility after 15 and 45 minutes rest are shown; after 45 minutes resistances are lower and extensibilities higher. Apparently, the shaping of the test-piece brings the dough into a condition of high resistance against deformation; this condition gradually disappears. The phenomenon bears some resemblance to stress relaxation; it differs from stress relaxation in that it proceeds while the test-piece is not under stress. Moreover, the decrease in resistance that is dealt with in this paragraph proceeds in the course of one or two hours, whereas true stress relaxation proceeds on a time scale of minutes. Under other conditions, changes of dough properties in the opposite direction, that is a stiffening of dough with time, can also be observed.

Because of the effects of strain and time, one can only obtain reproducible results in rheological tests, if the deformation history of the test-piece, including the mixing process, is rigidly standardized. Apart from its mechanical effect, mixing accelerates chemical reactions in dough which may affect dough properties.

14.5 The role of cross-links in gluten

In § 14.2 it has been explained that the properties of dough depend upon those of its continuous gluten phase. This gluten phase is considered to be a network of protein molecules with water in between. § 14.3 points to hydrogen bonds and disulphide bonds as the cross-links that most probably give this network its coherence. In this section we shall discuss how far one can understand dough properties on the basis of this hypothesis.

For the importance of the disulphide bond direct evidence is available. In § 14.2 it has already been mentioned that disulphide groups act as links between the constituents of the giant glutenin molecules. In addition, the rheological properties of dough are very sensitive to oxidation and reduction. Figs. 14.5 and 14.6 illustrate that oxidation leads to stiffer doughs. Fig. 14.5

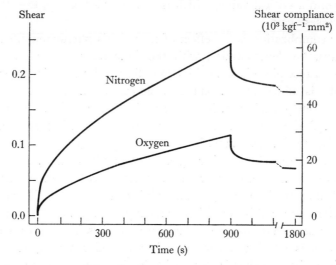

Fig. 14.5. Creep curves of doughs after mixing in nitrogen or oxygen. Shear stress 3.96×10^{-6} kgf mm^{-2}. After 900 s the stress was removed.

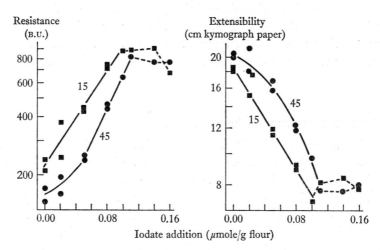

Fig. 14.6. Resistance to extension, and extensibility of doughs with various amounts of potassium iodate. Measurements were performed by means of the Brabender extensigraph. Squares represent measurements after 15 min rest, circles after 45 min. 1 B.U. $\equiv$ 1.24 gram-force.

refers to creep experiments with doughs of identical composition, mixed in nitrogen and oxygen, respectively. Fig. 14.6 shows the effect of potassium iodate on load–extension tests. Oxidizing agents such as potassium bromate, potassium iodate, azodicarbonamide, or acetone peroxide are intentionally added by the miller or baker to improve dough properties. Conversely, the addition of thiol compounds or sulphite softens the dough.

Analytical determinations confirm that atmospheric oxygen, as well as the reagents mentioned in the preceding paragraph, cause a decrease in the number of thiol groups in dough. They are oxidized to disulphide groups, according to the scheme in Fig. 14.1, but higher oxidation products, such as cysteic acid residues, are also possible. The consumption of bromate, iodate, and azodicarbonamide in dough can, for the greater part, be explained by their reaction with thiol groups.[6]

Blocking of the thiol groups, without formation of disulphide cross-links, causes similar changes in dough properties as does oxidation. Apparently, the effects are due to a reduction in thiol content, rather than to an increase in the number of cross-links. This can be explained by the model in Fig. 14.7, in which the deformation of gluten proceeds by the way of thiol–disulphide exchange reactions. The thiol compound XSH in this model may be the amino acid cysteine, the tripeptide glutathione, or a protein with thiol groups. The actual occurrence of thiol–disulphide exchange reactions in dough has been demonstrated by the addition of various thiol compounds, which were incorporated into the gluten proteins.[7]

In Fig. 14.7 it is assumed that between reactions (1) and (2) the conformation of the lower protein molecule changes as a result of Brownian motion; under stress its direction will be biased. The upper chain is arbitrarily considered to stay in position. Reaction (3) is the net result of reactions (1) and (2); the compound XSH is not consumed and becomes available again for another cycle of exchange reactions. Each cycle may contribute to the viscous deformation of the test-piece; any deformation, however large it may be, can be explained by a sufficient number of reaction cycles. If the elasticity of gluten is similar to that of rubber, its elastic deformation will, at a given stress, be smaller the more disulphide cross-links it contains. The model in Fig. 14.7 explains why a limited elastic deformation can be accompanied by a very large viscous one. Fig. 14.3 shows that, except at low stress and after a very short time, the deformation of dough is predominantly viscous.

The model in Fig. 14.7 predicts that oxidation of thiol groups to disulphide bonds reduces both its viscous and elastic deformation. Generally, the reduction in the viscous part is relatively more important; this is illustrated in

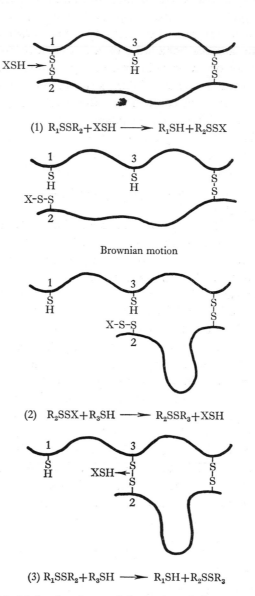

(1) $R_1SSR_2 + XSH \longrightarrow R_1SH + R_2SSX$

Brownian motion

(2) $R_2SSX + R_3SH \longrightarrow R_2SSR_3 + XSH$

(3) $R_1SSR_2 + R_3SH \longrightarrow R_1SH + R_2SSR_3$

Fig. 14.7. Model for the viscous deformation of gluten as a result of thiol–disulphide exchange reactions.

Fig. 14.5. This is in qualitative agreement with the fact that the relative decrease in thiol content is 20–40 times larger than the relative increase in disulphide groups. A quantitative explanation of the changes in dough properties as a result of oxidation is not yet possible. The model might be improved by the distinction between rheologically effective and ineffective thiol and disulphide gropus.[8]

Our knowledge about the hydrogen bonds is much less detailed. The most specific indication of their importance in rheological properties is the observation that dough or gluten with deuterium oxide instead of water shows a higher resistance to extension.[9]

Hydrogen bonds are easily broken and re-formed. This suggests that they play a role in temporary changes in properties, such as the increase in resistance against deformation of test-pieces that are shaped in the Brabender extensigraph (see §13.4). Rounding and rolling possibly cause an increased alignment of the protein molecules, and consequently, more hydrogen bonds. After that, Brownian motion leads again to more random conformations with less hydrogen bonds. This information is, however, unspecific; the argument might refer to hydrophobic bonds as well.

Summarizing, one can say that there is a qualitative understanding of one detail of dough behaviour, that is, its response to oxidation, thiol blocking, or disulphide cleaving reagents. We can also explain why gluten behaves as an elastic as well as a viscous material. However, a quantitative explanation is still beyond our reach. The same can be said of the explanation of general phenomena like the occurrence of a yield value, the particular shape of creep curves, and changes of dough properties with time. An important question, the answer to which is as yet unknown, is the relation between the properties of the gluten phase, the starch content, and dough properties; the possible interaction between starch kernels, that was mentioned in §14.2, forms part of this problem. A wide field awaits exploration.

References to chapter 14

1 A more extensive discussion of dough properties with many references has been published by A. H. Bloksma and I. Hlynka in *Wheat, chemistry and technology* (Ed. I. Hlynka, Amer. Assoc. Cereal Chemists, St Paul 1964), pp. 465–526. In the present chapter reference will only be made to papers that are not cited in this earlier discussion.

2 For a more comprehensive description of the breadmaking process see E. B. Bennion, *Breadmaking, its principles and practice* (3rd ed.; London, 1954).

3 For reviews see J. W. Pence, C. C. Nimmo and F. N. Hepburn in *Wheat, chemistry*

and technology (Ed. I. Hlynka, Amer. Assoc. Cereal Chemists, St Paul 1964), pp. 227–76; and J. S. Wall and A. C. Beckwith, *Cereal Sci. Today* **14**, 16 (1969).

4   Data are taken from papers cited in ref. 1, and in addition from A. H. Bloksma, *Rheol. Acta* **2**, 217 (1962); J. Glucklich & L. Shelef, *Kolloid-Z.u.Z. Polymere* **181**, 29 (1962); H. G. Muller, M. V. Williams, P. W. Russell Eggitt & J. B. M. Coppock, *J. Sci. Food Agr.* **13**, 572 (1962); C. H. Lerchenthal & H. G. Muller, *Cereal Sci. Today* **12**, 185 (1967).

5   G. E. Hibberd & W. J. Wallace, *Rheol. Acta* **5**, 193 (1966).

6   F. J. R. Hird & J. R. Yates, *Biochem. J.* **80**, 612 (1961). C. C. Tsen & W. Bushuk, *Cereal Chem.* **40**, 399 (1963). A. H. Bloksma, *J. Sci. Food Agr.* **15**, 83 (1964).

7   E. E. McDermott and J. Pace, *Nature* **192**, 657 (1961). D. G. Redman & J. A. D. Ewart, *J. Sci. Food Agr.* **18**, 15 (1967). T. Kuninori & B. Sullivan, *Cereal Chem.* **45**, 486 (1968).

8   A. H. Bloksma, Soc. Chem. Ind. (London), Monograph no. 27, 153–166, 1968.

9   V. L. Kretovich & A. B. Vakar, *J. Sci. Food Agr.* **15**, ii 147 (1964). A. B. Vakar, A. Ya. Pumpyanskii & L. V. Semenova, *J. Sci. Food Agr.* **18**, i 139 (1967). R. Tkachuk & I. Hlynka, *Cereal Chem.* **45**, 80 (1968).

15 Paints and lacquers

M. N. M. Boers

15.1 Introduction

The terms 'paints' and 'lacquers' are used in general to refer to a large number of materials of very different composition. It is often impossible to show any chemical relation among them: the one factor they have in common is their use.

They consist of a solution of a polymer in which is dispersed a pigment. The polymer is the vehicle which forms the body of the dry coat. The mechanism of the solidification process depends on the type of vehicle. Oleo-resins and alkyd resin paints, for instance, harden under the influence of oxygen which causes an interlinking of fatty acid chains resulting in a three-dimensional molecular network. Stoving lacquers like alkyd resin/melamine resin combinations are hardened by a condensation reaction at elevated temperature. Polyurethanes and epoxy resin/amine combinations are examples of vehicles which harden at room temperature. In this case the cross-linking is achieved by an addition reaction.

Hardening of the coat may also be a purely physical process, consisting of evaporation of the solvent only. Examples of this type of drying are the nitrocellulose lacquers. Similarly, the drying process of emulsion paints is also purely physical. The film formed after evaporation of the water then consists of sintered polymer particles.

There are three stages at which the rheological properties of coating materials have practical implications. The first is on application by brushing, spraying or any other technique. The second stage comprises the drying period beginning just after the application of the wet film and ending when the film is considered dry enough to handle. Then the final stage is where the elastic properties of the coat determine its performance in practice. Some of the rheological aspects of these three stages will be discussed in the following sections.

15.2   Rheology of liquid paints

Essential to the practical performance of a paint is its rheological behaviour on spreading to the substrate and during the period directly following, that is, during drying. Easy spreading on the substrate requires a low viscosity and the same holds for the levelling out of striations or other marks in the wet film.

On the other hand a wet film of too low a viscosity may run down from vertical surfaces and form 'tears' along edges and borders. Moreover, lowering the viscosity implies a further dilution of the pain with solvent which is altogether inefficient. A paint satisfactory overall is therefore a kind of compromise between opposing rheological requirements.

Paints or varnishes to be applied by spraying, dipping or rolling are usually near-Newtonian liquids with a viscosity of 0.5 to 1.5 poise. They generally contain a blend of low and high volatile solvents to obtain a proper balance between the initial viscosity and the increase of viscosity after application.

Good brushing paints nearly always show a yield value or are thixotropic in nature. It is the experience that purely Newtonian liquids have poor brushing properties. In the case of paints showing a yield value the Casson equation (15.1) appears to hold well for describing the relation between shear rate, D, and shear stress, τ.

$$\sqrt{\tau} = K_0 + K_1\sqrt{D}. \qquad (15.1)$$

In this equation K_0 and K_1 are constants representing respectively the yield value and the viscosity at infinite shear rate. Fig. 15.1 gives an example of the application of this equation to some pigment dispersions.[1]

On brushing, the shear rate caused in the wet film is of the order of $10^4 \ \text{s}^{-1}$ and it is generally found that viscosity values measured at such high shear rates correlate well with the ease of brushing experienced by painters. Fig. 15.2 illustrates the relationship between brushability and high shear viscosity for some types of paint.

Brushing paints with poor flow properties show the well-known phenomenon of brush marks, caused by insufficient levelling of the striations made by the brush. The levelling process has been analysed by several investigators.[3, 4] The relaxation time, t, has been found to depend on the size of the brush marks x, the wet film thickness h, the surface tension σ, and the viscosity η, according to equation (15.2) in which K is a constant

$$t = K\eta x^4/\sigma h^3. \qquad (15.2)$$

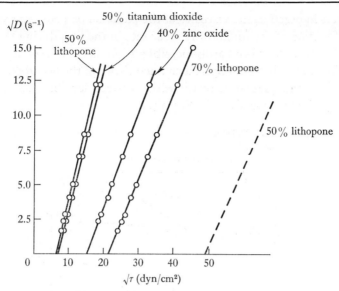

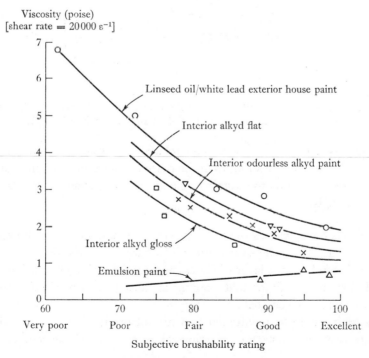

Fig. 15.1. Linear relationship between $\sqrt{\tau}$ and $\sqrt{D}$ according to the Casson equation for five pigment dispersions (ref. 1). — dispersions in linseed oil; - - - dispersion in paraffin oil.

Fig. 15.2. Correlation between subjective brushability rating and viscosity at a high shear rate (ref. 2).

However, in practice, the viscosity of the paint is not a constant value but a complex function of the (unknown) shear rate in the film during the levelling and of time. It is therefore hardly possible to predict the levelling properties of a certain paint only from physical characteristics of the liquid paint as such. An example of the qualitative relationship between levelling and viscosity in the case of thixotropy is given in Fig. 15.3.

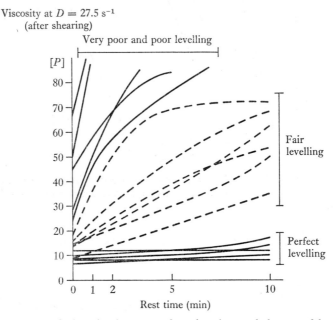

Fig. 15.3. Rate of viscosity increase after shearing and degree of levelling of brush marks for a number of thixotropic paints (ref. 5).

Sagging is a process more or less related to levelling but acting in the opposite direction and leading to an increase of unevenness of the paint film. For Newtonian liquids the sagging velocity at the surface of the layer on a vertical plane is

$$v = \rho g h^2 / 2\eta, \tag{15.3}$$

in which ρ is the density, g the gravitational constant, h the film thickness and η the viscosity. It can be seen from this equation that the degree of sagging strongly depends on the layer thickness and confirms the observation of every unskilled amateur painter that sagging is most pronounced with too thick and, moreover, uneven coats.

Finally, it has to be mentioned that not all profile defects of a paint coat originate from unsatisfactory rheological properties of the wet coat only.

Poor wetting of the substrate by the wet paint, for instance, may lead to the formation of craters in the coat notwithstanding good flow properties. Surface irregularities may also arise from a partial deflocculation of vehicle components during the evaporation of the solvents caused by an improperly balanced composition.

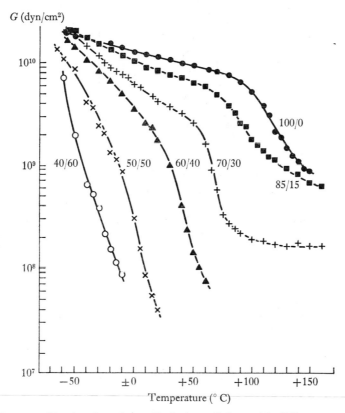

Fig. 15.4. Torsional modulus G of nitrocellulose with different amounts of plasticizer as a function of temperature (ref. 6).

15.3 Mechanical properties of paint coats

Paint coats may differ widely in mechanical behaviour. They can be hard and brittle or soft and extensible, depending on the nature of the vehicle and the kind of pigmentation. This wide range is necessary to meet the variety of substrates as different as, for instance, steel structures and rubber sheet.

One of the rheological characteristics of a paint coat is its elastic modulus as a function of temperature. The graphs in Figs. 15.4, 15.5 and 15.6 show this

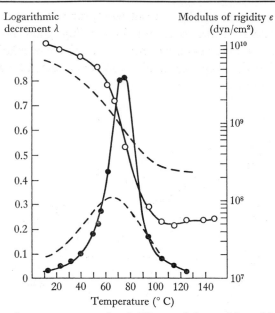

Fig. 15.5. Effect of temperature on the rigidity modulus and logarithmic decrement of an alkyd-melamine resin (- - -) and an epoxy-triamine resin (—) (ref. 7).

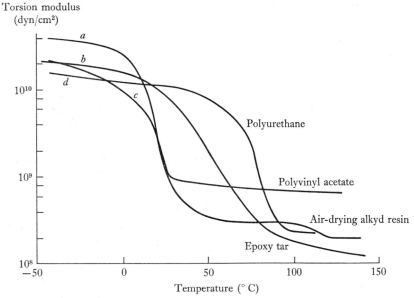

Fig. 15.6. Torsional modulus of four types of protective coats versus temperature (ref. 8).

relationship for various types of coats currently used. They show that most of the polymers applied in the coating technique are either glassy or visco-elastic at room temperature. Alkyd-melamine resins, for instance, have all the characteristics of a material in the glassy state, whereas in air-drying alkyd resin paint coat is a typical visco-elastic material. Coats which are nearly rubbery-elastic at room temperature are rare; some highly extensible types of polyurethane resins approximate to this behaviour.

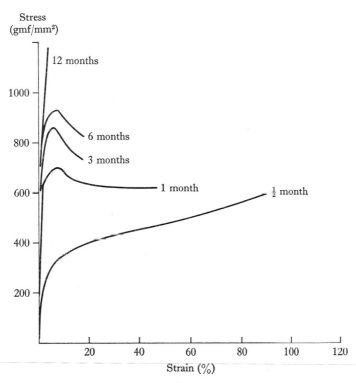

Fig. 15.7. Stress–strain curves of an alkyd resin paint coat weathered outdoors during various periods (ref. 8).

For the protective action of a paint coat over a prolonged period, not only the mechanical properties in its initial state of existence are of importance, but also to what extent they keep their strength with time. In outdoor conditions especially, a coat may deteriorate rapidly in stength and flexibility due to decay in the polymer structure. Fig. 15.7 gives an example of the influence outdoor weathering may have on the mechanical properties of a paint coat, in this case a moderately durable type of alkyd resin paint.

The greater part of the mechanical and thermal behaviour of a paint coat is determined by the nature of the vehicle. Pigment particles dispersed in the vehicle increase its modulus but usually do not significantly influence the modulus–temperature relation and therefore act as a neutral filler. Fig. 15.8 illustrates the stiffening effect of titanium dioxide. Sometimes, however, a pigment may have a more intensive influence on the mechanical properties of the coat. This will be the case if the pigment interferes with the drying

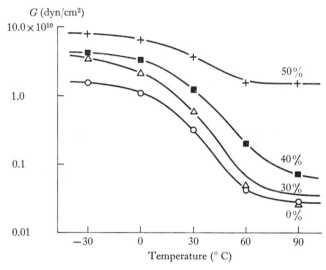

Fig. 15.8. Relation between torsional modulus G, temperature T and pigment volume concentration for an alkyd-melamine stoving resin pigmented with an inactive pigment (titanium dioxide) (ref. 9).

process by absorption of the drying agents and thus retards hardening of the coat, a phenomenon not unknown with oil-based carbon black paints, for instance.

In the foregoing, attention was paid to the mechanical character of paint coats in the form of free films and irrespective of the role of the substrate. However, what actually counts in practice is the performance of a coat on its substrate. Then only the properties of the coat as such but also its thickness and adhesive strength are of importance in the total protective action. Therefore the practical performance of a coat is actually evaluated more directly by testing practical criteria like the resistance to scratching, chipping, impact, wear and such-like damaging actions by making use of test panels coated under specified conditions. To achieve reproducible results it is important that such tests are carried out at a constant temperature and relative humidity of the air.

References to chapter 15

1 W. Heinz, *Materialprüf* **1**, 311 (1959).
2 Temple C. Patton, *Paint flow and pigment dispersion*, pp. 112 (New York, 1964).
3 R. J. Blackington, *Official Digest Federation Soc. Paint Technol.* **25**, 205 (1953).
4 N. D. P. Smith, *J. Oil Colour Chemists Assoc.* **44**, 618 (1961).
5 H. J. Freier, *Rheologica Acta* **3**, 254 (1964).
6 K. Wolf, FATIPEC II Congress, Noordwijk-Holland, Congress Book, 1953, p. 90.
7 Y. Inoue & Y. Kobatake, *Koll. Zeitschr.* **150**, 18 (1958).
8 Data from Paint Research Institute TNO, Delft-Holland.
9 U. Zorll, *Farbe u. Lack* **73**, 200 (1967).

16 Clay†

U. Daum and J. L. den Otter (§16.3)

Clay belongs to a large group of minerals having a layered silicate structure, and a particle size normally less than 2 μm. Natural clay deposits cover large parts of the earth's surface and, therefore, they are important in geology, soil science and engineering. Several clay minerals have found industrial applications, for example in the ceramic, paint and paper-coating industries, and in oil-well drilling. In all these applications, the rheological properties of clay–water or clay–organic solvent mixtures play an important role. Most applications involve clay–water mixtures, which possess complex properties due to the ionic nature of the clay mineral surface.

In the first section of this chapter, the structure of individual clay particles will be briefly reviewed, and correlated with forces acting between the particles. In the second section, the structure of clay–water mixtures will be discussed. The relation between this structure and the microrheology of clay will also be investigated.

In the third section, the macroscopic rheological behaviour of clay will be discussed from a phenomenological point of view. Unfortunately, no detailed correlation can at present be given between the microstructure of concentrated suspensions and their macroscopic rheological properties. It is possible, however, to state the rheological implications of the desired plasticity of clays.

16.1 Structure of clay particles

16.1.1. Crystal structure

Most clays minerals are composed of two types of layers.[1–3]

(1) Tetrahedral layers, consisting of SiO_4 tetrahedra.

(2) Octahedral layers, composed of closely packed oxygen and hydroxyl ions, and containing aluminium, magnesium or other cations in octahedral holes.

The unit layer of each clay mineral is composed of two or three of these layers.

An ubiquitous two-layer clay mineral is kaolinite, which crystallizes in the form of hexagonal platelets, 0.05–2 μm thick and 0.3–3 μm in diameter. Its

† The authors are highly indebted to Dr H. van Olphen, National Academy of Sciences, Washington, D.C., for his critical remarks.

structure is schematically shown in Fig. 16.1 (left). Each unit layer consists of one tetrahedral and one octahedral layer, connected by shared oxygen ions. The unit layers are kept together by van der Waals' forces, and probably by hydrogen bonds between the hydroxyl groups exposed on an octahedral surface and the oxygen atoms exposed on the adjacent tetrahedral surface.

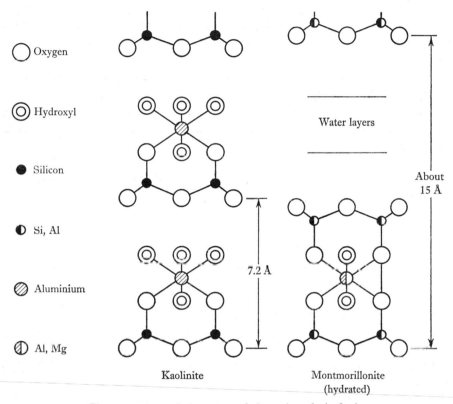

Fig. 16.1. Crystal structure of clay minerals (ref. 1).

Examples of three-layer clay minerals are montmorillonite and illite. They have unit layers consisting of one octahedral layer sandwiched between two tetrahedral layers (Fig. 16.1, right). Since only oxygen atoms are exposed, no hydrogen bonds can be formed between adjacent unit layers.

The crystal lattices of all clay minerals carry a negative charge due to iso-morphous substitution of Al for Si, or Mg for Al. This negative charge is compensated for by adsorption of cations on the crystal surfaces. In aqueous clay suspensions, these cations can be exchanged for other cations, if the latter are present in sufficiently high concentration.

In three-layer clay minerals, exchangeable cations are not only present on external crystal faces, but also between the unit layers, where they form regions of positive charge between the negative unit layers. This arrangement adds to the stability of the crystal at small layer distances, when the cations occupy midway positions between the unit layers, for which X-ray evidence can only be obtained in specimens with high cation populations, for example in vermiculites. At large interlayer distances, however, the cations form a diffuse electrical double layer, causing a repulsion between adjacent unit layers.

When a clay is suspended in dilute acid, the crystal edges acquire a positive charge owing to adsorption of protons by unsaturated bonds on the crystal surfaces. Under these conditions, the crystals show a slight anion-exchange capacity[4]. However, the net charge of each clay particle usually remains negative.

When the acidic suspension is neutralized, for example by addition of an alkali hydroxide, the crystal edges lose their positive charge as a result of deprotonation, and the anion-exchange capacity is lost.

16.1.2   Forces between clay particles

Carrying an electric charge, clay particles exert electrostatic forces on each other. Among other forces acting between clay particles, the following may be distinguished:

(1) Van der Waals' attractive forces.

(2) Repulsive forces caused by adsorption of solvent (water) on the clay surface.

The actual force between two clay particles depends on the following factors:

(1) The negative surface-charge density of the crystals caused by lattice substitutions.

(2) The water content of the clay–water mixture.

(3) The kind of exchangeable cations and anions present on the crystal surfaces.

(4) The pH of the medium.

(5) The kind and concentration of electrolytes in the aqueous phase. Of course, the force between particles always depends on their distance and mutual orientation. As a result, a preferential interparticle structure is frequently formed.

A well-known type of structure is that of swollen montmorillonite, in which water has penetrated between unit layers, without disturbing their parallelity. The amount of water taken up has been carefully studied as a function of the above-mentioned factors. Therefore, the swelling of three-layer clay minerals can serve as a useful model for the description of forces between clay particles in general.

16.2 Structure and microrheology of clay–water mixtures

16.2.1 The swelling of three-layer clay minerals

When lattice substitutions and exchangeable cations are neglected, the crystallographic unit cell of the structure shown in Fig. 16.1 (right) has the ideal composition, $Si_8Al_4O_{20}(OH)_4$, which corresponds to a 'molecular' weight of 721. Its surface area in the unit layer plane is 45.8 Å[2]. The lattice-charge densities of three-layer clay minerals range from 0 to 2 electron charges per unit cell (e.c./u.c.). Of course, both the cation-exchange capacity of the mineral and the surface-charge density of the unit layers are proportional to the lattice-charge density. For 1 e.c./u.c., the cation-exchange capacity is 140 meq/100 g, and the surface-charge density (on each side of the unit layer) 5.25×10^4 e.s.u./cm². The surface area available per monovalent exchangeable cation between units layers is inversely proportional to the lattice-charge density. For 1 c.c./u.c. it is, of course, exactly equal to 45.8 Å².

When the charge density is greater than 1.6 e.c./u.c., the compensating cations usually are potassium ions, that can only be exchanged with great difficulty, and must be considered to belong to the crystal structure, which does not allow penetration of water between its unit layers. This situation is found in the minerals muscovite and phlogopite.

At intermediate charge densities (from 0.5 to 1.6 e.c./u.c.), water can enter the space between unit layers, causing a swelling of the mineral. Hydration of the unit-layer surfaces and of the exchangeable cations gives a positive contribution to the energy, counteracted by work performed against the electrostatic forces between unit layers and cations.

At zero charge density, no cations are present between unit layers, and the latter are kept together by van der Waals' forces only. In spite of this, no water can penetrate between the unit layers. Apparently, the hydration energy of the unit-layer surfaces alone is not sufficient to overcome the van der Waals' forces which at small distances, are very strong. This situation obtains in the minerals pyrophyllite and talc.

Swelling three-layer clay minerals are called montmorillonoids or smectites. When equilibrated with water vapour at different pressures, these minerals form discrete hydrates, each containing an integral number of water mono-layers between a pair of unit layers. The number of monolayers taken up from saturated water vapour increases with decreasing lattice-charge density. At 1.6 e.c./u.c. (vermiculite), it is two.[5] The same number of monolayers is indicated by data obtained[6] on saponite and batavite (charge density: 1.1 and 1.4 e.c./u.c.[see 7]). For montmorillonite (0.5–0.8 e.c./u.c.), the number of monolayers adsorbed from saturated water vapour is found to be three.[6] When in contact with liquid water, montmorillonite sometimes shows a discrete unit-layer spacing corresponding to four monolayers of water.[8]

The range of water vapour pressures in which a hydrate is stable depends on the kind of exchangeable cation present,[9] but the unit-layer spacing for a constant number of monolayers depends only slightly on lattice-charge density and kind of cation. For a high charge-density vermiculite, X-ray analysis of the hydrates reveals that the exchangeable cations take a position midway between the unit layers,[10, 5] and the entropy of the absorbed layer of water is found to be less than that of normal liquid water.[5] In certain swelling clays, the interlayer water has been shown to be birefringent[11]. Thus, the exchangeable cations and the adsorbed water may be considered as part of the crystal structure of the swollen clay mineral. The range of swelling in which up to four monolayers of water are adsorbed is called the range of crystalline swelling. At maximum crystalline swelling, with four monolayers of water, the spacing of unit layers is about 22 Å, which corresponds to a water layer thickness of about 13 Å (see Fig. 16.1, right). The mineral then contains about 0.45 g of water per g of dry clay.

Saponite, batavite and vermiculite show no further swelling when they are immersed in water. Montmorillonite, however, continues to swell, provided that the exchangeable cations are alkali-metal ions. In the resulting highly swollen montmorillonite, the cations redistribute to form diffuse electrical double layers on the unit layer surface. According to the theory of diffuse electrical double layers,[e.g. 12] there is now an electrostatic repulsion between unit layers. This repulsion is suppressed by an increase in concentration of electrolytes in the surrounding water.

For a particular sodium montmorillonite, Norrish[8] found various stages of crystalline swelling at external sodium chloride concentrations greater than 0.3 mol/l. For lower salt concentrations, he found a wide distribution of layer spacings, but up to spacings of 130 Å (14 times that in dry montmoril-lonite) the position of the distribution maximum was inversely proportional to the square root of the salt concentration. Here, the sodium chloride con-centration was 0.01 mol/l. At still lower electrolyte concentrations, the parallelity of unit layers is lost, and the suspension consists of essentially free montmorillonite particles, composed of only one or a very few unit layers.[e.g. 13]

The swelling of montmorillonite beyond the crystalline range is called osmotic swelling. This designation is based on the fact that the double-layer repulsion may also be seen as the result of the osmotic activity of ions in the diffuse cation layer. For a given equilibrium spacing in the range of osmotic swelling, the electrostatic repulsion must be exactly balanced by attractive forces. At the large interlayer distances in question, van der Waals' forces are too weak to provide this balance. Both van Olphen[14] and Norrish and Rausell-Colom[15] assume that some montmorillonite layers are arranged in non-parallel directions, cross-links between parallel unit layers being provided by electrostatic attraction between positive layer edges and negative layer faces.

Many authors report anomalous properties for the interlayer water at unit-layer distances greater than 22 Å.[e.g. 16, 11] They claim that some form of crystalline ordering of the water layer may persist even in the range of osmotic swelling.

Polyvalent cations prevent the swelling of a montmorillonite beyond a unit-layer spacing of 20 Å, that is, beyond the crystalline range.[e.g. 7]

Illites are three-layer clay minerals with a lattice-charge density comparable to that of vermiculite (1.4 e.c./u.c.).[17] In montmorillonite, lattice substitutions occur mainly in the octahedral layers. In vermiculite and illite, by contrast, lattice substitutions predominate in the tetrahedral layers. As a consequence, the electrostatic attraction between unit layers and interlayer cations is stronger in illites than in montmorillonites. In illites, as in other non-swelling three-layer minerals, the interlayer cation usually is potassium. In natural illites, true illite layers often alternate randomly with smectite layers.[2] These mixed-layer illites are capable of swelling in water.

Van Amerongen[18] found a proportionality between specific surface area and cation-exchange capacity for thirty clays, chiefly composed of mixed-layer illites. From this proportionality, an available surface area of 90 Å^2 per monovalent exchangeable cation can be calculated, which is within the range found for montmorillonite (i.e. 57–92 Å^2 per monovalent cation, according to data given earlier in this paragraph). It may be concluded that the compensating potassium ions are not exchanged during measurement of the cation-exchange capacity, and that the illite particles are surrounded by montmorillonite layers.

When a montmorillonite, which has been equilibrated with water vapour of a certain pressure p, is immersed in pure water, a swelling pressure ΔP is developed with is related to p by the equation:[e.g. 19]

$$\Delta P = (RT/v) \ln (p_0/p), \qquad (16.1)$$

where R is the molar gas constant, T the absolute temperature, v the molar volume of liquid water and p_0 the vapour pressure of pure water. It is found from equation (16.1) that, in the range of osmotic swelling, ΔP has values of several atmospheres, which is sufficient to lift buildings. In the range of crystalline swelling, ΔP amounts to thousands of atmospheres. Pressures of this order of magnitude can occur under geological conditions, and are required to remove by compression the last monolayer of adsorbed water.[5]

16.2.2   The surface-charge density of kaolinite particles

According to data given in the preceding paragraph, the cation-exchange capacity of montmorillonite varies from 70 to 110 meq/100 g. The cation-exchange capacity of kaolinites usually is between 1 and 10 milliequiv./100 g. Since, in kaolinite, the exchangeable cations are located on the external crystal surfaces only, the surface-charge density of kaolinite particles can be shown to be of the same order of magnitude as that of the unit layers of montmorillonite.

For five kaolinites of different origin, Weiss[20, 21] found a proportionality between specific surface area (determined by nitrogen gas adsorption) and cation-exchange capacity. From this proportionality, and assuming that the exchangeable cations are uniformly distributed over the particle surface, an available surface area of 50 Å² per monovalent cation can be calculated. This is somewhat less than in montmorillonite (57–92 Å² per cation).

According to Weiss,[20] the exchangeable cations in kaolinite are mainly located on the tetrahedral surface of the particles. If this assumption is correct, the available surface area per monovalent cation in kaolinite is much less than in montmorillonite, corresponding with a higher surface-charge density for kaolinite.

16.2.3   Consistency of clay–water mixtures

When increasing amounts of water are added to a dry clay powder, a typical sequence of rheological conditions, or consistencies, is usually traversed. Three principal stages can be distinguished.

In the first, clay particles aggregate to form crumbs, or a mass that crumbles when it is deformed. The mass behaves more or less like a brittle solid.

In the second stage, the mass will sustain large deformations without rupturing. After removal of the stress, the mass retains its deformed shape, that is, it behaves like a plastic solid.

In the third stage, the mass is so easily deformed that it cannot retain its shape against the forces of gravity or gentle shaking. It behaves like a liquid.

The water content where a clay changes from one consistency to another depends more or less on the experimental procedure for discriminating between two consistencies. The weight percentage of water, referred to dry clay, at which the clay changes from a solid to a plastic consistency is called the upper plastic, or plastic limit (w_P). The water content at the transition from plastic to liquid behaviour is called the lower plastic, or liquid limit (w_L). The two limits define the range of water content in which the clay shows plastic behaviour.

Experimental procedures to determine these limits have been standardized, and are described for instance in ref. 19. Since these procedures are based on the early work of Atterberg,[22] these limits are sometimes referred to as Atterberg limits.

16.2.4   Interpretation of Atterberg limit values

Values of Atterberg limits of various clays, as determined by White,[23] have been tabulated by Grim.[24] Part of Grim's table is reproduced in Table 16.1.

(a) Kaolinite

From the data in Table 16.1 it appears the kaolinite may be plastic at water contents up to 75 % by weight. It is probable that at such relatively low water contents most kaolinite particles are bonded to one another.

TABLE 16.1. *Atterberg limit values* (ref. 23)

Clay		Na^+		K^+		Mg^{++}		Ca^{++}	
		w_P	w_L	w_P	w_L	w_P	w_L	w_P	w_L
Kaolinite	(1)	26	52	38	69	30	60	36	73
	(2)	28	29	28	35	28	39	26	34
Illite	(1)	34	61	43	81	39	83	40	90
	(2)	34	59	40	72	35	71	36	69
	(3)	41	75	41	72	43	98	42	100
Montmorillonite	(1)	93	344	57	161	59	158	65	166
	(2)	89	443	57	125	51	199	65	155
	(3)	97	700	60	297	53	162	63	177
	(4)	86	280	76	108	73	138	79	123

If, in plastic clays, permanent bonds exist between particles, these must be broken when a large deformation is applied. They are re-formed shortly after the deformation is discontinued. The resistance to deformation will primarily be determined by the number and strength of interparticle bonds. At the liquid or plastic limit, a clay mass has a certain rigidity, the actual value of which depends on the particular testing method by which it is determined. For high values of the limits, that is when the clay content is low, fewer interparticle contacts would be expected to exist than for low limit values, with a high clay content. The fact that the rigidity is the same in both cases can be explained by assuming that, for high limit values, the bonds between particles are stronger than for low limit values.

Since the surface-charge densities are similar in both cases, forces between kaolinite particles may be of similar magnitude as those between montmorillonite unit layers. Hofmann and co-workers[25] suggest that, between parallel surfaces of adjacent kaolinite particles, exchangeable cations take fixed positions, as in the crystalline swelling of montmorillonite. The same authors also suggest that a similar kind of bonding exists between different types of surfaces (edges and faces), giving rise to a house-of-cards structure.

Calcium montmorillonite does not swell beyond the crystalline region, in contrast to sodium montmorillonite. This implies that crystalline bonding is more pronounced in calcium than in sodium montmorillonite. This might explain the higher limit values found for calcium and magnesium kaolinite as compared with sodium kaolinite (see Table 16.1). At low electrolyte concentrations, no crystalline swelling is observed in sodium montmorillonite. Thus, crystalline bonding is not likely to occur between sodium kaolinite particles in (distilled) water. A discussion of the forces acting under these conditions is deferred to §16.2.6. We note that, for some sodium kaolinites (e.g. kaolinite (2) in Table 16.1), the range of plastic behaviour is very narrow, which indicates that the attractive forces between the clay particles are weak.

The limit values of potassium kaolinite are similar to those of calcium and magnesium kaolinite. This may be due to a particular bonding property of potassium ions, which is also demonstrated by the predominance of potassium as the bonding cation in non-swelling three-layer clay minerals.

Plastic calcium kaolinite masses can be liquefied by addition of sodium carbonate or sodium silicate, when, on the kaolinite surface, calcium is exchanged for sodium, and, in the interparticle liquid, calcium carbonate or calcium silicate is precipitated.[e.g. 26, 17] This process is applied in the preparation of ceramic casting slips.

(b) Montmorillonite and illite

In plastic montmorillonite and illite, water is present between the clay particles, as well as between the unit layers inside the particles themselves. In order to obtain the same amount of interparticle water, more has to be added to illite than to kaolinite, and to montmorillonite than to illite.

When calcium is exchanged for sodium, the limit values tend to increase by virtue of increased intracrystalline swelling, and to decrease as a result of decreased interparticle bonding. Apparently, the influence of intracrystalline swelling predominates in montmorillonite, and that of interparticle bonding in illite.

Due to their high water content, plastic montmorillonites suffer high shrinkage on drying. For this reason, pure montmorillonites are not suitable for ceramic applications. Nevertheless, some montmorillonite is often added to kaolinite in order to widen the range of plastic behaviour of the latter.[24]

(c) Influence of external electrolytes

Hofmann and co-workers[25] studied the influence of external electrolyte concentration on the consistency of kaolinite and montmorillonite gels. They selected gels of equal rigidity by imposing the condition that these do not flow from a test tube inverted after six seconds, during which time the gel is formed. As in the interpretation of Atterberg limits, the amount of electrolyte solution in such a gel is taken as a measure of the strength of bonds between particles. To prevent simultaneous cation-exchange, sodium chloride and hydroxide solutions were added to sodium clays, and calcium chloride and hydroxide solutions to calcium clays. Some of the results are shown in Table 16.2.

Kaolinite gels of equal rigidity always show a uniform increase in fluid content with increasing electrolyte concentration. In sodium kaolinite, the only attractive forces between particles are van der Waals' forces, which are independent of electrolyte concentration. However, electrostatic repulsion between particles, due to the interpenetration of electrical double layers, diminishes with increasing electrolyte concentration, resulting in a greater net attraction. Crystalline bonding between sodium kaolinite particles may occur at high sodium ion concentrations.

In calcium kaolinite, crystalline bonds may exist even at very low electrolyte concentrations, as is borne out by the fact that calcium kaolinite contains

TABLE 16.2. *Fluid content of thixotropic gels (ref. 25) (in ml of solution per 100 g of clay)*

Electrolyte concentration (equiv./l)	Na-kaolinite		Na-montmorillonite		Ca-kaolinite		Ca-montmorillonite	
	NaCl	NaOH	NaCl	NaOH	$CaCl_2$	$Ca(OH)_2$	$CaCl_2$	$Ca(OH)_2$
0	117	117	1270	1270	237	237	340	340
0.002	137	124	1000	900	307	267	350	330
0.02	173	120	1430	700	317	407	380	380
0.04	—	—	—	—	—	437	—	470
0.2	204	173	1235	1970	320	—	350	—
0.3	214	190	1030	2300	—	—	—	—
0.4	230	237	965	1770	—	—	—	—
0.5	264	274	765	1470	—	—	—	—
1.0	324	317	565	665	—	—	—	—
2.0	357	320	500	565	317	—	330	—
3	—	324†	—	—	—	—	—	—
5	500†	—	234†	500†	—	—	—	—
10	—	—	—	—	317†	—	247†	—

† Behaviour uncertain because of foaming and high viscosity.

more water than the corresponding sodium kaolinite gel. It was found[25] that calcium montmorillonite shows a uniform decrease in unit-layer spacing with increasing calcium ion concentration. From this, it may be inferred that, in calcium kaolinite, besides the decreased double-layer repulsion, there is an increased strength of crystalline bonding, when the calcium ion concentration is increased. One would, therefore, expect a higher fluid content for kaolinite gels at high calcium than at high sodium ion concentrations. The fact that the reverse is sometimes true (cf. $CaCl_2$ versus NaCl) might point to a different geometrical arrangement of particles in the two gels. The strong bonding forces in the presence of calcium chloride would, for instance, promote more face-to-face association of particles than would a high concentration of sodium chloride.[21]

The behaviour of montmorillonite gels is more complex. At high electrolyte concentrations (above 0.3 equiv./l), there is always a decrease in fluid content with increasing electrolyte concentration. This indicates a predominant influence of the deswelling of montmorillonite crystals. At lower electrolyte concentrations, there is always an increase of fluid content with increasing electrolyte concentration, sometimes preceded by a region of decreasing fluid content. This may show an alternating predominance of deswelling and increased interparticle bonding. A different explanation of a similar behaviour of dilute montmorillonite suspensions will be given in §16.2.5*b*.

16.2.5   Dilute suspensions

(*a*) *Kaolinite*

At a pH of 10, the viscosity of a sodium kaolinite suspension, containing 15.5% by weight (or 6.5% by volume) of clay, increases with increasing sodium chloride concentration.[27] This is in qualitative agreement with the finding that, under similar conditions, the fluid content of thixotropic gels increases (see Table 16.2).

On the contrary, in the range of pH between 6.5 and 7, the same kind of sodium kaolinite[27] shows a decrease in viscosity with increasing sodium chloride concentration. This is in qualitative agreement with results obtained by van Olphen,[28] and Schofield and Samson.[4]

The last-mentioned authors observed a positive adsorption of chloride ions from a 0.005 N sodium chloride solution by a sodium kaolinite that had been prepared at pH 3. When 1 milliequiv. of sodium hydroxide is added per 100 g of kaolinite, its chloride-ion adsorption capacity disappears. When the same amount of sodium hydroxide is added to a salt-free suspension of the same kaolinite, a 1000-fold drop in viscosity is observed. According to data of Michaels and Bolger,[27] this drop occurs at pH 7.1, and it is accompanied by an inflection in the titration curve of sodium kaolinite.

Schofield and Samson ascribed the adsorption of chloride ions at pH values below 7.1 to the presence of a positive charge on the edges of the kaolinite crystals. At higher pH values, they found a negative adsorption of chloride ions, which they ascribed to electrostatic repulsion of these ions by the negative basal faces of the crystals. The high viscosity of the slightly acidic, salt-free suspensions was ascribed to attraction between positive crystal edges and negative crystal faces, possibly giving rise to a cubic card-house structure throughout the suspension. This is in agreement with the fact that such a suspension shows a Bingham yield stress.[28] The decrease in viscosity attending the addition of sodium chloride is explained as a result of the compression of electrical double layers on both positive edges and negative faces, which reduces the attraction between the two. Complete deflocculation cannot be produced by the mere addition of sodium chloride, since, at higher salt concentrations, the double-layer repulsion between negative faces is so much reduced that van der Waals' attraction between faces begins to predominate, giving rise to card-pack rather than to card-house structures. In the salt-flocculated condition, suspension viscosity and sediment volume are less than in the acid-flocculated condition, but greater than in the completely deflocculated state, produced by small amounts of sodium hydroxide. With acid sodium kaolinite, the gradual transition from acid-flocculation to salt-flocculation causes a uniform decrease in viscosity, when salt is added.

When deflocculated by addition of sodium hydroxide, a sodium kaolinite suspension remains completely deflocculated up to pH 11, where, according to Michaels and Bolger,[27] the concentration of added sodium ions may be 0.05 equiv./l. This concentration is sufficient to induce salt-flocculation providing the counter-ion is chloride (see Table 16.2). In the presence of hydroxyl ions, salt-flocculation is prevented by adsorption of hydroxyl ions on the crystal edges. When sodium chloride is added at pH 10, salt-flocculation occurs. This also happens when more sodium hydroxide is added.

At a pH of about 7, calcium or barium hydroxide causes a much smaller drop in viscosity than does sodium hydroxide.[4, 29] Apparently, the clay remains flocculated owing to the persistence of crystalline bonding between kaolinite particles. For the same reason, a calcium kaolinite is not deflocculated by sodium hydroxide.[26]

Sodium or calcium hydroxide sometimes produces an increase in viscosity of acid-flocculated kaolinites, at least for pH values below 7.[4, 29] Michaels and Bolger[27] acribe this effect to the presence, in acid-treated kaolinite, of adsorbed positive hydrolyzed alumina on the basal faces. On the addition of alkali, the alumina would be neutralized and desorbed before the neutralization of positive crystal edges is complete. The resulting increase in particle repulsion would allow the existing card-house flocs to swell before ultimately breaking down.

Direct evidence for the existence of positive edge-charges in acid kaolinites had earlier been obtained by Thiessen,[30] who showed by means of electron micrographs that the particles of a negative gold sol are adsorbed exclusively on the edges of kaolinite crystals. The concept of positive edge-charge was first introduced into clay colloid-chemistry by van Olphen,[31] in an investigation of sodium montmorillonite.

(b) Montmorillonite

The behaviour of a dilute, salt-free sodium montmorillonite suspension upon the addition of sodium chloride is similar to that of a sodium kaolinite suspension, with the exception that, after an initial drop in viscosity and yield stress, these quantities rise again with addition of more sodium chloride.[32, 33] The minimum yield stress of a gel with a montmorillonite concentration of 3.22 % by weight was found at a sodium chloride concentration of about 5 milliequiv./l. Only at concentrations above 100 milliequiv./l, the yield stress tends to decrease again, as it does in suspensions of sodium kaolinite.

Van Olphen explained the high yield value and viscosity of the salt-free suspension by assuming that it was flocculated by electrostatic attraction between positive crystal edges and negative crystal faces. He attributes the complex behaviour, described above, to the existence of units consisting of two parallel montmorillonite crystals held together by one crystal, perpendi-

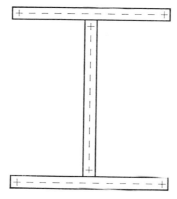

Fig. 16.2. Double-T unit of montmorillonite crystals (ref. 3).

cular to both (see Fig. 16.2). In such units, there is a delicate balance between edge-to-face attraction and face-to-face repulsion. The former is supposed to predominate at zero electrolyte concentration, and the latter at a concentration of about 5 millequiv./l.

At still higher electrolyte concentrations, electrostatic edge-to-face attraction would once more predominate, as a result of diminished face-to-face repulsion. Only at high electrolyte concentrations (above 100 millequiv./l) would flocculation be governed by van der Waals' attraction, resulting in the formation of card-pack flocs, and a low suspension viscosity and sediment volume.

The changes observed when sodium hydroxide is added to a suspension of sodium montmorillonite are qualitatively similar to those following the addition of sodium chloride.[34]

The same author[34] found a gradual decrease in Bingham yield stress and plastic viscosity of salt-free suspensions when sodium was exchanged for calcium. This would be expected to result from a gradual deswelling of the mineral.[35]

Calcium chloride does not reduce the viscosity and yield stress of a salt-free calcium montmorillonite suspension, which is in agreement with our assumption that calcium clays always flocculate as a result of the formation of crystalline bonds.

16.2.6   The origin of plasticity of deflocculated clay–water systems

In our discussion of the Atterberg limits of sodium kaolinite we observed that, at low electrolyte concentrations, no crystalline bonds can be formed between particles of this clay type. Two alternative explanations can be given for the plastic behaviour of sodium clays.

(1) Flocculation by van der Waals' forces.

(2) Flocculation by electrostatic attraction between positive edges and negative faces.

Van der Waals' attraction will predominate at high electrolyte concentrations. Under these conditions, suspensions of sodium kaolinite and montmorillonite are reported to be thixotropic.[4, 32–34] This means that, even at these high salt concentrations, double-layer repulsion is high enough to retard flocculation. It can be concluded that, at low salt concentrations, double-layer repulsion will effectively prevent flocculation by van der Waals' forces alone.

In publications concerning the flow behaviour of clays, reference to relevant pH values is often omitted. For instance, in publications on sodium montmorillonite, no mention is made of the pH at which it changes from the edge-to-face flocculated to the deflocculated state. However, when authors explicitly state that a particular gel of a sodium clay was not edge-to-face flocculated,[e.g. 25] other explanations for its finite rigidity have to be found.

Two such explanations have been forwarded. Warkentin[36] and Warkentin and Yong[37] ascribed the rigidity of deflocculated suspensions of sodium kaolinite and montmorillonite near the liquid limit to electrostatic double-layer repulsion between particles. This repulsion defines an interaction volume around each particle. When the interaction volumes of different particles overlap (i.e. at water contents below the liquid limit), the particles are kept in a fixed configuration, which prevents their free movement. At water contents above the liquid limit, on the other hand, interaction between particles is weak enough to permit easy movement.

Another possible explanation of the plastic behaviour of clays is that there exists a region of structured water around each clay particle, possessing a certain rigidity of its own. A similar region of structured water, referred to in § 16.2.1, is present between the unit layers of swollen montmorillonite. A great deal of evidence for the existence of such regions was forwarded by early workers in the field,[e.g. 38] but some controversy still persists over the relevant data.[39, 16] The present state of the subject has recently been reviewed by Low and White.[40] In view of accumulating evidence, it seems likely that structured water regions are in some way essential to the plastic behaviour of clays.

16.2.7 Deflocculation by alkali polyphosphates

Flocculated clay suspensions are deflocculated by small amounts of alkali salts containing certain polyvalent anions, particularly polyphosphates.

Polyphosphate anions are linear polymers of the structure given in Fig. 16.3:[41]

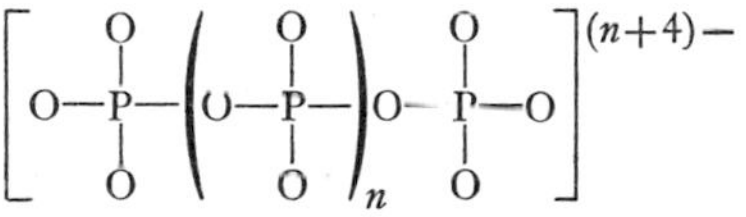

$$\left[O-\underset{\underset{O}{|}}{\overset{\overset{O}{\|}}{P}}-\left(O-\underset{\underset{O}{|}}{\overset{\overset{O}{\|}}{P}}- \right)_n O-\underset{\underset{O}{|}}{\overset{\overset{O}{\|}}{P}}-O \right]^{(n+4)-}$$

Fig. 16.3. Structure of linear polyphosphate anions.

Cyclic polyphosphates are also known. Polyphosphate anions form soluble complexes with many cations, such as sodium, calcium and aluminium. Hence, it is understandable that these anions improve the stability of negative particles towards the flocculating action of salts.[31] Moreover, the formation of crystalline bonds can be prevented by exchange of calcium for alkali ions.[31]

Polyphosphates produce an increase in the electrophoretic mobility of clay particles[31] as well as an increase in their cation-exchange capacity.[42] These findings are satisfactorily explained by the assumption that clay particles adsorb polyphosphate. Quantitative chemical studies of this adsorption were made by van Olphen,[31] Michaels[42] and Lyons.[43]

By virtue of the repulsive forces between them, adsorption of negative polyphosphate ions on negative crystal faces is, of course, not likely to occur. The amounts of polyphosphate adsorbed are in agreement with the assumption that adsorption takes place on the crystal edges only. Of low molecular weight polyphosphates ($n = 1$ or 2, see Fig. 16.3), the amounts adsorbed by kaolinite are such that each poly-anion could be accomodated, in a plane configuration, on the octahedral parts of the edge-surface only.[42, 43] It is generally assumed that these octahedral surfaces are the sites of the positive edge-charge at low pH, by comparison with the surface potential of hydrous alumina sols.[3] Therefore, from the adsorption of low molecular weight polyphosphates, it might concluded that this adsorption is purely electrostatic in nature. However the adsorption of high molecular weight anions (n = about 30) on montmorillonite is so high that only one or two monomeric units per anion can be directly attached to an octahedral edge-surface. Here, adsorption occurs far in excess of the amount required for compensation of the positive edge-charge. Possibly, monomeric groups of the polyphosphate anions form chemical complexes with aluminium ions exposed on the edge-surfaces.

As a result of the high adsorption described, the sign of the positive edge-charge is reversed. This explains the large drop in viscosity and yield value in acid-flocculated clay suspensions.

Alkali polyphosphates are extensively used as deflocculants for montmorillonite in oil-well drilling, and for kaolinite in the paper-coating industry.[24] In both these applications, clay slips are desired with a low viscosity and a clay concentration as high as possible. Polyphosphates slowly hydrolyze in aqueous solution. The rate of hydrolysis is increased in the presence of kaolinite.[42, 43] For this reason, polyacrylate deflocculants, which are not prone to hydrolysis, are coming more and more into use.

16.3   Phenomenological rheology of clays (by J. L. den Otter)

In investigating the deformation behaviour, structural details of a material usually are disregarded, that is it is regarded as a continuum. However, it should eventually be possible to correlate the microstructure described in the preceding paragraphs with rheological properties. At present this is possible only to a very limited extent.

16.3.1   Theories

A large number of theories have been proposed, describing elastic and flow properties of dilute and concentrated suspensions. The most successful theories describe simple systems, for instance flow of dilute suspensions of particles with a simple shape. Examples of these are the well-known theories for spheres[45] and for ellipsoids,[46, 47] where Brownian motion is taken into account. The flow behaviour of even dilute suspensions of irregularly-shaped particles cannot be satisfactorily predicted. Experimental and theoretical results for very dilute suspensions are given by Goldsmith and Mason.[48] Nearly all experimental work is limited to the region of Newtonian viscosity, at low shear rates. For the flow behaviour of moderately concentrated (up to 15 volume per cent) suspensions, various theories give conflicting predictions,[49–52] even for the simple case of weakly interacting spheres. Theories describing the flow behaviour of concentrated suspensions (greater than 15 volume per cent) of strongly interacting particles have been advanced by some authors.[53–56] However, since the flow behaviour of concentrated suspensions of particles even with the simplest possible shape (spheres) cannot be described satisfactorily, more complicated systems at present appear to be beyond the grasp of theory. This applies even more to clay–water mixtures, which are extremely complex.

For this reason, the best that now can be tried is to perform measurements on different types of clay–water suspensions, and to try to correlate differences in flow behaviour with changes in such parameters as temperature, magnitude of deformation, and concentration of clay or electrolytes, preferably over a small range.

An experimental difficulty in investigating clay–water suspensions is their instability. They are almost always found subject to ageing, that is their (flow) properties change with time. Experiments of some duration may give wrong results, because the structure of the material at the end of the experiment differs from that at the beginning. This may be due to irreversible coagulation of particles, or sometimes, to chemical reactions.

16.3.2   Elastic properties

Most experimental techniques for investigating elastic or viscous properties of matter involve shear. In some experiments in soil mechanics the compression modulus is measured. All clay–water mixtures, except extremely dilute ones, show both viscous and elastic properties at fairly low rates of deformation. These can best be investigated by imposing a sinusoidal shear on the material. If the (small) deformation is given by $\gamma_0 \sin \omega t$ (γ_0 representing amplitude of deformation, ω circular frequency, and t time), the shear stress τ can be expressed as:[57,58]

$$\frac{\tau}{\gamma_0} = (G' \sin \omega t + G'' \cos \omega t), \tag{16.2}$$

where G' is the storage modulus, G'' the loss modulus. For most materials, G' and G'' are found to depend upon ω and γ_0. In the linear region, for small deformations, the ratio τ/γ_0 is independent of γ_0 for any given value of ω. For deformations larger than γ_c this is no longer true, and for most materials discussed in this chapter the moduli, and especially G', decrease progressively with increasing amplitude of deformation. Essentially the same occurs when the deformation is not sinusoidal, but a step function. In this case the experiments are more difficult to perform, however. An example of the measurement of the moduli G' and G'' as functions of frequency is given in Fig. 16.4, for a suspension of an English paper coating clay (kaolinite) at 40° C. The clay concentration was 30.3 % (vol./vol.). The clay had been deflocculated by the addition of 0.4 % (wt./wt. of clay) of a sodium polyphosphate (Calgon) and the suspension adjusted to pH 7 with sodium hydroxide. The suspension

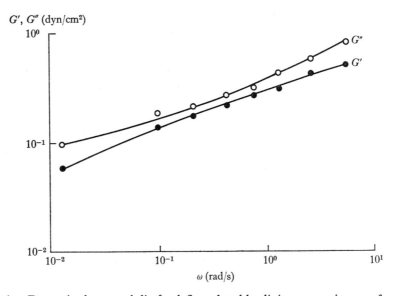

Fig. 16.4. Dynamic shear moduli of a deflocculated kaolinite suspension as a function of circular frequency (shear amplitude about 0.14).

obtained was a free-flowing liquid. The measurements were performed with a concentric-cylinder viscometer developed by one of the authors.[59]

A characteristic of the visco-elastic properties of clay–water suspensions is that they depend very much on the time-scale in which the experiments are performed.

The suspension in Fig. 16.4 becomes non-linear at small deformations. The storage and loss modulus are plotted in Fig. 16.5 as functions of the amplitude of deformation at a constant value of ω. It is clear that G'', and even more so G' decrease with increasing amplitude of deformation. As discussed before, the magnitude of the critical deformation γ_c, above which the material exhibits non-linear behaviour, depends very much on the microstructure of the material, and also on the applied frequency.

Dynamic shear experiments on clay–water systems of compositions similar to those used in the ceramic industry were reported by Astbury, Moore and Lockett.[60] These authors found that all systems investigated were non-linear in the entire experimental deformation range (2.5×10^{-3} up to about 0.1). It was found that the experimental results could not be accommodated by

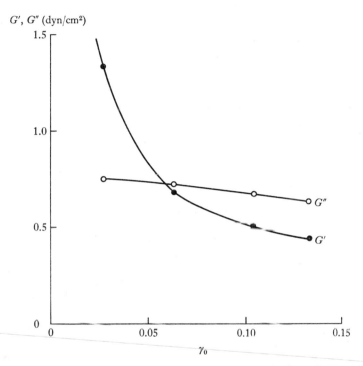

Fig. 16.5. Dynamic shear moduli of the same suspension as in Fig. 16.4, as a function of shear amplitude at a constant circular frequency of 2.62 rad/s.

equation (16.2), but that higher harmonics were necessary to describe the relation between stresses and deformation (except perhaps for a bentonite clay).

Large numbers of careful rheological measurements on clay soils have been reported. These experiments were performed for civil engineering purposes. Much of this work consists of measurements on clay soils of compression moduli, also in dynamic testing.[61–66]

All measurements were performed in the non-linear region, although deformations as small as 10^{-5} were reported in compression testing.[61, 65] Unfortunately, many experiments are reported without a clear estimate of the applied deformation. Another difficulty in the interpretation of experiments in the non-linear region is that the deformation in the material may well be inhomogeneous. This means that a large part of the total deformation may be concentrated in a small area, so that the sample behaves in a non-linear way, although in a homogeneous deformation of the same magnitude the sample might still be in the linear region.

Related to the elastic properties are normal stress differences in steady shear flow, which were investigated by Beazley in deflocculated clay suspensions.[67]

16.3.3 Plasticity

Many clay–water mixtures are plastic during deformation. This means that, when the material is deformed under stress, it will readily assume a new shape. This new shape will be retained nearly unaltered when the stress is taken away. To obtain the desired plasticity a previous kneading is sometimes necessary.

During the whole deformation process, the material does not crack and cohesion is retained, so that the material flows fairly easily. On the other hand, for most industrial applications it is desirable that the material has a considerable rigidity almost immediately after the deformation, enabling it to retain its new shape without support. To some extent the last two requirements would seem to exclude each other. Using the treatment of elasticity given previously, the mechanism of plasticity can be explained and expressed in terms of fundamental rheological properties of the material. The material should have a rigidity (shear modulus) which is very non-linear, so that the modulus falls off very rapidly with increasing deformation, but it should reach a fairly large value rapidly once the deformation is stopped. In other words, the structure resulting from interactions between clay particles is broken down at very small deformations, and it is reformed quickly afterwards.

The plastic flow behaviour can therefore be predicted from systematic measurements of the moduli as functions of deformation. The consistency after deformation can be investigated by measuring the dynamic moduli as function of time for small deformations (preferably within the linear region) after a large preceding deformation.

16.3.4   Flow properties

The flow properties of a material subjected to simple shear can be represented by a flow curve, which gives the relation between shear stress τ and shear rate D. As an example, the flow curve for a flocculated clay–water suspension is shown in Fig. 16.6. The clay was the same as that of Figs. 16.4 and 16.5. It was dispersed, without Calgon, in distilled water at a concentration of 25.0 % (vol./vol.). The pH of the suspension was 4.6. Measurements were

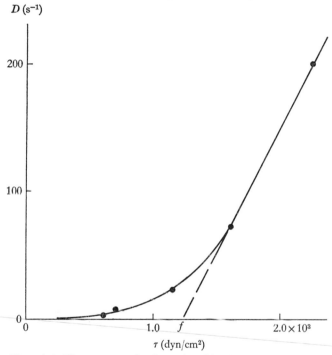

Fig. 16.6. Flow curve of a flocculated kaolinite suspension.

made at 22° C, using a cone-and-plate viscometer. The suspension was a stiff paste, although its water content (115 g H_2O per 100 g clay) might still be above the liquid limit (compare Table 16.1). To avoid crowding, measuring points at low shear stress are not shown in Fig. 16.6.

The flow behaviour of clay suspensions is usually interpreted in terms of a Bingham liquid, which has a flow curve described by:

$$\left.\begin{array}{rcll} \eta_b D & = & \tau - f & \text{for} \quad \tau > f, \\ D & = & 0 & \text{for} \quad \tau < f, \end{array}\right\} \tag{16.3}$$

where η_b is a constant viscosity (see also equation (1.6)).

In Fig. 16.7 an alternative plot to that of Fig. 16.6 is given, where the viscosity $\eta = \tau/D$ is plotted as a function of shear rate. Measuring points at low shear stress are included in this figure.

The shear stress f of equation (16.3) is called the yield value. For fairly high shear rates, results are sometimes plotted as in Fig. 16.6. If the flow curve in Fig. 16.6 were based on measurements at shear rates higher than, say,

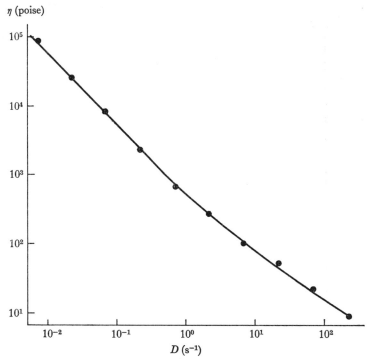

Fig. 16.7. Double-logarithmic representation of the flow curve given in Fig. 16.6.

70 s^{-1}, a yield value of about $f = 1.2 \times 10^3$ dyn/cm^2 would be obtained. Experiments at lower shear stresses show quite clearly that, in contradiction to equation (16.3), the shear rate in this region is not zero. This shows that in many cases the concept of a yield value is an over-simplification of physical reality.

It is doubtful whether materials exist that do not deform permanently, even under small stresses. In practice, however, when used on a small time scale, a yield value can give a useful description of the experimental results.

It is to be expected that a correlation exists between yield value and shear modulus G of a non-linear material:

$$f \approx G\gamma_c \qquad (16.4)$$

where γ_c is the deformation at structural breakdown.

16.3.5 Technical testing methods for clays

Technical testing methods usually are not intended to measure a physical property in a scientific manner. They are invariably meant to predict the suitability of a material for some technical application. Many of them are quite satisfactory in this respect, whereas more scientific rheological measurements would often be too expensive or time-consuming. On the other hand, fundamental rheological measurements often give a better understanding of the relation between structure and mechanical properties for a far wider range of variables than would be possible with technical testing methods. Ideally a technical testing method would be developed out of a more extensive investigation using scientific rheological measurements.

References to chapter 16

1 G. W. Brindley (Ed.), *The X-ray identification and crystal structures of clay minerals* (2nd Ed.; London, 1961).
2 R. A. Grim, *Clay mineralogy* (2nd. Ed.; New York, 1968).
3 H. van Olphen, *An introduction to clay colloid chemistry* (New York, 1963).
4 R. K. Schofield & H. R. Samson, *Disc. Faraday Soc.* **18**, 135 (1954).
5 H. van Olphen, *J. Colloid Sci.* **20**, 822 (1965).
6 A. Scholz, *Kolloid-Z.* **173**, 61 (1960).
7 U. Hofmann, A. Weiss, G. Koch, A. Mehler & A. Scholz, *Proc. Natl. Conf. Clays Clay Minerals* **4**, 273 (1956).
8 K. Norrish, *Disc. Faraday Soc.* **18**, 120 (1954).
9 R. W. Mooney, A. G. Keenan & L. A. Wood, *J. amer. Chem. Soc.* **74**, 1371 (1952)
10 W. F. Bradley, E. J. Weiss & R. A. Rowland, *Proc. Natl. Conf. Clays Clay Minerals* **10**, 117 (1963).
11 B. V. Derjaguin & R. Greene-Kelly, *Trans. Faraday Soc.* **60**, 449 (1964).
12 J. Th. G. Overbeek in *Colloid science* (H. R. Kruyt, Ed.; Amsterdam, 1952), volume 1, chapter 6.
13 H. van Olphen, *Rec. Trav. Chim.* **69**, 1309 (1950).
14 H. van Olphen, *J. Colloid Sci.* **17**, 660 (1962).
15 K. Norrish & J. A. Rausell-Colom, *Proc. Natl. Conf. Clays Clay Minerals* **10**, 123 (1963).
16 P. F. Low, *Adv. Agron.* **13**, 269 (1961).
17 U. Hofmann, *Angew. Chem.* **80**, 736 (1968).
18 H. van Amerongen in *Science of ceramics* (G. H. Stewart, Ed.; London, 1965), volume 3, p. 53.

19   R. N. Yong & B. P. Warkentin, *Introduction to soil behaviour* (New York, 1966).
20   A. Weiss, *Z. Anorg. Allgem. Chem.* **299**, 92 (1959).
21   A. Weiss, *Rheol. Acta* **2**, 292 (1962).
22   A. Atterberg, *Intern. mitt. boden* **1**, 4 (1911).
23   W. A. White, Ph.D. Thesis, University of Illinois, 1955.
24   R. A. Grim, *Applied clay mineralogy* (New York, 1962).
25   U. Hofmann, R. Fahn & A. Weiss, *Kolloid-Z.* **151**, 97 (1957).
26   A. L. Johnson & F. H. Norton, *J. Amer. Cer. Soc.* **24**, 189 (1941).
27   A. S. Michaels & J. C. Bolger, *Ind. Eng. Chem. Fundamentals* **3**, 14 (1964).
28   H. van Olphen, *Clay colloid chemistry* (New York, 1963), p. 99.
29   N. Street, *Aust. J. Chem.* **9**, 467 (1956).
30   P. A. Thiessen, *Z. Elektrochem.* **48**, 675 (1942).
31   H. van Olphen, *Rec. Trav. Chim.* **69**, 1313 (1950).
32   H. van Olphen, *Disc. Faraday Soc.* **11**, 82 (1951).
33   H. van Olphen, *J. Colloid Sci.* **19**, 313 (1964).
34   K. H. Hiller, *Rheol. Acta* **3**, 132 (1964).
35   H. van Olphen, *Proc. Natl. Conf. Clays Clay Minerals* **6**, 196 (1958).
36   B. P. Warkentin, *Nature* **190**, 287 (1961).
37   B. P. Warkentin & R. N. Yong, *Proc. Natl. Conf. Clays Clay Minerals* **9**, 210 (1962).
38   W. A. Weyl & W. C. Ormsby in *Rheology, theory and applications* **3** (F. R. Eirich, Ed.; New York, 1960).
39   R. T. Martin, *Proc. Natl. Conf. Clays Clay Minerals* **9**, 28 (1962).
40   P. F. Low and J. L. White, 'hydrogen bonding and polywater in clay–water systems', note accepted for publication in *Proc. Natl. Conf. Clays Clay Minerals* **18**.
41   J. R. van Wazer, *Phosphorus and its compounds* **1** (New York, 1958).
42   A. S. Michaels, *Ind. Eng. Chem.* **50**, 951 (1958).
43   J. W. Lyons, *J. Colloid Chem.* **19**, 399 (1964).
44   J. R. van Wazer & E. Besmertnuk, *J. Phys. & Colloid Chem.* **54**, 89 (1950).
45   A. Einstein, *Ann. Physik* **19**, 289 (1906). **34**, 591 (1911).
46   W. Kuhn & H. Kuhn *Helv. Chim. Acta* **28**, 97 (1945).
47   H. Giesekus, *Rheol. Acta* **2**, 50 (1962).
48   H. L. Goldsmith & S. G. Mason in *Rheology, theory and applications* **4** (F. R. Eirich, Ed.; New York, 1967).
49   R. Simha, *J. Appl. Phys.* **23**, 1020 (1952).
50   V. Vand, *J. Phys. & Colloid Chem.* **52**, 277 (1948).
51   H. L. Frisch & R. Simha, in *Rheology, theory and applications* **1** (F. R. Eirich, Ed.; New York, 1956).
52   Do Ik Lee, *Trans. Soc. Rheology* **13**, 273 (1969).
53   A. S. Michaels & J. C. Bolger, *Ind. Eng. Chem Fundamentals* **1**, 153 (1962).
54   M. M. Cross, *J. Colloid Sci.* **20**, 417 (1965).
55   H. D. Weymann, *Proc. Intern. Congr. Rheology* 4th (E. H. Lee, Ed.; New York, 1965), part 3, p. 573.
56   R. J. Hunter & S. K. Nicol, *J. Colloid Interface Sci.* **28**, 250 (1968).
57   J. D. Ferry, *Viscoelastic properties of polymers* (New York, 1961).
58   A. J. Staverman & F. R. Schwarzl in *Die Physik der Hochpolymeren* (H. A. Stuart, Ed.; Berlin 1956), volume 4.
59   J. L. den Otter, *Rheol. Acta* **8**, 355 (1969).
60   N. F. Astbury, F. Moore & J. A. Lockett, *Trans. Brit. Ceram. Soc.* **65**, 435, (1966).

61 R. J. Krizek & R. L. Kondner, *Int. Symp. Rheol. Soil Mech.* (Grenoble, 1964).
62 A. G. Franklin & R. J. Krizek, *Proc. Natl. Conf. Clays Clay Minerals* **16**, 353 (1968).
63 Tan Tjong Kie, Thesis, Delft, 1954.
64 R. L. Kondner & R. J. Krizek, *J. Franklin Inst.* **279**, 366 (1965).
65 R. L. Kondner & R. J. Krizek, *Proc. ASTM* **64**, 944 (1964).
66 K. H. Roscoe, *Proc. Int. Conf. Soil Mechan. Found. Eng.* 3, **1**, 186 (1953).
67 K. M. Beazley, *Tappi* **50**, 151 (1967).

17 Miscellaneous substances

H. K. de Decker and R. Houwink (§17.4)

17.1 Gelatin and glue[14]

17.1.1 Description and general behaviour

Gelatin and glue are both obtained from the same material, the 'collagen' in animal bones and skins, by extraction with water. Depending on the conditions of the aqueous extraction process, the properties of gelatin can vary widely. Glue can be regarded as one of the gelatin types, usually of lower molecular weight than ordinary gelatin.

The raw product, usually called 'gelatine', contains a few per cent of various salts in addition to the real '*gelatin*', a derived protein with a polypeptide chain. Its molecular weight (average and distribution) depends on its history with respect to temperature and contact with water. One can assume that the collagen base material has very high molecular weight and is cross-linked; it swells in water but is not soluble. Gelatine is derived from collagen by hydrolytic scission of the chain and the cross-links of the collagen. There is no definite point where this process terminates; hence the indefinite structure of gelatin. The molecular weight average is around 50000–100000, but extremes up to 400000 occur in the molecular weight distribution of many types of gelatin.

A great number of amino acids have been found to be structural components of the gelatin chain; glycine, proline, and alanine are some of the most frequently occurring. The peptide bond—$CO.NH$— linking two amino acid units, is the most common feature in the gelatin structure (about one in every 100 molecular weight units). Due to the fact that some of the amino acid units carry reactive groups in the side chain (for example, glutamic acid carries a —$COOH$ group and lysine carries a —NH_2), there are considerable numbers of free amino and carboxyl groups on the chain (about one of each in every 1000 molecular weight units); see Fig. 17.1.

Gelatin is completely miscible with water from 0 to 100%. A solution containing a low proportion of water behaves as solid gelatin. Solutions with less than 10% gelatin behave as other solutions of high polymers, at least above 40° C; but below that temperature they gradually solidify, becoming stiffer and more elastic. The technically amazing phenomenon of water

438

containing only 1 % solid material and yet behaving as a solid in several properties is a well-known fact that can be observed in dilute gelatin solutions at room temperature.

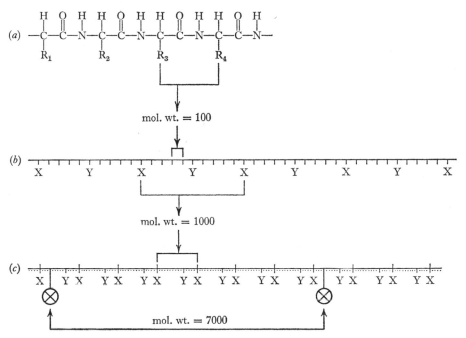

Fig. 17.1. Polypeptide chain molecule for gelatin. (*a*) Detailed chemical structure: R, etc. are side groups of many types; one link averages mol. wt. = 100; segment shown mol. wt. = 400. (*b*) Chain segment of mol. wt. = 4000 showing ionized groups X = COOH and Y = NH_2. (*c*) Chain segment of mol. wt. = 10000 showing cross-linking loci at ⊗.

17.1.2 Structure

Gelatin molecules are shown schematically in Fig. 17.1 as long flexible chains, having many (1:1000 mol. wt. units) 'ionic' sites which in water are positively or negatively charged and orient substantial numbers of water molecules in their vicinity (cf. the swelling of bentonite clay in water). Some sites on the chain are loci for possible cross-linking (estimates indicate one such locus every 7000 mol. wt. units). Cross-links are formed when an ionic site is in a situation where it would become more stable by associating with a site of opposite sign, than by polarizing water molecules. Only some of the ionic sites are prone to this situation, and then only under certain conditions of temperature, pH, etc. This explains why the gel formation and the stiffness are both dependent on these variables.

Thus, gelatin molecules in water tend to form a network in which (*a*) water molecules are polarized and immobilized in a structure as in a bentonite gel (Chapter 16); (*b*) elasticity is caused by the flexible cross-linked gelatin chains in the same way as in vulcanized rubber (Chapter 9).

Assuming that the same 'random coil' theory of elasticity applies to gelatine gels as to vulcanized rubber, a comparison between the two materials can be made and the differences in the Young's modulus explained in terms of

(1) water effect by volume dilution of the network;

(2) water effect by dissociation of cross-links;

(3) number of cross-link sites per molecule.

There is approximate equivalence between gelatin and rubber if the effect (2) causes 80% dissociation of cross-links at 10% concentration, and 50% dissociation at 30% gelatine concentration. As an example, the modulus of a 10% gelatine gel is 0.2 kgf/cm^2; rubber under similar conditions of dilution and cross-linking site density would show 1.0 kgf/cm^2. The factor of 5 is probably due to dissociation of gelatin cross-links.

An empirical equation (17.1) links Young's modulus M, and the concentration C, of gelatine gels (K and n are constants)

$$M = KC^n. \qquad (17.1)$$

The exponent n is less than or equal to 2 and always above 1. This can be understood in terms of the effects (1) and (2). Effect (1) produces a linear concentration dependence and (2) a further concentration factor with an exponent m approaching 1 under some conditions; the resulting exponent is

$$n = 1 + m \leqslant 2.$$

The structure of gelatin solutions is also indicated by their compressibility. In the gel state, values around 10×10^{-6} are found, against about 1×10^{-6} for most solids.

When the gel is transformed into a sol by heating, the compressibility increases to 48×10^{-6}, a value similar to that for water (see Table 6.1). This indicates that the water in the gel is denser (more closely packed by ionic attraction?) then in liquid water or gelatin sol, but that even in the gel the dense packing of a solid is not reached.

17.1.3 Rheology

From the above picture of the structure, it can be understood that the rheology of gelatin solutions is strongly dependent on concentration, molecular weight, temperature, time, shear stress, and pH. This dependence is similar to that for other polymer solutions, but is also related to the instability and sensitivity of the ionic cross-links. A brief discussion of each of the variables follows.

(a) Temperature

Viscosity of the solutions increases as temperature decreases, till a point is reached where solid elasticity is observed. This is usually around 40° C, but will be observed at a lower temperature if the criterion for elastic reaction is too high and the concentration is too low. For example, tiny nickel splinters suspended in a 0.5 % gelatin solution at room temperature can be displaced by a magnet, but they will spring back when the magnet is removed. The same solution shows liquid behaviour under larger stresses. In the 'gel' state, the modulus increases as the temperature is lowered further.

(b) Concentration

As would be expected, the modulus of elasticity is higher for more concentrated gelatin gels as shown in Table 17.1.

TABLE 17.1. *Elasticity of gelatin gels (at room temperature)*

Gelatin concentration (%)	Young's modulus (kgf/cm^2)
2.5	1×10^{-2}
5	4×10^{-2}
10	2×10^{-1}
20	1
30	1.5
45	3
87	5×10^3

(c) Molecular weight

The modulus and tensile strength, as well as the viscosity of a gelatin solution, all decrease with molecular weight. Gel formation does not take place at all if the molecular weight is below about 15 000, presumably because less than two cross-link sites per average molecule are then available.

15

Hydrolytic action causes the above mentioned properties to decrease with time when gelatin solutions are boiled, due to gradual molecular degradation.

(*d*) Time

The most striking effect of time is that on the rigidity (modulus) of gels. Fig. 17.2 shows some examples. The increase of stiffness with time appears to be limitless, although 18 hours is usually sufficient to establish that it does not change strongly any more.

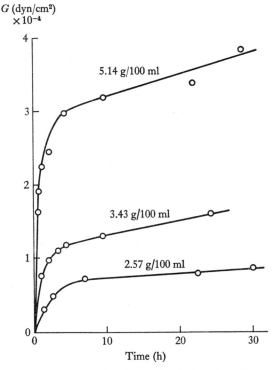

Fig. 17.2. Increase of rigidity of gelatin gels with time.

(*e*) Shear stress

The rather loose structure of a gelatin gel is destroyed by shearing. The modulus drops to zero and viscous behaviour is observed, but after prolonged standing the gel structure reforms. The viscosity of the solution is subject to thixotropy.

(f) pH

Both below and above a certain point (the isoelectric point, usually around pH = 5) the viscosity of a sol increases. This can be understood by taking into account the observation of the 'isoelectric opalescence' which indicates the presence of aggregates of gelatin molecules of 40 million molecular weight units at the isoelectric point. This kind of 'concentration' of gelatin in aggregates diminishes the interaction with the solvent and therefore the total volume of bound water decreases, leading to a lower viscosity. By the same

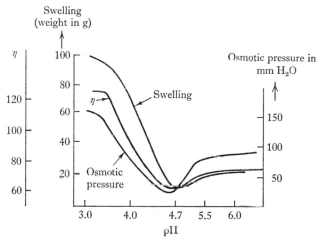

Fig. 17.3. Influence of pH on viscosity, swelling power and osmotic pressure of a 1 % gelatin sol in water.

token, this concentration will cause greater stiffness in the gel structure, as is sometimes observed at the isoelectric point.

Data for the viscosity as a function of the pH are given in Fig. 17.3.

17.2   Creams, pastes and jellies[1, 2]

A number of materials used as food, cosmetics, or medicines, are neither liquids nor solids in the usual sense; they are creamy or jelly-like.

These materials consist of several components and have a structure that remains elastically rigid up to the yield value, and above which the structure collapses and viscous flow occurs. Both yield value and elastic modulus are low, but the viscosity is many times higher than that of water. This feature means the materials stay in place under limited pressure: a layer of butter on bread does not flow away; a gelatine pudding will 'stay'. Nevertheless, both must be easy to chew and swallow in the case of food, or to apply and remove in the case of unguents and cosmetics, and therefore the viscosity cannot be too high.

In a subjective characterization of these materials, some are said to be creamy, others jelly-like, and many somewhere in between. Recent investigations have shown that more 'creaminess' means a lower elastic modulus, and more 'jelly-like' corresponds to a higher elastic modulus. Usually the viscosity is higher for a cream or paste than for a jelly, but even at equal viscosity the criterion based on the elastic modulus correlates with the subjective classification of cream versus jelly. When using 'viscosity' in this context, and remembering that these complex materials have a D–τ relationship with more than one constant (see Chapter 4), the 'viscosity' selected for comparison should be at a shear rate comparable to that occurring in spreading, swallowing, etc. In rotation viscometers with a 'ring' width of 1 cm at a radius of 4 cm, a rotation speed of 2–5 rev/min is appropriate; that is a shear rate D of between 50 and 125 cm/cm min. Tongue movements during eating generally have a frequency of about 30 per minute and amplitudes of the order of 1 cm.

Viscosity determinations can be meaningful in the assessment of smooth liquid foods by human perception.[3] The Weber–Fechner law relates the sensory response to a stimulus (in this case the resistance against deformation) by a power law of the type:

$S_r = kS_t^n$ where S_r is sensory response, S_t is stimulus and k is a constant.

For a series of 5 soups, ranging from watery-thin to very thick, the latter 50 times thicker than the former, a panel assessed the consistency by tasting, and the results were plotted against the shearing stress measured. A value of 1.28 was found for n. Thus, the sensory response, in this case, is a sharper tool than the physical evaluation, a fact not uncommon in empirical psychophysical comparisons, where n is usually between 0.5 and 2.

The overall rheology of these various materials can be described roughly by a Bingham type equation:

$$\tau - \tau_0 = \eta D \text{ (see Chapter 4).}$$

However, correction terms are needed to express the exact D–τ relationship. Moreover, thixotropic breakdown of the structure shifts the D–τ relationship with time and stress.

Table 17.2 shows various classes of materials, with some examples, and

TABLE 17.2. *Survey of creams, pastes and jellies*

Class	Examples	Yield value (gf/cm²)
Syrups	Honey, sugar syrup	0
Unguents	Chocolate (45° C), Vaseline	
Creams	Facial cream	0–10
Pastes	Mayonnaise, porridge, butter, toothpaste	
Jellies	Pectin solution, gelatine solution	1–100
Solids	Cheese, chocolate (20° C)	Over 100

indicates for each a typical yield value at room temperature. The viscosities vary within each class; they are generally around 1–10 poise at room temperature and at typical rates of deformation. The elastic modulus in the area of elastic deformation is of the order 10^4 gf/cm² (10^7 dyn/cm²) for butter at 10° C when stressed for 1 s, but it is considerably lower for most of these materials under prolonged stress, since they are subject to creep. Generally, the elastic modulus decreases at higher temperature.

The elastic modulus decreases under heavier loads, that is, the ratio of the elastic compliance of the material to the stress is not constant. This is one reason why the yield value is ill-defined; the modulus approaches zero gradually at a stress where compliance is mostly viscous.

For a 10 % gelatin gel in water the elastic modulus under small stresses for periods of a few minutes can be measured reproducibly; it is about 200 gf/cm² or 2×10^5 dyn/cm². Measured under the same conditions, the elastic modulus of the other classes will be found to diminish when going from the bottom to the top of the table.

The phenomenon of 'stringiness' will be more obvious as the yield value is lower and the viscosity higher.

From the structural viewpoint, these products can be subdivided into four types:

(1) Emulsions, i.e. fine dispersions of one liquid in another, e.g. mayonnaise.

(2) Dispersions of fine crystalline or amorphous solid particles, e.g. Vaseline, yogurt, chocolate.

(3) Dispersions of somewhat larger solid particles, e.g. apple sauce, porridge.

(4) Polymer solutions, e.g. starch, custard, pectin or gelatine jellies.

In addition, there are variations in the nature of the continuous or liquid phase leading to substantial differences in its viscosity. It is high for Vaseline or chocolate where the continuous phase is a lubricating oil or cacao butter, and low for gelatine solution. The continuum in type (3) is usually a structured liquid with non-Newtonian flow; these systems are less subject to thixotropy than the others. The jelly condition occurs only in type (4). Creamy behaviour is found in all three of the other types.

17.3 Rheology of soaps[14]

17.3.1 Definitions

'Soaps' (or 'surfactants' or 'detergents') are chemicals which have a tendency to accumulate at the boundary or interface of two phases. By their presence at the interface they reduce the interfacial tension and promote 'wetting'.

An example is ordinary household soap which concentrates at the water–air interface, reducing the surface tension and making the water foamy; the substantial increase in surface from the air bubbles in the foam is stabilised by the soap. Another example is the detergent used in lubricating oils. Carbon particles are kept in suspension by the adsorbed layer of detergent molecules which reduce their tendency to agglomerate and prevent their forming a cake in the combustion engine.

Chemically the structure of a soap has two parts, one hydrophilic and one hydrophobic. In ordinary soap, for example, sodium stearate, the stearate hydrocarbon chain is compatible with oil but not with water, but the COONa part is hydrophilic and in water dissociates into COO^- and Na^+. Many soaps have ionizing end groups; others are non-ionic.

17.3.2 Pure soaps

As a pure substance, sodium stearate is a solid at room temperature, reasonably well crystallized, but having a 'soapy' or 'smectic' touch. Furthermore, a sharp melting point cannot be determined. Both characteristics are due to the tendency of the soap molecules to orient parallel to each other in layers, so that the polar end of one layer faces the polar end of the next layer, and the hydrocarbon chains are also adjacent. This orientation persists at relatively high temperatures. It also persists and is even promoted under great shearing stress. It is the 'smectic' arrangement of liquid crystals discussed in Chapter 6. The viscosity of sodium stearate decreases as the temperature rises, but three sharp transitions are found. Discontinuities in the viscosity curve at $225°$ C and 262° C indicate transition stages from the true crystalline solid to the smectic liquid crystal and the final jump in the curve at 290° C indicates the transformation of the liquid crystal to the true (isotropic) liquid.

17.3.3 Solutions

Concentrated solutions of soaps in non-polar solvents have a tendency to form liquid crystals and this tendency increases as the concentration increases. A striking illustration of this was shown by J. J. Hermans using poly-γ-benzyl-

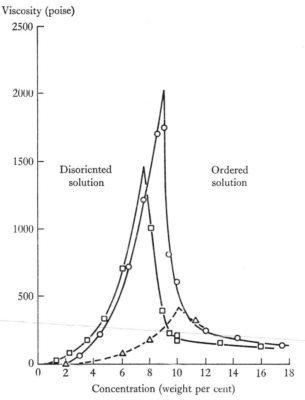

Fig. 17.4. Viscosity change with concentration for poly-γ-benzyl-L-glutamate in *m*-cresol. Molecular weights: □, 342 000; ○, 270 000; △, 220 000. Stress < 100 dyn/cm².

L-glutamate, a polymer that has surfactant properties, in *m*-cresol solution. Higher concentrations led to greater mutual orientation and sliding of the oriented bundles of molecules, and hence a *lowering* of viscosity instead of the increase usually connected with more concentrated solutions. These observations are represented in Fig. 17.4.

As indicated in Chapter 6, the shearing stress also has an orienting effect, and will therefore lower viscosity in sufficiently concentrated soap solutions.

The viscosity of dilute soap solutions reflects the tendency of soap to form micelles, or packages of molecules oriented so that the end most compatible with the solvent is adjacent to the solvent (Fig. 17.5). This arrangement enables the micelle to create a 'hydrocarbon' medium surrounded by water. This is utilized in the production of synthetic rubber and other emulsion polymers which are formed inside the soap micelles.

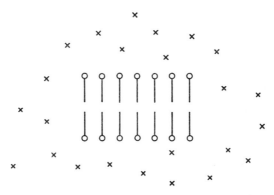

Fig. 17.5. Soap micelle in water. ○ represents COO^-Na^+; | represents $CH_3(CH_2)_n$; × represents H_2O.

At first sight, micelle formation should not influence the solution viscosity as long as the micelles do not grow together into extensive networks since this would lead to smectic behaviour (see above). However, at concentrations well below that at which liquid crystals are formed, there is a definite viscosity increment above the value normally expected, for the given volume concentration, in accordance with Einstein's law. This increment is due to the immobilization of solvent around the micelles.

This is true for aqueous soap solutions and also for soap solutions in oil. These latter systems find practical use as greases. The 'structure' built up in the oil by the soap to form a grease is destroyed by shearing stresses and results in the smectic behaviour essential for these materials. The complete structure can be re-established if sufficient time or a temperature increase is allowed.

A very special example of a 'liquid crystal' of the smectic type is the double layer on the inner and outer surfaces of a soap bubble.

17.4 Sulphur† (by R. Houwink)

17.4.1 Constitution[4–6]

Sulphur is known to exist in a great number of allotropic modifications. Of the crystalline and amorphous forms which exist at room temperature, all eventually revert to an eight-membered ring in an orthorhombic crystalline modification. When orthorhombic sulphur is heated it changes to a monoclinic crystalline structure at temperatures above 95° C and melts at about 113° C. The melt is a pale yellow fluid with a viscosity of 6–10 centipoise, only slightly higher than that of water. The exact melting point varies between 113° C and 119° C depending upon the allotropic composition. As the temperature of the

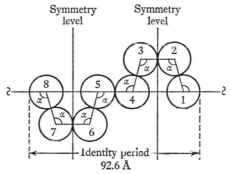

Fig. 17.6. Part of a polymeric sulphur chain.

sulphur is raised, the melt consisting primarily of S_8 rings changes to polymeric chains above 159° C. The polymeric chains can be preserved by quenching the liquid sulphur and holding it at temperatures between −30° C and −10° C.

Elastic sulphur can be considered as a solution of polymeric sulphur S_8^{Ch} (Ch stands for chain; this is also known as S_μ sulphur) in a monomer of S_8^R (R stands for ring; this is also known as S_λ sulphur). Both are amorphous and weight fractions of polymer as high as 40 % have been observed. The monomer can be considered to act as a plasticizer lowering T_g of the pure polymer from 75° C to −30° C. Chain molecules of length 10^5 S_8 units have been reported.[7] In accordance with this structure the modulus–temperature curve of elastic sulphur shows the characteristic behaviour of a plasticized amorphous polymer (see Fig. 3.8). There is a glassy region of high modulus, a sharp transition region starting at T_g (−30° C) and a rubbery plateau region. The latter can, however, only be observed under special experimental conditions.

† The author is much indebted to The Sulfur Institute, Washington, D.C. for constructive comments concerning this section.

Our knowledge concerning the strength of the S—S bonds is still limited.[8] The S—S bond energy in compounds like $CH_3.S—S.CH_3$, estimated at 60–70 kcal/mol is almost twice as high as that of the O—O bond; in S_8 rings it is estimated at 33 kcal/mol, that is of the same order of magnitude as in the O—O bond. This may still be an attractive basis for the development of sulphur polymers of special technical value. To achieve this the crystallization phenomena should be suppressed and a convenient cross-linking mechanism found, in order to effect 'inverse vulcanization';[9] that is instead of vulcanizing a hydrocarbon polymer by means of sulphur, 'vulcanizing' sulphur by means of hydrocarbons.

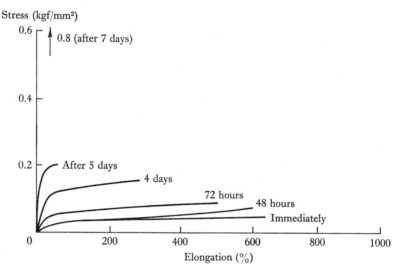

Fig. 17.7. Stress–strain curves for elastic sulphur prepared at 350 °C, at various intervals after preparation (ref. 10).

17.4.2   Elastic behaviour

Quenching of sulphur from temperatures of 200° C down to −30° C results in a yellow translucent glass which shows no evidence of crystallization. When heated above −30° C it first becomes rubbery elastic and then crystallisation sets in, becoming increasingly rapid above −10° C. However, it is possible by working very quickly in the range between −30 and −10° C to study the properties of elastic sulphur. When drawn out it will crystallize into fibres. The shape of the stress–strain diagram for highly elastic sulphur is largely dependent on two factors: (*a*) the temperature at which the molten sulphur is poured, and (*b*) the time lapse since pouring the sulphur.

Fig. 17.7 shows a set of curves for elastic sulphur prepared at 350° C (the temperature at which the best quality is obtained) illustrating the gradual stiffening which takes place with time.[10] Fig. 17.8 shows the remarkable

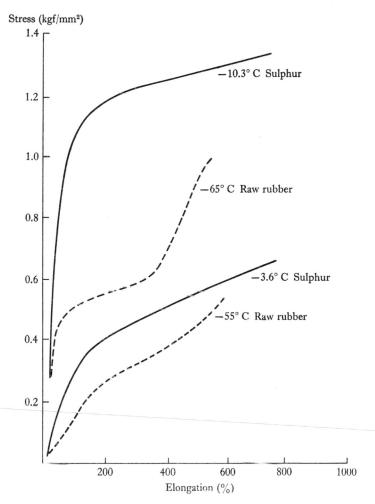

Fig. 17.8. Stress–strain curves for elastic sulphur and raw rubber (ref. 11).

similarity between the stress–strain curves for elastic sulphur and unvulcanized (raw) rubber at appropriately selected temperatures. This similarity may lend weight to the search for 'inverse vulcanization'.

The tensile strength of crystallized sulphur after repeated stretching,[12] may be as high as 11 kgf/mm², approaching that of cellulose (20 kgf/mm²) and unvulcanized rubber at room temperature ($14\,\mathrm{kgf/mm^2}$). That this value was attained, indicates a very thorough orientation of the chain molecules. However, this high tensile strength decreases over several days as a result of the transition $S_8^{Ch} \rightarrow S_8^{R}$.

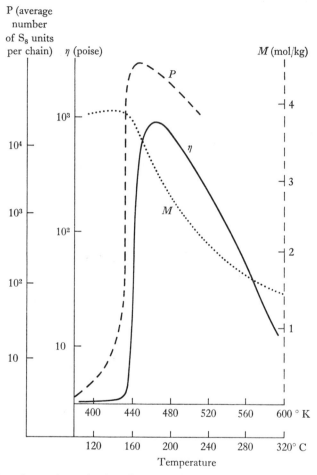

Fig. 17.9. Viscosity, polymerization degree and concentration of monomeric sulphur *M* in liquid sulphur as function of temperature (ref. 4).

17.4.3   Viscous behaviour

Several endeavours have been made[12] to explain the rheological behaviour of liquid sulphur up to 159° C, on the basis of the equilibrium between the two molecular species, S_8 rings and long polymer chains. The liquid shows a normal viscosity up to the transition temperature, 159° C, then follows the anomaly of an abrupt viscosity increase followed by a gradual decrease at still higher temperatures, and all these changes are perfectly reversible.

Fig. 17.9 shows that the changes in viscosity are parallelled by changes in the chain length and inversely by the amount of S_8 in the polymer chains.

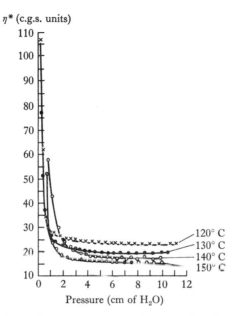

Fig. 17.10. Relation between apparent viscosity η^* and pressure in the viscometer for molten sulphur.

The presence of polymer chains in the solution is further supported by the fact[13] that over the temperature interval 120° C to 150° C, visco-elastic flow can be observed. This is shown in Fig. 17.10 where the viscosity η^* is represented as a function of the applied pressure in the viscometer. This shearing-stress dependence of η^* is presumably due to orientation of the longer molecules.

References to chapter 17

1 G. W. Scott Blair, *Foodstuffs, their plasticity, fluidity and consistency* (Amsterdam, 1953).
2 *Proceedings Symposium Foodstuffs* (Kon. Ned. Chemische Vereniging, 1962).
3 F. W. Wood, Soc. Chem. Ind. Monograph no. 27 (1968).
4 R. F. Bacon & R. Fanelli, *J. Amer. Chem. Soc.* **65**, 639 (1943).
5 B. Meyer (Ed.), *Elemental sulfur* (Interscience, 1965).

6   A. V. Tobolsky, *The chemistry of sulfides* (Interscience, 1968).
7   F. J. Touro & T. K. Wiewirowski, *J. Phys. Chem.* **70**, 239 (1966).
8   Prior, *Mechanics in sulfur reactions* (McGraw-Hill, 1962).
9   M. D. Barnes in ref. 5, p. 365. Vulcanizing agents might be divinylbenzene or 1,5-cyclooctadiene.
10   K. Sakurada & H. Erbring, *Koll. Z.* **72**, 129 (1935).
11   J. D. Strong, *J. Phys. Chem.* **32**, 1225 (1928).
12   See e.g. A. Eisenberg, *Macromolecules* **2**, 44 (1969).
13   W. Ostwald & H. Maless, *Koll. Z.* **63**, 305 (1933).
14   F. R. Eirich, *Rheology* (New York, 1965/7).

Index